Contents

C000213577

Preface

This is the first of a series of books written for students preparing for A level Mathematics. Books 1 and 2 of the series cover the work required for a single-subject A level in Mathematics or in Pure and Applied Mathematics. Book 1 covers the essential core of sixth-form mathematics now accepted by the GCE Boards, while Book 2 covers the applied mathematics, i.e. the numerical methods, mechanics and probability, contained in most single-subject syllabuses. Book 3 covers the additional pure mathematics needed by students taking the double-subject Mathematics and Further Mathematics, and by those taking Pure Mathematics as a single subject.

Vector notation and vector techniques are used wherever appropriate, particularly where these methods illuminate or simplify the work, but their use is avoided whenever they appear likely to confuse the student. Set language is employed wherever it is considered helpful, but it is not introduced at all times as a matter of principle.

The material is arranged under well-known headings and is organised so that the teacher is free to follow his or her own preferred order of treatment. The chapter contents are listed and an index is also provided to make it easy for both the teacher and the student to refer back rapidly to any particular topic. For ease of reference, a list of the notation used is given at the back of the book together with a list of formulae.

Each topic is developed mainly through worked examples. There is a brief introduction to each new piece of work followed by worked examples and numerous simple exercises to build up the student's technical skills and to reinforce his understanding. It is hoped that this will enable the individual student working on his own to make effective use of the books, and the teacher to use them with mixed ability groups. At the end of each chapter there are many miscellaneous examples, taken largely from past A level examination papers. In addition to their value as examination preparation, these miscellaneous examples are intended to give the student the opportunity to apply the techniques acquired from the exercises throughout the chapter to a considerable range of problems of the appropriate standard.

We are most grateful to the University of London University Entrance and School Examinations Council (L), the Associated Examining Board (AEB), the University of Cambridge Local Examinations Syndicate (C) and the Joint Matriculation Board (JMB) for giving us permission to use questions from their past examination papers.

We are also grateful to the staff of Macmillan, particularly Mrs Janet Hawkins and Mr Tony Feldman, for the patience they have shown and the help they have given us in the preparation of these books.

C. W. Celia
A. T. F. Nice
K. F. Elliott

1 Sets, functions and operations

1.1 Sets

All manner of people collect things ranging from match-boxes and beer mats to works of art of great value. Many collectors sort their collections into categories, e.g. stamps sorted by country of origin.

Mathematicians have this habit of collecting and sorting. From earliest time men have been interested in numbers and have arranged them in sets, e.g. even numbers, prime numbers, etc. Indeed the idea of a set is so important in mathematics that a notation and a language have been developed for dealing with sets.

A set is a well-defined collection of objects called elements or members of the set.

Well-defined means that objects which are members of a set are distinguishable from objects which are not, and each member of a set is different from every other member of the set. The fair-haired people in a room would not be a well-defined collection since in some cases it might be a matter of opinion whether a particular person was fair or not. Also the numbers 1, 1, 2 would not constitute a set.

The elements of a set may be named in a list or may be given by a description, and the list or description is enclosed in braces { }. For instance, the set of days of the week may be given as

{Sunday, Monday, Tuesday, Wednesday, Thursday, Friday, Saturday}

or as   {the days of the week}.

Sets may be finite, i.e. having a finite number of elements, or they may be infinite. The days of the week form a finite set and the set of even numbers is an infinite set.

In addition to the braces used above, other symbols have been devised to enable statements about sets to be made with brevity and precision. For instance, let $E = \{$even numbers$\}$. The statement '2 is an element of the set of even numbers' or '2 belongs to the set of even numbers' or indeed '2 is an even number' may be abbreviated to

$2 \in E$, $\in$ standing for 'is an element of' or 'belongs to'.

Similarly the statement '3 is not an element of the set of even numbers' or '3 is not an even number' may be abbreviated to

$3 \notin E$, $\notin$ standing for 'is not an element of' or 'does not belong to'.

The colon : is used as an abbreviation for 'such that', e.g. $\{x : x^2 - 9 = 0\}$ which is read as 'the set of all values of the number x such that $x^2 - 9 = 0$'.

If each member of set A belongs to set B and also each member of set B belongs to set A, then A and B are the same set.

Exercises 1.1

1 Define (i) by listing the elements, (ii) by describing the elements, four different sets of
   (a) numbers   (b) shapes   (c) non-mathematical objects.
2 (a) Give two examples of a finite set of numbers.
   (b) Give two examples of an infinite set of numbers.
3 List the elements of the sets
   (a) {prime numbers less than thirty}
   (b) {letters of the alphabet consisting of straight lines only}
   (c) {perfect squares between 2 and 50}.
4 Describe in words the sets
   (a) $\{2, 4, 6, 8\}$ (d) $\{x : x > 0\}$
   (b) $\{2, 3, 5, 7, 11\}$ (e) $\{x : x \leqslant 0\}$
   (c) $\{x : x^2 = 1\}$ (f) $\{x : 0 < x < 3\}$.
5 If $A = \{1, 2, 3, 4, 5\}$ and $B = \{8, 9, 10\}$, complete the following statements by inserting $\in$, $\notin$, A, B or elements of the sets A and B:
   (a) $3 \ldots A$ (c) $2 \in \ldots$ (e) $1 \notin \ldots$ (g) $5 \ldots B$ (i) $\ldots \notin A$
   (b) $4 \ldots B$ (d) $9 \in \ldots$ (f) $8 \notin \ldots$ (h) $8 \ldots A$ (j) $\ldots \notin B$.
6 Given that $A = \{$factors of 30$\}$, express in set notation the statements
   (a) 2 is a factor of 30   (b) 4 is not a factor of 30.
   Express in words the statements (c) $5 \in A$ (d) $7 \notin A$.

1.2   Intersection and union of sets

It is possible for an element to belong to two or more sets.

Let $A = \{1, 2, 3, 4, 5\}$ and $B = \{2, 4, 6\}$. Then $2 \in A$ and $2 \in B$. This is illustrated in Fig. 1.1 in a diagram which is known as a Venn diagram. The numbers 2 and 4

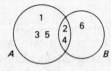

Fig. 1.1

belong to both set A and set B. $\{2, 4\}$ is the set of numbers belonging to both set A and set B. This set is represented by $A \cap B$ and is called the intersection of sets A and B. $A \cap B$ is read as 'A intersection B' or sometimes as 'A cap B'. The set $A \cap B$ consists of those elements which belong to both set A and set B,

i.e.
$$2 \in A \text{ and } 2 \in B \quad \text{and so} \quad 2 \in (A \cap B),$$
$$1 \in A \text{ but } 1 \notin B \quad \text{and so} \quad 1 \notin (A \cap B).$$

Let $P = \{x : x > 1\}$ and $Q = \{x : x < 4\}$.

Then $P \cap Q = \{x : 1 < x < 4\}$.

Two sets which have no common elements are said to be disjoint. Consider the disjoint sets $E = \{$even numbers$\}$ and $D = \{$odd numbers$\}$. There are no numbers which belong to both sets and so there are no elements in the set $E \cap D$. The set with no elements is called the empty set and is represented by $\varnothing$ or $\{\ \}$. Thus $E \cap D = \varnothing$ or $E \cap D = \{\ \}$.

Consider again the sets $A = \{1, 2, 3, 4, 5\}$ and $B = \{2, 4, 6\}$. The numbers 1, 2, 3, 4, 5, 6 all belong to either the set A or the set B or to both set A and set B. This set $\{1, 2, 3, 4, 5, 6\}$ is known as the union of the sets A and B and is represented by $A \cup B$. $A \cup B$ is read as 'A union B' or sometimes as 'A cup B'.

In Fig. 1.2 the shaded region in Venn diagram (a) represents the set $A \cup B$ and the shaded region in Venn diagram (b) represents the set $A \cap B$.

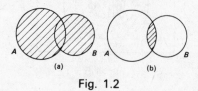

(a)   (b)

Fig. 1.2

Figure 1.3 shows the Venn diagram for the disjoint sets E and D.

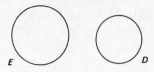

E D

Fig. 1.3

Sometimes all the elements of one set are members of another set. For instance, let $\mathbb{N} = \{0, 1, 2, 3, \ldots\}$. Then every element of the set $E = \{$positive even numbers$\}$ belongs to the set $\mathbb{N}$. The set E is then said to be a subset of the set $\mathbb{N}$ and this is written as $E \subset \mathbb{N}$ and is illustrated in the Venn diagram in Fig. 1.4.

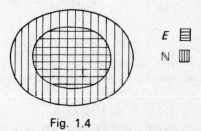

Fig. 1.4

Every set is considered to be a subset of itself. The empty set is a subset of all sets.

It is sometimes necessary to limit one's consideration to the elements of a particular set so that all sets considered will be subsets of this set. This set is called the universal set and is represented by the symbol $\mathscr{E}$. For instance, we may consider only elements which belong to $\mathbb{R}$, the set of real numbers. Subsets of $\mathbb{R}$ include $\mathbb{N}$, the set $\{0, 1, 2, \ldots\}$; $\mathbb{Z}$, the set $\{0, \pm 1, \pm 2, \ldots\}$; and P, the set of prime numbers. Then $\mathbb{R}$ is said to be the universal set for the subsets $\mathbb{N}$, $\mathbb{Z}$ and P. In a Venn diagram the universal set is represented by a rectangle with the subsets shown as rings within this rectangle. The Venn diagram for the universal set $\mathbb{R}$, represented by $\mathscr{E}$ and the subsets $\mathbb{N}$, $\mathbb{Z}$ and P is shown in Fig. 1.5.

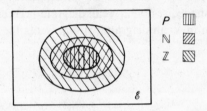

Fig. 1.5

If the set A is a subset of the universal set $\mathscr{E}$, the set consisting of those elements of $\mathscr{E}$ which are not members of A is called the complement of A and is represented by A'. Consider the set of natural numbers to be the universal set and the subset E the set of positive even numbers. Then E', the complement of E, will consist of the. positive odd numbers.

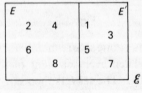

Fig. 1.6

This is illustrated in Fig. 1.6. Note that $A \cup A' = \mathscr{E}$ and $A \cap A' = \varnothing$.

Exercises 1.2

1   Give three examples of pairs of sets which are
    (a) disjoint   (b) not disjoint.
2   Give three examples of sets which are the same as the empty set.
3   Find $A \cap B$ and $A \cup B$ when
    (a) $A = \{1, 2, 3, 4, 5\}$, $B = \{4, 6, 8\}$
    (b) $A = \{$positive even numbers less than 12$\}$, $B = \{$factors of 12$\}$

(c) $A = \{x \in \mathbb{N}: x > 3\}$, $B = \{x \in \mathbb{N}: x < 7\}$
(d) $A = \{x \in \mathbb{Z}: x > 0\}$, $B = \{x \in \mathbb{Z}: x > 4\}$.

4 Illustrate in Venn diagrams the sets A, B, $A \cap B$ and $A \cup B$ of question 3(a), (b), (c) and (d).

5 If $A = \{1, 2, 3\}$, how many subsets does A have? List the subsets.

6 If $A = \{2, 4, 6\}$, $B = \{\text{factors of } 12\}$, $C = \{\text{odd numbers between } 0 \text{ and } 8\}$, state which of the following statements is true:
$A \subset B$, $B \subset A$, $A \subset C$, $C \subset B$, $B \subset C$, $C \subset A$, $A \cap C = \varnothing$,
$A \cup C = \{\text{natural numbers less than } 8\}$, $B \cap C = \varnothing$.

7 If $\mathscr{E} = \{x \in \mathbb{N}: 1 \leqslant x \leqslant 12\}$, $A = \{2, 4, 6, 8, 10\}$, $B = \{3, 6, 9, 12\}$, list the elements of the sets
(a) A' (c) $A \cap B$ (e) $A \cup B$
(b) B' (d) $(A \cap B)'$ (f) $(A \cup B)'$.

1.3 Ordered pairs

The order in which the elements of a set are listed does not change the set, so that $\{a, b, c\} = \{c, a, b\}$. However, there are situations where the order in which numbers appear is important. For instance, the height and weight of a person may be 175 cm and 70 kg respectively, and this could be represented by (175, 70). This is an example of an 'ordered pair'. It is clearly ordered since (175,70) is different from (70, 175). The position of a point P in the plane of coordinate axes Ox, Oy can be given by the ordered pair (x, y), where x and y are the distances OQ and QP shown in Fig. 1.7. The ordered pair (y, x) gives the position of the point P' so that the ordered pairs (x, y) and (y, x) give different points.

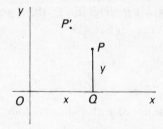

Fig. 1.7

A set of ordered pairs may be formed from two sets A and B by forming all the ordered pairs (a, b) such that $a \in A$ and $b \in B$. This set is called the product set (sometimes the cartesian product) of A and B and is denoted by $A \times B$, so that $A \times B = \{(a, b): a \in A, b \in B\}$. If $A = \{1, 2, 3\}$ and $B = \{2, 4\}$, then

$$A \times B = \{(1, 2), (1, 4), (2, 2), (2, 4), (3, 2), (3, 4)\}.$$

The product set $A \times A$ is sometimes denoted by A^2. An example of such a product set is $\mathbb{R} \times \mathbb{R}$, where $\mathbb{R}$ is the set of real numbers, i.e. $\mathbb{R} \times \mathbb{R} = \{(x, y): x \in \mathbb{R}, y \in \mathbb{R}\}$. Each element of this product set can be represented by one point in the x–y

plane, i.e. the plane of the coordinate axes Ox, Oy. The product set $\mathbb{R} \times \mathbb{R}$ is then represented by the x–y plane.

Exercises 1.3

1 Represent the following ordered pairs in a diagram of the x–y plane:
$(0, 3)$, $(1, 2)$, $(-2, 1)$, $(4, -3)$, $(-5, -2)$.
2 Write down the ordered pairs represented by the points A, B, C and D in the following diagram.

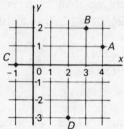

3 If $A = \{1, 2, 3\}$, $B = \{4, 5\}$, $C = \{1, 3\}$, list the elements of the sets
(a) $A \times B$ (d) $C \times A$ (g) $A \times A$
(b) $B \times A$ (e) $B \times C$ (h) $B \times B$
(c) $A \times C$ (f) $C \times B$ (i) $C \times C$.
4 If there are m elements in set A and n elements in set B, find the number of elements in the sets
(a) $A \times B$ (b) $B \times A$ (c) $A \times A$ (d) $B \times B$.
5 If $A = \{x \in \mathbb{R} : 0 \leqslant x \leqslant 5\}$, $B = \{x \in \mathbb{R} : x \geqslant 3\}$ and $C = \{x \in \mathbb{R} : -2 \leqslant x \leqslant 1\}$, illustrate the following sets in diagrams of the x–y plane:
(a) $A \times B$ (d) $C \times A$ (g) $A \times A$
(b) $B \times A$ (e) $B \times C$ (h) $B \times B$
(c) $A \times C$ (f) $C \times B$ (i) $C \times C$.

1.4 Functions

Elements of one set may be linked with elements of another set. For example, if

$$F = \{\text{league club footballers}\}$$
and $$C = \{\text{football league clubs}\},$$

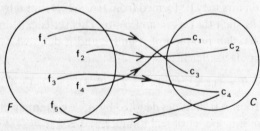

Fig. 1.8

each member of the set F can be linked with one member of the set C, this member being the club for which the footballer plays. This situation is illustrated in Fig. 1.8.

Figure 1.8 is an example of a mapping diagram. The elements of set F have been mapped to the elements of set C. Each element of the set F has been associated with just one element of the set C. The way in which the elements of set F have been linked with the elements of set C is that the elements of set F 'play for' the elements of set C, i.e. 'plays for' is the relation which associates each element of set F with just one element of set C.

A way or rule or method of associating each element of one set with a unique element of another set is called a *function*. A function maps the elements of set A to the elements of set B. The set A is called the *domain* of the function and the set B is called the *codomain* of the function. Note that not every element of the codomain need have a corresponding element in the domain but every element of the domain must correspond to some element in the codomain.

Let $A = \{1, 2, 3, 4\}$ and $B = \{1, 2, 3, 4, 5, 6, 7, 8\}$. Let us map each element, x, of the set A to its double, $2x$, in the set B and denote this function by the letter d.

We write d$: A \rightarrow B$ to indicate that d is a function which maps the elements of the set A to the elements of the set B. This is illustrated in the mapping diagram in Fig. 1.9. We say that 2 is the image of 1 under the function d since the function d maps 1 to 2 and we write this as d$(1) = 2$. Similarly d$(2) = 4$, d$(3) = 6$, d$(4) = 8$.

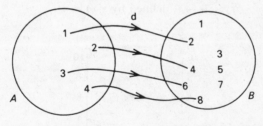

Fig. 1.9

Only the elements 2, 4, 6, 8 in the set B are images of elements of the set A. The set $\{2, 4, 6, 8\}$ is a subset of the set B. This subset is known as the image of the set A under the function d and is called the range (or range set) of the function.

In this example $A = \{1, 2, 3, 4\}$ is the domain of the function d, $B = \{1, 2, 3, 4, 5, 6, 7, 8\}$ is the codomain of the function d and the set $\{2, 4, 6, 8\}$ is the range of the function d.

A function relates each element of the domain to one and only one element of the codomain.

It is possible of course to relate the elements of one set to the elements of another set without the relation being a function. As an example of such a relation we take the domain to be the set $A = \{1, 4, 9\}$ and the codomain to be $\mathbb{Z}$, the set of integers, and relate the elements of A to their square roots in $\mathbb{Z}$. Then 1 is related to both $+1$ and -1, 4 is related to both $+2$ and -2 and 9 is related to both $+3$ and -3. This relation is not a function. It is just a relation.

To define a function we need to state the domain and codomain and to state the relation which links each element of the domain with one and only one element of the codomain. This relation is sometimes referred to as the *rule* of the function. The notation used is illustrated in the following examples.

The function s maps elements of the domain $\mathbb{R}$, the set of real numbers, to their squares in the codomain, also $\mathbb{R}$. This is written as

$$s:\mathbb{R} \to \mathbb{R} \text{ defined by } s(x) = x^2$$

or $\qquad\qquad s:\mathbb{R} \to \mathbb{R} \text{ defined by } s:x \mapsto x^2.$

Consider another function r which maps elements of the domain $\mathbb{Z}^+$, the set of positive integers, to their reciprocals in the codomain $\mathbb{Q}$, the set of rational numbers. The notation to express this function would be

$$r:\mathbb{Z}^+ \to \mathbb{Q} \text{ defined by } r(x) = 1/x,$$

or $\qquad\qquad r:x \mapsto 1/x, \qquad \text{where } x \in \mathbb{Z}^+$

Functions may be represented diagrammatically in a variety of ways. The mapping diagram in Fig. 1.9 is one way. Another way of representing a function which maps one set of numbers to another set of numbers is shown in Fig. 1.10. The function illustrated is

$$s:\mathbb{R} \to \mathbb{R} \text{ defined by } s(x) = x^2.$$

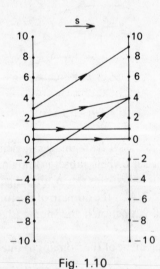

Fig. 1.10

The domain is represented by the left-hand number line and the codomain is represented by the right-hand number line.

Functions have traditionally been represented in a graphical way with which the reader will be familiar. To illustrate this method we consider the function

$$f: \mathbb{R} \to \mathbb{R} \text{ defined by } f(x) = x + 1.$$

Each element x of the domain and its image $f(x)$ form an ordered pair $(x, f(x))$. This ordered pair can be represented by the coordinates (x, y) of one point in the plane of coordinate axes Ox, Oy. Thus the ordered pairs

$$\ldots, \ (-1, f(-1)), \ (0, f(0)), \ (2.1, f(2.1)), \ (\sqrt{5}, f(\sqrt{5})), \ \ldots$$

are represented by the points with coordinates

$$\ldots, \ (-1, 0), \ (0, 1), \ (2.1, 3.1), \ (\sqrt{5}, 1 + \sqrt{5}), \ \ldots \text{ respectively.}$$

Thus the infinite set of ordered pairs $(x, f(x))$, where $x \in \mathbb{R}$, is represented by the infinite set of points whose y-coordinates exceed their x-coordinates by 1, i.e. points on the line $y = x + 1$. This set of points is shown in Fig. 1.11 which is another way of illustrating the function f diagrammatically.

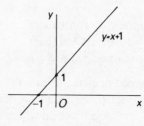

Fig. 1.11

The function f is thus illustrated in a graph by representing each element x of the domain and its image $f(x)$ in the codomain by the point with coordinates (x, y), where $y = f(x)$. The resulting graph is often called the graph of $y = f(x)$.

Just as numbers can be sorted into sets so can functions. For instance, there are odd and even functions.

A function f is said to be an even function if $f(-x) = f(x)$ for all values of x. A function f is said to be an odd function if $f(-x) = -f(x)$ for all values of x.

The function $s: \mathbb{R} \to \mathbb{R}$ defined by $s(x) = x^2$ is an even function since $(-x)^2 = x^2$ and so $s(-x) = x^2 = s(x)$.

The function $d: \mathbb{R} \to \mathbb{R}$ defined by $d(x) = 2x$ is an odd function since $2(-x) = -2x$ and so $d(-x) = -2x = -d(x)$.

Exercises 1.4

1   Illustrate in three different diagrams each of the following functions:

$f: \mathbb{R} \to \mathbb{R}$ is defined by (a) $f(x) = 2x$ (b) $f(x) = x^2$
(c) $f(x) = 3x + 4$ (d) $f(x) = x^2 - 3x + 2$.

2 By choosing your own domains, codomains and rules, give two examples of relations which (a) define functions, (b) do not define functions.

3 Determine whether the function f is odd, even or neither when

$f: \mathbb{R} \rightarrow \mathbb{R}$ is defined by (a) $f(x) = 3x$
(b) $f(x) = 2x^2$
(c) $f(x) = 2x + 1$
(d) $f(x) = (x + 1)^2$
(e) $f(x) = x^2 + 1$
(f) $f(x) = (x - 1)(x + 1)$
(g) $f(x) = (x - 1)/(x + 1)$.

1.5 Composition of functions

Functions may be combined.

Consider the sets

$$F = \{\text{league club footballers}\},$$
$$C = \{\text{football league clubs}\},$$
$$D = \{\text{football league divisions 1, 2, 3, 4}\}.$$

Elements of set F can be mapped to elements of set C and then elements of set C can be mapped to elements of set D. This is illustrated in the mapping diagram shown in Fig 1.12. Here we have one function which maps set F to set C, and a second function which maps set C to set D, together giving a third function which maps set F to set D. This third function is called the composite function of the other two functions.

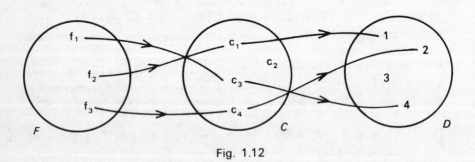

Fig. 1.12

Consider the functions $f: \mathbb{R} \rightarrow \mathbb{R}$ defined by $f(x) = x^2$ and $g: \mathbb{R} \rightarrow \mathbb{R}$ defined by $g(x) = x + 1$. Let us denote by F the composite function of the function f, which maps x to x^2, and the function g, which then maps x to $x + 1$.

f maps 1 to 1 and then g maps 1 to 2 and so F maps 1 to 2,

i.e.   f(1) = 1   and g(1) = 2   giving F(1) = 2
and   f(2) = 4   and g(4) = 5   giving F(2) = 5
       f(3) = 9   and g(9) = 10   giving F(3) = 10

and in general

$$f(x) = x^2 \quad \text{and } g(x^2) = x^2 + 1 \quad \text{giving } F(x) = x^2 + 1,$$

i.e. $$F(x) = g[f(x)].$$

The function F is said to be the composite function gf, sometimes written g o f.
    If we denote the composite function fg by G, then

$$G(x) = f[g(x)] = f(x+1) = (x+1)^2.$$

It will be seen that $G(x) \neq F(x)$ and so the composite functions fg and gf are not the same.

    It is not always possible to form a composite function. To form a composite function fg, the image set of the function g must be the domain of the function f or a subset of this domain.

Exercises 1.5

1   For the functions f and g, where

$$f : \mathbb{R} \to \mathbb{R} \text{ is defined by } f(x) = 2x$$
and $$g : \mathbb{R} \to \mathbb{R} \text{ is defined by } g(x) = x - 3,$$

find fg(1), fg(2), gf(−1), gf(−3), fg(x), gf(x).

2   For the functions p and q, where

$$p : \mathbb{R} \to \mathbb{R} \text{ is defined by } p(x) = 2x + 1$$
and $$q : \mathbb{R} \to \mathbb{R} \text{ is defined by } q(x) = (x+1)^2,$$

find pq(−1), pq(0), qp(−2), qp(4), pq(x), qp(x).

3   If $f : \mathbb{R} \to \mathbb{R}$ is defined by $f(x) = x + 1$
    and $g : \mathbb{R} \to \mathbb{R}$ is defined by $g(x) = x^2$,

determine whether each of the composite functions gf, fg, ff, gg is odd, even or neither.

4   If f is an odd function and g is an even function, determine whether each of the composite functions fg, gf, ff, gg, fgf and gfg is odd, even or neither.

1.6 Inverse functions

For a given function f, can a function g be found such that $gf(x) = x$? If one can be found and if also $fg(x) = x$, then g is denoted by f^{-1}. The function f^{-1} is said to be the inverse function of f.

    Consider the function $f : \mathbb{R} \to \mathbb{R}$ defined by $f(x) = 2x$.

If $g(x) = \frac{1}{2}x$, then $g[f(x)] = g(2x) = x$,
and $f[g(x)] = f(\frac{1}{2}x) = x$.

Hence $\qquad g[f(x)] = f[g(x)] = x,$

i.e. $\qquad gf(x) \quad = fg(x) = x$

and so g is f^{-1}, the inverse function of f. For this particular function f, the inverse function f^{-1} exists. These functions f and f^{-1} are illustrated in Fig. 1.13. It can be seen that each element of the domain maps to one and only one element of the codomain, and no two elements of the domain map to the same element in the codomain. (Such a mapping is said to be one–one.) Also to every element of the codomain there is an element of the domain which maps to it. This last property means that the range equals the codomain.

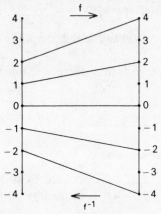

Fig. 1.13

Next consider the function $g : \mathbb{R} \to \mathbb{R}$ defined by $g(x) = x^2$. This function is illustrated in the mapping diagram shown in Fig. 1.14. The function g maps both $+1$ and -1 to 1, i.e. $g(1) = 1$ and $g(-1) = 1$. Mapping from the codomain to the domain, 1 would be linked to both $+1$ and -1 so that this relation linking the

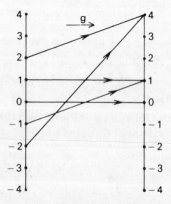

Fig. 1.14

codomain with the domain would not be a function. Consequently, for this function g an inverse function g^{-1} does not exist.

In general, an inverse function f^{-1} will exist for a function f if

(i) the range of f = the codomain of f

(ii) $x_1 = x_2$ implies that $f(x_1) = f(x_2)$

(iii) $f(x_1) = f(x_2)$ implies that $x_1 = x_2$ } for all x_1, x_2 in the domain of f

The second condition can be written as $(x_1 = x_2) \Rightarrow (f(x_1) = f(x_2))$, where the symbol $\Rightarrow$ means 'implies that'. Also the third condition can be written as $(f(x_1) = f(x_2)) \Rightarrow (x_1 = x_2)$. We can combine these two statements into one

$$(x_1 = x_2) \Leftrightarrow (f(x_1) = f(x_2)),$$

where the symbol $\Leftrightarrow$ means 'implies that and is implied by'.

Sometimes it is possible to obtain a function that possesses an inverse by reducing the domain of a given function. It has just been seen that the function $g: \mathbb{R} \to \mathbb{R}$ defined by $g(x) = x^2$ does not have an inverse function. However consider the function $f: \mathbb{R}^+ \to \mathbb{R}^+$ defined by $f(x) = x^2$, where $\mathbb{R}^+$ is the set of positive real numbers. It can easily be seen that, with this domain, the conditions for the existence of an inverse function are satisfied and so an inverse function f^{-1} does exist.

Exercises 1.6

1   Prove that an inverse function f^{-1} exists and find $f^{-1}(x)$ when
$f: \mathbb{R} \to \mathbb{R}$ is defined by (a) $f(x) = 3x$
(b) $f(x) = 2x + 1$
(c) $f(x) = 1 - 4x$.

2   Suggest a suitable domain for the relation r to define a function when the relation r associates x with

(a) $\dfrac{1}{x}$ (b) $\dfrac{1}{x-1}$ (c) $\dfrac{1}{x^2+1}$ (d) $\dfrac{1}{x^2-1}$.

3   Suggest suitable domains and codomains for the following relations to define functions with inverse functions:

(a) $r: x \mapsto 4x^2$ (c) $r: x \mapsto \dfrac{1}{1+x}$

(b) $r: x \mapsto \dfrac{1}{x^2}$ (d) $r: x \mapsto x^2 - 9$.

In each case find $r^{-1}(x)$.

1.7   Operations

One element of a set may be combined with another element of the set to produce a third element. For instance, one element of $\mathbb{N}$, the set $\{0, 1, 2, \ldots\}$, may be added to another element of that set to produce a third number, e.g. $3 + 4 = 7$.

Here the mathematical operation of addition has been performed on the numbers 3 and 4 to produce the third number 7. Addition is said to be a binary operation on the set because it combines *two* elements of the set.

Addition is but one of many binary operations which may be performed on a set of numbers. Other well-known binary operations are subtraction, multiplication and division. It is easy to devise many other binary operations. For instance, we might define a binary operation o on $\mathbb{Q}$, the set of rational numbers, by $a \text{ o } b = ab + a$ so that $\frac{1}{2} \text{ o } \frac{1}{3} = \frac{2}{3}$ and $\frac{1}{3} \text{ o } \frac{1}{2} = \frac{1}{2}$.

When an operation o is defined on a set S a structure is produced which will have certain properties which can be investigated.

Consider the structure consisting of the operation o defined on the set S. If $a \in S$ and $b \in S$ and $a \text{ o } b = c$, then c may be, but need not be, a member of the set S.

If o is the operation of addition and if $S = \mathbb{N}$, the set of natural numbers, then, for all elements a, b of the set $\mathbb{N}$, c will also be a member of the set $\mathbb{N}$.

However if o is the operation of subtraction and if $S = \mathbb{N}$, then $a \text{ o } b = c$ will not always be a member of the set $\mathbb{N}$, e.g. if $a = 3$ and $b = 7$, then $a \text{ o } b = 3 - 7 \equiv -4$ and $-4 \notin \mathbb{N}$.

Furthermore c may not always be different from both a and b. For the operation of addition on the set $\mathbb{Z} = \{0, \pm 1, \pm 2, \dots\}$, let $a = 3$ and $b = 0$, then $a \text{ o } b = 3 + 0 = 3$, i.e. $c = a$.

Also $a \text{ o } b$ may, but need not, produce the same element as $b \text{ o } a$. With the operation o as addition on the set $\mathbb{Z}$, $a \text{ o } b = b \text{ o } a$, e.g. $3 + 4 = 7$ and $4 + 3 = 7$. However with the operation o as subtraction on the set $\mathbb{Z}$, then $a \text{ o } b \neq b \text{ o } a$, e.g. $3 - 4 = -1$ but $4 - 3 = 1$.

These ideas can be expressed in general terms.

Binary operation: A binary operation defined on a set S is a rule which assigns to each ordered pair (a, b), where a, $b \in S$, a unique element c.

Closure: A set S is said to be closed under an operation o if, for every ordered pair (a, b), where a, $b \in S$, the element $a \text{ o } b \in S$. The operation o is then said to be a closed binary operation on S.

Commutativity: An operation o defined on a set S is said to be commutative if, for every ordered pair (a, b), where a, $b \in S$, $a \text{ o } b = b \text{ o } a$.

In arithmetic we soon discover that $(3 + 4) + 5 = 3 + (4 + 5)$, i.e. in evaluating $3 + 4 + 5$ we may either add the sum of 3 and 4 to 5 or add 3 to the sum of 4 and 5. This property is called associativity.

Associativity: A closed operation o defined on a set S is said to be associative if, for every a, b, $c \in S$, $(a \text{ o } b) \text{ o } c = a \text{ o } (b \text{ o } c)$.

Identity element: For an operation o defined on a set S an identity element e is said to exist if, for every element $a \in S$, $a \text{ o } e = e \text{ o } a = a$, where $e \in S$.

For the operation of addition defined on $\mathbb{R}$, the set of real numbers, the identity element is 0 since $a + 0 = 0 + a = a$.

Similarly, for the operation of multiplication on the set $\mathbb{R}$, the identity element is 1 since $a \times 1 = 1 \times a = a$.

With the operation of multiplication on $\mathbb{R}$ we know that, if $a \neq 0$, $a \times \frac{1}{a} = \frac{1}{a} \times a = 1$ and we say that $\frac{1}{a}$ is the multiplicative inverse of a.

Inverse elements: For an operation o defined on a set S which has an identity element e, an element $a \in S$ has an inverse element a^{-1} if $a \circ a^{-1} = a^{-1} \circ a = e$ and $a^{-1} \in S$.

For the operation of addition defined on $\mathbb{Z}$, the set of integers, $-a$ is the inverse of a since $a + (-a) = (-a) + a = 0$, the identity element.

On the other hand, for the operation of addition defined on $\mathbb{N}$, the set of natural numbers, a has no inverse since $-a \notin \mathbb{N}$.

Exercises 1.7

1   Consider the operation $\cap$ (intersection) defined on the set S of sets $\varnothing$, $\mathscr{E}$, A, B, C, ...

(a) Find the identity element of the set S under the operation $\cap$.

(b) Find the element of the set S which is the inverse of the set A under the operation $\cap$.

(c) Using Venn diagrams, or otherwise, show that the operation $\cap$ on S is (i) commutative, (ii) associative.

2   Consider the operation $\cup$ (union) on the set S of sets $\varnothing$, $\mathscr{E}$, A, B, C, ...

(a) Find the identity element of the set S under the operation $\cup$.

(b) Find the element of the set S which is the inverse of the set A under the operation $\cup$.

(c) Using Venn diagrams, or otherwise, show that the operation $\cup$ on S is (i) commutative, (ii) associative.

3   The operation o is defined on $\mathbb{R}$, the set of real numbers, by

$$a \circ b = \frac{a}{a+b}, \quad \text{where } a + b \neq 0.$$

(a) Is the set $\mathbb{R}$ closed under the operation o?

(b) Is the operation o (i) associative, (ii) commutative?

(c) For the operation o on $\mathbb{R}$, is there an identity element and if there is, give it.

(d) Does each element of $\mathbb{R}$ have an inverse element under the operation o?

4   The operation $*$ is defined on $\mathbb{R}$, the set of real numbers, by

$$a * b = a + ab.$$

(a) Is the set $\mathbb{R}$ closed under the operation $*$?

(b) Is the operation $*$ (i) associative, (ii) commutative?

(c) For the operation $*$ on $\mathbb{R}$, is there identity element? If there is, give it.

(d) Does each element of $\mathbb{R}$ have an inverse element under the operation $*$?

5   The operation o is defined on the set $S = \{0, 1, 2, 3, 4\}$ by

$$a \circ b = |a - b| \quad (|a - b| \text{ denotes the magnitude of } a - b).$$

(a) Is the set S closed under the operation o?

(b) Is the operation o (i) associative, (ii) commutative?

(c) For the operation o on S, is there an identity element? If there is, give it.

(d) Does each element of S have an inverse element under the operation o?

Miscellaneous exercises 1

1 Define (a) an odd function, (b) an even function. For each of the following two functions, decide whether it is odd or even (or neither) and justify your answers

$$f(x) = \begin{cases} x^2 \text{ for } x > 0, \\ -x^2 \text{ for } x \leqslant 0. \end{cases} \qquad g(x) = \begin{cases} \sqrt[3]{x} \text{ for } x > 0, \\ -x^3 \text{ for } x \leqslant 0. \end{cases}$$

Sketch the graphs of these two functions and explain how you recognise, from the graphs, that $f(x)$ possesses an inverse function whereas $g(x)$ does not. Write down a formula (or formulae) for the inverse function of $f(x)$ and sketch the graph of this inverse function. [L]

2 If $f: x \mapsto x/(1+x)$ is defined on the domain of positive real numbers, find the range of f.

Find the inverse relation. State the condition for an inverse relation to be a function and determine whether the condition is satisfied in this case. [C]

3 The function f is defined by $f: x \mapsto x - (1/x)$ over the domain of all non-zero real numbers. Sketch the graph of f and state the range of f.

Explain carefully how you can tell from the graph that, if the domain is restricted to either positive real numbers or negative real numbers, then f is one-one, whereas if the domain of f is not so restricted, then f is not one-one. Find an explicit formula for the inverse function f^{-1} in the case where the domain of f is the set of positive real numbers.

State the range of f when the domain is the set

$$\{x : 0 < x^2 < 1\}.$$

Determine whether f is one-one in this case. [C]

4 The domain of the function f is the set

$$D = \{x : x \in \mathbb{R}, \ -2 < x < 3\}.$$

The function $f: D \to \mathbb{R}$ is defined by

$$f(x) = \begin{cases} 2x - 1 & \text{for} \quad -2 < x \leqslant 1, \\ x^2 & \text{for} \quad 1 < x \leqslant 2, \\ 10 - 3x & \text{for} \quad 2 < x < 3. \end{cases}$$

Find the range of this function and sketch a graph of this function. Explain why there is no inverse function to f. Suggest an interval such that f, restricted to this interval, will have an inverse function. Give an expression for the inverse function in this case. [L]

5   The domain of f is the set $\{x: -3 < x < 3$, where x is a real number$\}$. Sketch the graph of the function f defined by

$$f:x \mapsto -5-x \quad (-3 < x \leqslant -1),$$
$$f:x \mapsto 2x^3 \quad (-1 < x \leqslant 1),$$
$$f:x \mapsto 5-x \quad (1 < x < 3).$$

State the range of $f(x)$.

Define the inverse relation in similar form and determine whether this relation is a function. [C]

6   $\mathbb{R}$ is the set of all real numbers. A function f from $\mathbb{R}$ into $\mathbb{R}$ is given by

$$f:x \mapsto \frac{x}{x^2+1}.$$

Find the image of $1/a$ under f, where $a \in \mathbb{R}$ and $a \neq 0$. Deduce that f is not one-one.

Show that if a and b are real numbers with $a > b \geqslant 1$, then

$$f(b) - f(a) = \frac{(a-b)(ab-1)}{(1+a^2)(1+b^2)}$$

and deduce that $f(b) > f(a)$. Deduce that if the domain of f is restricted to the subset of $\mathbb{R}$ given by $\{x:x \in \mathbb{R}, x \geqslant 1\}$, then f is one-one, and state in this case the range of f. Sketch the graph of the function f over the domain $\mathbb{R}$. [C]

7   The domain of a mapping m is the set of all real numbers except 1. The mapping is given by

$$m:x \mapsto \left| \frac{x}{x-1} \right|.$$

(i) State with a reason whether or not m is a function.

(ii) State the range of m.

(iii) Show that

$$(x > 1) \Rightarrow (m(x) > 1) \text{ and that } (x < 0) \Rightarrow (m(x) < 1).$$

(iv) Sketch the graph of $y = m(x)$. [C]

8   (i) If $u(x)$ is an odd function and $v(x)$ is an even function, determine whether $u(u(x))$, $u(v(x))$, $v(u(x))$ are necessarily odd, necessarily even, or neither.

(ii) Sketch the graph of the function $x|x|$ and state the feature of this graph that enables one to say that the function has an inverse function. Sketch the graph of this inverse function. [L]

9   (i) If $f(x)$ is a non-vanishing even function and $g(x)$ is an odd function, determine whether $g(x)/f(x)$, $(g(x))^3$, $g(x^2)$ are necessarily odd, necessarily even, or neither.

(ii) Sketch the graph of the function

$$f(x) = \frac{x}{x^2+1}$$

(continued)

and explain why this function does not possess an inverse function.

Is it possible, by changing the values of f(x) for negative values of x, to produce a new function which does possess an inverse? If so, do so; if not, explain why it cannot be done. [L]

10 The operation o is defined on $\mathbb{R}^+$, the set of positive real numbers, by

$$a \circ b = \frac{a^2 + b^2}{a + b}.$$

(a) Is the set $\mathbb{R}^+$ closed under the operation o?

(b) Is the operation o (i) associative, (ii) commutative?

(c) For the operation o on $\mathbb{R}^+$, is there an identity element? If there is, give it.

(d) Does each element of $\mathbb{R}^+$ have an inverse element under the operation o? Give reasons for your answers.

11 The operation * is defined on $\mathbb{R}$, the set of real numbers by

$$x * y = 3xy, \quad (x \neq 0, \ y \neq 0).$$

(a) Does the set $\mathbb{R}$ under the operation * have the properties of (i) closure, (ii) associativity, (iii) commutativity?

(b) For the operation * on $\mathbb{R}$ is there an identity element? If there is, give it.

(c) Find the inverse of x under the operation *.

2 Algebra 1

2.1 The graph of $y = ax^2 + bx + c$

The reader will have already drawn graphs of algebraic functions by tabulating values and plotting points and will know that the points whose coordinates satisfy the equation $y = ax + b$ lie on a straight line. This line crosses the y-axis at the point where $y = b$ and makes an angle with the x-axis whose tangent equals a so that a is the gradient of the line.

What form is taken by the points in the x–y plane whose coordinates satisfy the equation $y = ax^2 + bx + c$?

Putting $x = 0$ gives $y = c$ showing that the curve $y = ax^2 + bx + c$ crosses the y-axis where $y = c$. Thus

if $c = 0$,   the curve goes through the origin,
if $c > 0$,   the curve cuts the y-axis above the origin,
if $c < 0$,   the curve cuts the y-axis below the origin.

From drawing graphs of particular quadratic functions the reader may have noticed that, when a is positive, the curve is $\cup$ shaped and, when a is negative, the curve is $\cap$ shaped.

When $a > 0$: ax^2 is positive for all values of x, and for numerically large positive or negative values of x the value of ax^2 will be large and positive and so will the value of y.

When $a < 0$: similarly in this case, when x is numerically large, y will be negative and numerically large.

When $a = 0$: then $y = bx + c$ so that in this special case the graph is a straight line.

Provided that a is not zero,

$$y = ax^2 + bx + c = a\left[x^2 + \frac{b}{a}x + \frac{c}{a} \right]$$

$$= a\left[\left(x + \frac{b}{2a} \right)^2 + \frac{c}{a} - \frac{b^2}{4a^2} \right]$$

$$= \left(\frac{4ac - b^2}{4a} \right) + a\left(x + \frac{b}{2a} \right)^2.$$

If $a > 0$, y will have a least value of $\dfrac{4ac - b^2}{4a}$ when $\left(x + \dfrac{b}{2a} \right)^2 = 0.$

If $a < 0$, (say $a = -p$, where p is positive),

$$y = \frac{4ac - b^2}{4a} - p\left(x + \frac{b}{2a}\right)^2,$$

and clearly y will have a greatest value of $\dfrac{4ac - b^2}{4a}$ when $\left(x + \dfrac{b}{2a}\right)^2 = 0$.

Thus y will have either a greatest or a least value when $x = -\dfrac{b}{2a}$.

Whether a is positive or negative the curve whose equation is $y = ax^2 + bx + c$

is a parabola and the point $\left(-\dfrac{b}{2a}, \dfrac{4ac - b^2}{4a}\right)$ is called the vertex of the parabola.

When $x = -\dfrac{b}{2a} + k$, $\quad y = \left(\dfrac{4ac - b^2}{4a}\right) + a\left[-\dfrac{b}{2a} + k + \dfrac{b}{2a}\right]^2$

$$= \left(\frac{4ac - b^2}{4a}\right) + ak^2.$$

When $x = -\dfrac{b}{2a} - k$, $\quad y = \left(\dfrac{4ac - b^2}{4a}\right) + a\left[-\dfrac{b}{2a} - k + \dfrac{b}{2a}\right]^2$

$$= \left(\frac{4ac - b^2}{4a}\right) + ak^2.$$

Thus y has the same value for $x = -\dfrac{b}{2a} + k$ as for $x = -\dfrac{b}{2a} - k$ and so the curve

is symmetrical about the line $x = -\dfrac{b}{2a}$. This line is known as the axis of the

parabola.

Example

Find the greatest or least value of y when
(a) $y = 3x^2 - 2x + 1$ (b) $y = -x^2 + 2x - 3$.

(a) $y = 3x^2 - 2x + 1 = 3\left[x^2 - \dfrac{2}{3}x + \dfrac{1}{3}\right]$

$$= 3\left[\left(x - \frac{1}{3}\right)^2 + \frac{1}{3} - \frac{1}{9}\right]$$

$$= 3\left(x - \frac{1}{3}\right)^2 + \frac{2}{3}.$$

$\therefore y$ has a least value of $\dfrac{2}{3}$ when $x = \dfrac{1}{3}$.

(b) $y = -x^2 + 2x - 3 = -(x^2 - 2x + 3)$
$$= -[(x-1)^2 + 3 - 1]$$
$$= -(x-1)^2 - 2.$$
$\therefore$ y has a greatest value of -2 when $x = 1$.

Exercises 2.1(i)

1 State the coordinates of the points of intersection with the y-axis of the following graphs:

(a) $y = 3x + 4$ (e) $2x - 3y - 4 = 0$ (i) $y = 7 - x - x^2$
(b) $y = 5 - 2x$ (f) $y = x^2 - x + 3$ (j) $y = x^3 - x^2 + x - 1$
(c) $y + 2x = 0$ (g) $y = 2x^2 - 4x - 7$ (k) $y = x^4 + x^2 + 3$.
(d) $x + 2y + 3 = 0$ (h) $y = 3x^2 - 2x$

2 Find the least value of y when

(a) $y = x^2 - 2x + 3$ (e) $y = x^2 - 4x + 4$ (h) $y = 3x^2 - 6x - 7$
(b) $y = x^2 + 6x + 12$ (f) $y = x^2 + 8x + 16$ (i) $y = 4x^2 - 6x + 1$
(c) $y = x^2 - 4x - 3$ (g) $y = 2x^2 - 4x + 1$ (j) $y = 10x^2 + 4x - 3$.
(d) $y = x^2 + x - 4$

In each case state the value of x when y has its least value.

3 Find the greatest value of y when

(a) $y = 5 + 2x - x^2$ (e) $y = -x^2 - 6x - 9$ (h) $y = -2x^2 + 12x - 7$
(b) $y = 8 - 4x - x^2$ (f) $y = 10x - x^2 - 25$ (i) $y = 6 - x - 2x^2$
(c) $y = -4 + 6x - x^2$ (g) $y = -3x^2 - 12x + 8$ (j) $y = 4x - 5 - 3x^2$.
(d) $y = -1 + x - x^2$

In each case state the value of x when y has its greatest value.

4 State whether y has a greatest or least value and find this greatest or least value when

(a) $y = x^2 - 6x + 7$ (d) $y = 9 - 4x + x^2$ (g) $y = 6 + x - 2x^2$
(b) $y = 5 - 2x - x^2$ (e) $y = 11 - 6x - x^2$ (h) $y = 3x - 2 - 4x^2$
(c) $y = x^2 + 4x + 4$ (f) $y = 2x^2 - 8x - 3$ (i) $y = -4 - 12x - 9x^2$.

It has been shown that the value of c gives the value of y at the point where the curve $y = ax^2 + bx + c$ crosses the y-axis. How can we find where the curve crosses the x-axis? Indeed does the curve cross the x-axis at all?

The curve crosses the x-axis where $y = 0$, i.e. where $ax^2 + bx + c = 0$. Consider the case where $ax^2 + bx + c$ can be factorised by inspection and thus expressed in the form $a(x - \alpha)(x - \beta)$. Then $y = 0$ when $a(x - \alpha)(x - \beta) = 0$, i.e. the curve crosses the x-axis where $x = \alpha$ and $x = \beta$.

Now consider the case where values for α and β cannot readily be found. When $a > 0$, y has a least value $(4ac - b^2)/4a$. If this least value exceeds 0, then y will never equal or be less than 0, i.e. the curve will neither meet nor cross the x-axis. Thus there will be no point of intersection of the curve and the x-axis if $(4ac - b^2)/4a > 0$, i.e. if $b^2 - 4ac < 0$.

When $a < 0$, y has a greatest value $(4ac - b^2)/4a$. In this case there will be no point of intersection of the curve and the x-axis if this greatest value is less than 0, i.e. if

$b^2 - 4ac < 0$. Thus whatever the sign of a, the curve will not meet the x-axis if $b^2 - 4ac < 0$.

Consider the case of $b^2 - 4ac = 0$.

When $a > 0$, y has a least value $(4ac - b^2)/4a = 0$, i.e. the lowest point on the curve is on the x-axis so that the curve touches the x-axis.

When $a < 0$, y has a greatest value of 0 and so the highest point on the curve is on the x-axis and again the curve touches the x-axis. Thus when $b^2 - 4ac = 0$ the curve $y = ax^2 + bx + c$ touches the x-axis and the equation $ax^2 + bx + c = 0$ has two equal roots, $x = -\dfrac{b}{2a}$.

Consider the case of $b^2 - 4ac > 0$.

When $a > 0$, the least value of y, $(4ac - b^2)/4a$, is negative so that the lowest point is below the x-axis and the curve crosses the x-axis twice.

When $a < 0$, the highest point is above the x-axis and again the curve crosses the x-axis twice.

Thus whatever the sign of a, the curve will cross the x-axis twice if $b^2 - 4ac > 0$, and the equation $ax^2 + bx + c = 0$ will have two distinct real roots.

The following table illustrates the link between the curve $y = ax^2 + bx + c$ and the roots of the equation $ax^2 + bx + c = 0$ given by the formula

$$x = \frac{-b \pm \sqrt{(b^2 - 4ac)}}{2a}.$$

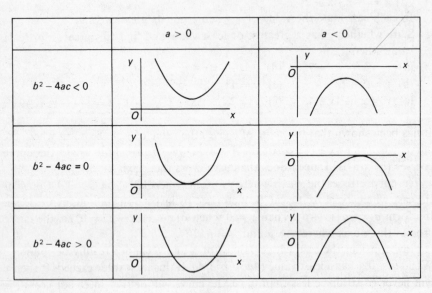

Fig. 2.1

Note that the greatest or least value of y occurs where $x = -\dfrac{b}{2a}$ and the curve cuts the x-axis where $x = -\dfrac{b}{2a} + \dfrac{\sqrt{(b^2-4ac)}}{2a}$ and $x = -\dfrac{b}{2a} - \dfrac{\sqrt{(b^2-4ac)}}{2a}$.

Thus if the curve crosses the x-axis at points P and Q, the greatest or least value of y occurs when x has the value of the x-coordinate of the mid-point of PQ.

Exercises 2.1(ii)

1 Which of the following curves do not meet the x-axis?
   (a) $y = 2x^2 + x + 5$ (c) $y = -2x^2 + 3x + 1$
   (b) $y = 3x^2 + 2x - 1$ (d) $y = -3 + 2x - x^2$
2 Which of the following curves touch the x-axis?
   (a) $y = x^2 - 4x + 4$ (f) $y = -4 - 4x - x^2$
   (b) $y = x^2 - x + 1$ (g) $y = 24x - 16 - 9x^2$
   (c) $y = x^2 + 6x + 9$ (h) $y = 9x^2 + 6x + 1$
   (d) $y = 5 - 4x - x^2$ (i) $y = 9 - 12x + 4x^2$
   (e) $y = 8 - 16x - x^2$ (j) $y = 9 - 4x^2$
3 Which of the following expressions are positive for all real values of x?
   (a) $x^2 - 2x + 5$ (d) $x^2 - 6x + 6$ (g) $3x^2 - 4x + 1$
   (b) $x^2 - 2x - 1$ (e) $2x^2 - 6x + 5$ (h) $4x^2 - 10x + 7$
   (c) $x^2 + 4x + 2$ (f) $5 - 3x - x^2$
4 Which of the following expressions are negative for all real values of x?
   (a) $1 - 4x - x^2$ (c) $-3 + 3x - x^2$ (e) $-7 - 4x - 3x^2$
   (b) $-1 - 3x - x^2$ (d) $-2 - 5x - 2x^2$ (f) $-2 - 7x - 4x^2$
5 Find the set of values of a for which $ax^2 + 3x + 4$ is positive for all real values of x.
6 Find the set of values of c for which $c + 2x - 3x^2$ is negative for all real values of x.
7 Show that $3x^2 + 4x + 2 > 0$ for all real values of x.
8 Show that $-2 + 3x - 2x^2 < 0$ for all real values of x.
9 Find the coordinates of the points of intersection with the coordinate axes of the following curves.
   (a) $y = 2x^2 - 8x + 6$ (c) $y = 3x^2 - 12x + 12$ (e) $y = -x^2 + x - 6$
   (b) $y = 4 + 3x - x^2$ (d) $y = 6x^2 + 7x - 5$
   Find also the greatest or least value of y in each case.
   Sketch the five curves.

The steps to be taken in drawing the curve $y = ax^2 + bx + c$ can now be listed.
(1) Plot the point of intersection of the curve and the y-axis at $(0, c)$.
(2) Note the sign of a: when $a > 0$ the curve has the form $\cup$, when $a < 0$ the curve has the form $\cap$.
(3) (i) If possible factorise $ax^2 + bx + c$ by inspection, expressing it in the form $a(x - \alpha)(x - \beta)$; then plot the points of intersection of the curve and the x-axis at $(\alpha, 0)$ and $(\beta, 0)$.

(ii) If this is not possible, find the value of $b^2 - 4ac$.

$b^2 - 4ac < 0$, the curve does not meet the x-axis,

$b^2 - 4ac = 0$, the curve touches the x-axis where $x = -\dfrac{b}{2a}$,

$b^2 - 4ac > 0$, the curve cuts the x-axis in two points; find these two points by solving the equation $ax^2 + bx + c = 0$.

(4) Find the highest or lowest point on the curve, i.e.

$$\left(-\frac{b}{2a}, \frac{4ac - b^2}{4a} \right).$$

(An alternative method of finding greatest and least values will be dealt with in chapter 9.)

Exercises 2.1(iii)
Sketch the following curves

(1) $y = x^2 + 2x + 3$ (6) $y = x^2 + 7x + 12$ (11) $y = 4x^2 - 12x + 9$
(2) $y = x^2 - 8x + 16$ (7) $y = -5 - 2x - x^2$ (12) $y = 6x^2 + x - 1$
(3) $y = x^2 - 5x + 6$ (8) $y = -4 - 4x - x^2$ (13) $y = 3x^2 + 4x + 5$
(4) $y = x^2 - 2x + 3$ (9) $y = 6 + x - x^2$ (14) $y = 24x - 9 - 16x^2$
(5) $y = x^2 + 6x + 9$ (10) $y = 2x^2 - 3x + 4$ (15) $y = 6 + 13x - 5x^2$.

2.2 Relations between the roots and the coefficients of equations

Let α and β be the roots of the quadratic equation $ax^2 + bx + c = 0$. Then the equation can be put in the form $a(x - \alpha)(x - \beta) = 0$,

i.e. $\qquad\qquad\qquad ax^2 - a(\alpha + \beta)x + a\alpha\beta = 0$.

Compare with the equation $ax^2 + bx + c = 0$.
Equating coefficients we get

$$-a(\alpha + \beta) = b \quad \text{giving} \quad \alpha + \beta = -\frac{b}{a}$$

and $\qquad\qquad a\alpha\beta = c \quad \text{giving} \quad \alpha\beta = \frac{c}{a}.$

Also if α and β are the roots of a quadratic equation, then

$$(x - \alpha)(x - \beta) = 0$$
$$x^2 - (\alpha + \beta)x + \alpha\beta = 0.$$

Thus a quadratic equation can be expressed as

$$x^2 - (\text{sum of the roots})x + (\text{product of the roots}) = 0.$$

Example 1
If α and β are the roots of the quadratic equation $ax^2 + bx + c = 0$,

find (a) $\alpha^2 + \beta^2$, (b) $\alpha^3 + \beta^3$, and form quadratic equations whose roots are
(i) α^2 and β^2, (ii) α^3 and β^3.

$$\alpha + \beta = -b/a \quad \text{and} \quad \alpha\beta = c/a$$

$$\alpha^2 + \beta^2 = (\alpha + \beta)^2 - 2\alpha\beta = \frac{b^2}{a^2} - \frac{2c}{a} = \frac{b^2 - 2ac}{a^2}.$$

$$\alpha^3 + \beta^3 = (\alpha + \beta)^3 - 3\alpha^2\beta - 3\alpha\beta^2$$
$$= (\alpha + \beta)^3 - 3\alpha\beta(\alpha + \beta)$$
$$= \left(-\frac{b}{a}\right)^3 - 3\left(\frac{c}{a}\right)\left(-\frac{b}{a}\right) = \frac{b(3ac - b^2)}{a^3}.$$

For the equation whose roots are α^2 and β^2,

$$\text{sum of the roots} = \frac{b^2 - 2ac}{a^2}$$

$$\text{product of the roots} = \alpha^2\beta^2 = (\alpha\beta)^2 = \frac{c^2}{a^2}$$

$\therefore$ the required equation is $x^2 - \left(\frac{b^2 - 2ac}{a^2}\right)x + \frac{c^2}{a^2} = 0,$

i.e. $\qquad\qquad\qquad a^2 x^2 - (b^2 - 2ac)x + c^2 = 0.$

For the equation whose roots are α^3 and β^3,

$$\text{sum of the roots} = \frac{b(3ac - b^2)}{a^3}$$

$$\text{product of the roots} = \alpha^3\beta^3 = (\alpha\beta)^3 = \frac{c^3}{a^3}$$

$\therefore$ the required equation is $x^2 - \left(\frac{b(3ac - b^2)}{a^3}\right)x + \frac{c^3}{a^3} = 0,$

i.e. $\qquad\qquad\qquad a^3 x^2 - b(3ac - b^2)x + c^3 = 0.$

Example 2
If α and β are the roots of the equation $x^2 + px + q = 0$, form the equation whose roots are $1/\alpha$ and $1/\beta$.

$$\alpha + \beta = -\frac{p}{1} = -p \quad \text{and} \quad \alpha\beta = \frac{q}{1} = q$$

For the required equation,

$$\text{sum of the roots} = \frac{1}{\alpha} + \frac{1}{\beta} = \frac{\beta + \alpha}{\alpha\beta} = \frac{-p}{q}$$

$$\text{product of the roots} = \frac{1}{\alpha}\frac{1}{\beta} = \frac{1}{q}$$

$$\therefore \quad \text{the required equation is} \quad x^2 - \left(-\frac{p}{q}\right)x + \frac{1}{q} = 0,$$

i.e.
$$qx^2 + px + 1 = 0.$$

An alternative method is to put $x = 1/y$ in the given equation which gives

$$\frac{1}{y^2} + \frac{p}{y} + q = 0, \text{ i.e. } 1 + py + qy^2 = 0,$$

where $y = 1/x$, and so the roots are the reciprocals of the roots of the given equation.

The relations between the roots and the coefficients of equations of higher degree can be found.

Let $\alpha_1, \alpha_2, \alpha_3$ be the roots of the cubic equation $ax^3 + bx^2 + cx + d = 0$.

Then
$$a(x - \alpha_1)(x - \alpha_2)(x - \alpha_3) = 0,$$

i.e.
$$a\left[x^3 - (\alpha_1 + \alpha_2 + \alpha_3)x^2 + (\alpha_1\alpha_2 + \alpha_2\alpha_3 + \alpha_3\alpha_1)x - \alpha_1\alpha_2\alpha_3\right] = 0.$$

Comparing coefficients we get

$b = -a(\alpha_1 + \alpha_2 + \alpha_3)$ i.e. $\alpha_1 + \alpha_2 + \alpha_3 = -\dfrac{b}{a}$

$c = a(\alpha_1\alpha_2 + \alpha_2\alpha_3 + \alpha_3\alpha_1)$ i.e. $\alpha_1\alpha_2 + \alpha_2\alpha_3 + \alpha_3\alpha_1 = \dfrac{c}{a}$

$d = -a\alpha_1\alpha_2\alpha_3$ i.e. $\alpha_1\alpha_2\alpha_3 = -\dfrac{d}{a}$

Similar results can be obtained for equations of higher degree.

Example 3

If α, β, γ are the roots of the cubic equation $x^3 + px^2 + qx + r = 0$, form the equation whose roots are $\alpha^2, \beta^2, \gamma^2$.

$$\alpha + \beta + \gamma = -\frac{p}{1} = -p, \ \alpha\beta + \beta\gamma + \gamma\alpha = \frac{q}{1} = q, \ \alpha\beta\gamma = -\frac{r}{1} = -r.$$

For the required equation

$$\text{sum of the roots} = \alpha^2 + \beta^2 + \gamma^2 = (\alpha + \beta + \gamma)^2 - 2\alpha\beta - 2\beta\gamma - 2\gamma\alpha$$
$$= p^2 - 2q$$

$$\text{product of the roots} = \alpha^2\beta^2\gamma^2 = (\alpha\beta\gamma)^2 = r^2$$

$$\alpha^2\beta^2 + \beta^2\gamma^2 + \gamma^2\alpha^2 = (\alpha\beta + \beta\gamma + \gamma\alpha)^2 - 2\alpha^2\beta\gamma - 2\alpha\beta^2\gamma - 2\alpha\beta\gamma^2$$
$$= (\alpha\beta + \beta\gamma + \gamma\alpha)^2 - 2\alpha\beta\gamma(\alpha + \beta + \gamma)$$
$$= q^2 - 2(-r)(-p) = q^2 - 2pr$$

∴ the required equation is

$$x^3 - (p^2 - 2q)x^2 + (q^2 - 2pr)x - r^2 = 0.$$

Exercises 2.2

1 Find the sum and product of the roots of the following equations.

(a) $2x^2 + 3x - 4 = 0$ (d) $5x^2 + 2x - 3 = 0$

(b) $3x^2 - x - 6 = 0$ (e) $\dfrac{x^2 + 3}{x} = 1$

(c) $9x^2 + 24x + 16 = 0$ (f) $1 + \dfrac{1}{x} = x$

In each case state the nature of the roots.

2 If α and β are the roots of the equation $2x^2 - x + 4 = 0$, find, without solving the equation,

(a) $\alpha + \beta$ (b) $\alpha\beta$ (c) $2\alpha + 2\beta$ (d) $\dfrac{1}{\alpha} + \dfrac{1}{\beta}$ (e) $\alpha^2 + \beta^2$.

3 If α and β are the roots of the equation $3x^2 - 6x - 7 = 0$, find, without solving this equation,

(a) $(\alpha + \beta)^2$ (b) $(\alpha - \beta)^2$ (c) $\alpha^2 - \beta^2$.

4 If α and β are the roots of the equation $x^2 + 3x - 5 = 0$, find, without solving this equation,

(a) $(\alpha + \beta)^3$ (b) $\alpha^2 - \alpha\beta + \beta^2$ (c) $\alpha^3 + \beta^3$.

5 If α and β are the roots of the equation $2x^2 - 5x - 4 = 0$, find, without solving this equation,

(a) $(\alpha - \beta)^2$ (b) $(\alpha - \beta)^3$ (c) $\alpha^3 - \beta^3$.

6 If α and β are the roots of the equation $2x^2 + x - 7 = 0$, form, without solving this equation, the equations whose roots are

(a) $2\alpha, 2\beta$ (b) $3\alpha, 3\beta$ (c) $\dfrac{1}{\alpha}, \dfrac{1}{\beta}$ (d) $\dfrac{\alpha}{\beta}, \dfrac{\beta}{\alpha}$ (e) $\alpha + 1, \beta + 1$.

7 If α and β are the roots of the equation $3x^2 - 2x + 1 = 0$, form, without solving this equation, the equations whose roots are

(a) α^2, β^2 (b) $\dfrac{1}{\alpha^2}, \dfrac{1}{\beta^2}$ (c) $(\alpha + \beta)^2, (\alpha - \beta)^2$.

8 If α and β are the roots of the equation $px^2 + qx + r = 0$, form, without solving this equation, the equations whose roots are

(a) $\dfrac{1}{\alpha^2}, \dfrac{1}{\beta^2}$ (b) α^2, β^2 (c) α^3, β^3.

9 If one root of the equation $px^2 + qx + r = 0$ is twice the other root, show that $2q^2 = 9pr$.

10 The roots of the equation $x^2 - ax + 4 = 0$ differ by 1. Find the possible values of a.

11 If p, q, r are the roots of the equation $x^3 - x^2 + 2x - 3 = 0$, find

(a) $p + q + r$ (b) $pq + qr + rp$ (c) pqr

(d) $p^2 + q^2 + r^2$ (e) $\dfrac{1}{p} + \dfrac{1}{q} + \dfrac{1}{r}$.

12 If p, q, r are the roots of the equation $2x^3 + 3x^2 - x - 4 = 0$, form the equation whose roots are

(a) p^2, q^2, r^2 (b) $\dfrac{1}{p}, \dfrac{1}{q}, \dfrac{1}{r}$ (c) $p - 2, q - 2, r - 2$.

13 If p, q, r, s are the roots of the equation $ax^4 + bx^3 + cx^2 + dx + e = 0$, show that

(a) $p + q + r + s = -\dfrac{b}{a}$ (b) $pqr + qrs + rsp + spq = -\dfrac{d}{a}$ (c) $pqrs = \dfrac{e}{a}$.

2.3 Linear inequalities

The reader will have manipulated equations and discovered that the balance of an equation is undisturbed by adding to or subtracting the same quantity from each side of the equation. Also the balance is not disturbed by multiplying both sides of an equation by the same number nor by dividing both sides by any number other than zero. Do these rules for manipulating equations apply to inequalities?

If John is older than Jane now, he will be older than Jane in two year's time and was older than Jane two years ago. Thus if John's age is a years now and Jane's age is b years, then

$$a > b \Leftrightarrow a + 2 > b + 2 \quad \text{and} \quad a - 2 > b - 2.$$

It can be shown that, for all values of a, b and c,

$$a > b \Leftrightarrow a + c > b + c \quad \text{and} \quad a - c > b - c$$

and

$$a < b \Leftrightarrow a + c < b + c \quad \text{and} \quad a - c < b - c.$$

In our example of the ages of John and Jane a, b, $a - c$ and $b - c$ are all positive. It is left as an exercise for the reader to show that these results are true in the cases in which a, b, $a - c$ and $b - c$ are negative.

When a photograph is enlarged all its lengths are multiplied by n. If one line of length x is longer in the original photograph than another line of length y, then the enlargement of the first line will be longer than the enlargement of the second line, i.e.

$$x > y \Leftrightarrow nx > ny.$$

If $nx = p$ and $ny = q$, then $p > q$ and $p/n > q/n$.

In general $a > b \Leftrightarrow na > nb$ and $a/n > b/n$ when $n > 0$

and $a < b \Leftrightarrow na < nb$ and $a/n < b/n$ when $n > 0$.

If $a > b$, by subtracting $(a + b)$ from both sides, we get $-b > -a$. Multiplying both sides by n, where $n > 0$, we get $-nb > -na$, i.e.

$$a > b \Leftrightarrow -na < -nb.$$

Similarly
$$a > b \Leftrightarrow \frac{a}{-n} < \frac{b}{-n}.$$

Thus multiplying or dividing both sides of an inequality by a negative number has the effect of reversing the inequality sign.

Now consider the inequality $a > b$ when $a > 0$ and $b > 0$. Multiplying both sides by a gives $a^2 > ab$ and multiplying both sides by b gives $ab > b^2$,

so that $$a^2 > ab > b^2,$$
i.e. $$a > b \Rightarrow a^2 > b^2.$$

Similarly, when $a > 0$ and $b > 0$, $a > b \Rightarrow a^n > b^n$, where n is a positive integer.

Summarising, the following rules for manipulating inequalities are obtained.
(1) Equal additions to and equal subtractions from both sides of an inequality may be made.
(2) Both sides of an inequality may be multiplied or divided by a positive number.
(3) If both sides of an inequality are multiplied by or divided by a negative number, then the inequality sign is reversed.
(4) Both sides of an inequality between positive numbers may be raised to a positive integral power.

Example 1
Find the set of values of x for which $3x + 4 > 5$.

$$3x + 4 > 5 \Leftrightarrow 3x > 1 \qquad \text{using } rule\ 1$$
$$3x > 1 \Leftrightarrow x > \tfrac{1}{3} \qquad \text{using } rule\ 2$$

i.e. the required set is $\{x \in \mathbb{R} : x > \tfrac{1}{3}\}$.

Example 2
Find the set of values of x for which $5 - 2x < 7$.

$$5 - 2x < 7 \Leftrightarrow 5 < 7 + 2x \qquad \text{using } rule\ 1$$
$$5 < 7 + 2x \Leftrightarrow -2 < 2x \qquad \text{using } rule\ 1$$
$$-2 < 2x \Leftrightarrow -1 < x \qquad \text{using } rule\ 2$$

or
$$5 - 2x < 7 \Leftrightarrow -2x < 2 \qquad \text{using } rule\ 1$$
$$-2x < 2 \Leftrightarrow x > -1 \qquad \text{using } rule\ 3$$

i.e. the required set is $\{x \in \mathbb{R} : x > -1\}$.

Example 3
Find the set of values of x for which $\dfrac{3x - 2}{x} < 1$.

When $x > 0$;
$$\frac{3x - 2}{x} < 1 \Leftrightarrow 3x - 2 < x \qquad \text{using } rule\ 2$$

$$3x - 2 < x \Leftrightarrow 3x < x + 2 \qquad \text{using } rule\ 1$$

$$3x < x + 2 \Leftrightarrow 2x < 2 \qquad \text{using } rule\ 1$$
$$2x < 2 \Leftrightarrow x < 1 \qquad \text{using } rule\ 2$$

giving $0 < x < 1$.

When $x < 0$; $\qquad \dfrac{3x-2}{x} < 1 \Leftrightarrow 3x - 2 > x \qquad \text{using } rule\ 3$

$$3x - 2 > x \Leftrightarrow 2x > 2 \qquad \text{using } rule\ 1$$
$$2x > 2 \Leftrightarrow x > 1 \qquad \text{using } rule\ 2$$

but here $x < 0$ and so the required set cannot contain any negative values of x.

$\therefore \qquad \dfrac{3x-2}{x} < 1$ when $0 < x < 1$,

i.e.   the required set is $\{x \in \mathbb{R} : 0 < x < 1\}$.

Exercises 2.3

1 Find the set of values of x for which

(a) $x + 3 > 5$ (f) $x + 7 < -2$ (k) $4x - 1 < -2$

(b) $x + 3 > -5$ (g) $5 - x < 3$ (l) $3 - 4x > 2$

(c) $4 - x > 3$ (h) $5 - x < -3$ (m) $4 - 5x > -1$

(d) $4 - x > -3$ (i) $2x + 1 > 7$ (n) $5 - 6x < 7$

(e) $x + 7 < 2$ (j) $3x - 2 < 5$ (o) $1 - 3x > -6$.

2 Find the set of values of x for which

$$\frac{3x+4}{x} < 1.$$

(a) when $x > 0$ (b) when $x < 0$.
State the complete set of values of x for which

$$\frac{3x+4}{x} < 1.$$

3 Find the set of values of x for which

$$\frac{4-2x}{x} > 1$$

(a) when $x > 0$ (b) when $x < 0$.
State the complete set of values of x for which

$$\frac{4-2x}{x} > 1.$$

4 Find the set of values of x for which

$$\frac{4x+8}{x-1} > 3$$

(a) when $x > 1$ (b) when $x < 1$.

State the complete set of values of x for which

$$\frac{4x+8}{x-1} > 3.$$

5   Find the set of values of x for which

(a) $\dfrac{3+4x}{x} < 3$ (c) $\dfrac{3x+4}{x+3} > 2$ (e) $\dfrac{1-4x}{x+2} > 4$

(b) $\dfrac{3-4x}{2-x} < 2$ (d) $\dfrac{2x-4}{x-1} < 3$ (f) $\dfrac{2-3x}{1-2x} < 5$.

2.4 Quadratic inequalities

Next, inequalities of the form $ax^2 + bx + c \gtrless 0$ are considered.

If $b^2 - 4ac \leqslant 0$, then $ax^2 + bx + c \geqslant 0$ for all values of x when $a > 0$, and $ax^2 + bx + c \leqslant 0$ for all values of x when $a < 0$.
These results follow from the work of section 2.1.

If $b^2 - 4ac > 0$, the equation $ax^2 + bx + c = 0$ can be solved.
Let the roots be α and β with $\beta > \alpha$. Then

$$ax^2 + bx + c = a(x-\alpha)(x-\beta).$$

	$x-\alpha$	$x-\beta$	$(x-\alpha)(x-\beta)$
$x < \alpha$	$-$ve	$-$ve	$+$ve
$\alpha < x < \beta$	$+$ve	$-$ve	$-$ve
$\beta < x$	$+$ve	$+$ve	$+$ve

When $a > 0$, $ax^2 + bx + c > 0$ when $x < \alpha$ and   when $x > \beta$,
$\qquad\qquad ax^2 + bx + c < 0$ when $\alpha < x < \beta$.

When $a < 0$, $ax^2 + bx + c > 0$ when $\alpha < x < \beta$,
$\qquad\qquad ax^2 + bx + c < 0$ when $x < \alpha$ and   when $x > \beta$.

The same results may be obtained by considering the graph of

$$y = ax^2 + bx + c = a(x-\alpha)(x-\beta).$$

Figure 2.2 overleaf shows the graph (a) when $a > 0$, (b) when $a < 0$. It is easy to see the values of x for which $y\ (= ax^2 + bx + c)$ is positive or negative.

Example 1

Find the set of values of x for which $2x^2 + 5x - 3 > 0$.

The roots of the equation $2x^2 + 5x - 3 = 0$ are $\frac{1}{2}$ and -3.

$\therefore \qquad\qquad 2x^2 + 5x - 3 = 2(x+3)(x-\frac{1}{2})$

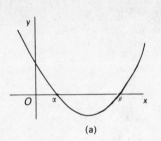

(a)

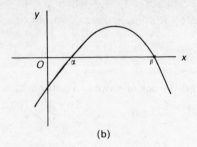

(b)

Fig. 2.2

	$x + 3$	$x - \frac{1}{2}$	$2(x+3)(x-\frac{1}{2})$
$x < -3$	$-$ve	$-$ve	$+$ve
$-3 < x < \frac{1}{2}$	$+$ve	$-$ve	$-$ve
$\frac{1}{2} < x$	$+$ve	$+$ve	$+$ve

Thus $\qquad 2(x+3)(x-\frac{1}{2}) > 0$ when $x < -3$ and when $x > \frac{1}{2}$,

i.e. $\qquad 2x^2 + 5x - 3 > 0$ when $x < -3$ and when $x > \frac{1}{2}$,

and the required set is the union of the sets $\{x \in \mathbb{R} : x < -3\}$ and $\{x \in \mathbb{R} : x > \frac{1}{2}\}$.

Alternatively, Fig. 2.3 shows a sketch of the graph of $y = 2x^2 + 5x - 3$. From the graph it is clear that y is positive when $x < -3$ and when $x > \frac{1}{2}$. Hence $2x^2 + 5x - 3 > 0$ when $x < -3$ and when $x > \frac{1}{2}$.

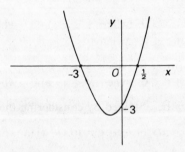

Fig. 2.3

Example 2

Find the set of values of x for which $1 + 2x - 3x^2 < 0$.

The roots of the equation $1 + 2x - 3x^2 = 0$ are $-\frac{1}{3}$ and 1.

$\therefore \qquad 1 + 2x - 3x^2 = -3(x + \frac{1}{3})(x - 1)$

	$x + \frac{1}{3}$	$x - 1$	$(x + \frac{1}{3})(x - 1)$	$-3(x + \frac{1}{3})(x - 1)$
$x < -\frac{1}{3}$	$-$ve	$-$ve	$+$ve	$-$ve
$-\frac{1}{3} < x < 1$	$+$ve	$-$ve	$-$ve	$+$ve
$1 < x$	$+$ve	$+$ve	$+$ve	$-$ve

Thus $\qquad -3(x + \frac{1}{3})(x - 1) < 0 \quad$ when $x < -\frac{1}{3}$ and when $x > 1$,

i.e. $\qquad 1 + 2x - 3x^2 < 0 \quad$ when $x < -\frac{1}{3}$ and when $x > 1$,

and the required set is the union of the sets $\{x \in \mathbb{R} : x < -\frac{1}{3}\}$ and $\{x \in \mathbb{R} : x > 1\}$.

Alternatively, Fig. 2.4 shows a sketch of the graph of $y = 1 + 2x - 3x^2$. From the graph it is clear that y is negative when $x < -\frac{1}{3}$ and when $x > 1$. Hence, as before, $1 + 2x - 3x^2 < 0$ when $x < -\frac{1}{3}$ and when $x > 1$.

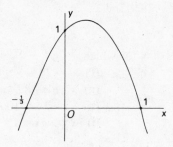

Fig. 2.4

Example 3

Find the set of values of x for which

$$\frac{4x + 8}{x - 1} > 3.$$

(Here we consider two further methods for dealing with this example which was question 4 in exercises 2.3.)

Multiplying both sides of the inequality by $(x - 1)^2$ and using *rule 2* of page 29 we get

$$(4x + 8)(x - 1) > 3(x - 1)^2$$
$$\therefore \qquad 4x^2 + 4x - 8 > 3x^2 - 6x + 3$$
$$\therefore \qquad x^2 + 10x - 11 > 0$$
and $\qquad (x + 11)(x - 1) > 0$
giving $\qquad x < -11 \quad$ or $\quad x > 1$,

i.e. the required set is the union of the sets $\{x \in \mathbb{R} : x < -11\}$ and $\{x \in \mathbb{R} : x > 1\}$.

Alternatively,

$$\frac{4x+8}{x-1} > 3 \Leftrightarrow \frac{4x+8}{x-1} - 3 > 0,$$

$$\frac{4x+8}{x-1} - 3 = \frac{4x+8-3x+3}{x-1} = \frac{x+11}{x-1} > 0.$$

	$x+11$	$x-1$	$\dfrac{x+11}{x-1}$
$x < -11$	$-$ve	$-$ve	$+$ve
$-11 < x < 1$	$+$ve	$-$ve	$-$ve
$1 < x$	$+$ve	$+$ve	$+$ve

Giving, as before, $x < -11$ or $x > 1$.

Exercises 2.4(i)

1 Find the set of values of x for which
 (a) $x^2 - 5x + 6 < 0$ (f) $x^2 - 2x + 5 > 0$
 (b) $x^2 + 7x + 12 < 0$ (g) $x^2 + x - 12 \geqslant 0$
 (c) $x^2 - x - 6 < 0$ (h) $x^2 - 3x \leqslant 10$
 (d) $x^2 - 3x > 0$ (i) $x^2 \geqslant 3x - 4.$
 (e) $x^2 + 2x - 1 > 0$

2 Find the set of values of x for which
 (a) $12 - 4x - x^2 < 0$ (f) $3x - 2 > x^2$
 (b) $-x^2 - 4x - 3 < 0$ (g) $x^2 \geqslant 4x - 4$
 (c) $-8 + 2x - x^2 < 0$ (h) $15 + 2x \leqslant x^2$
 (d) $4 - x^2 > 0$ (i) $x^2 - 9 \leqslant 0.$
 (e) $4x - x^2 > 0$

3 Find the set of values of x for which
 (a) $2x^2 - 11x + 12 < 0$ (g) $1 - x - 6x^2 < 0$
 (b) $6x^2 + 5x + 1 > 0$ (h) $3 + 5x - 2x^2 > 0$
 (c) $5x^2 + 3x - 2 \leqslant 0$ (i) $-4 + 2x - 3x^2 < 0$
 (d) $4x^2 - 9 \geqslant 0$ (j) $1 - 3x - 4x^2 \geqslant 0$
 (e) $3x^2 \leqslant 2x$ (k) $1 - 4x^2 \leqslant 0$
 (f) $3x^2 \geqslant x - 1$ (l) $6 + x \geqslant 12x^2$

4 Find the set of values of x for which
 (a) $\dfrac{x^2 - 2}{x} < 1$ (d) $\dfrac{3x+1}{x-1} < 2$

 (b) $\dfrac{6 - x^2}{x} > 1$ (e) $\dfrac{3x-2}{x+1} < 4$

 (c) $\dfrac{x-5}{x-1} > -3$ (f) $\dfrac{2x+3}{3-2x} > 1.$

The work of this section and section 2.1 can be used to find the set of values which can be taken by a rational expression such as

$$\frac{ex^2 + fx + g}{px^2 + qx + r} \quad (= y, \text{ say}).$$

We can express this in the form

$$(py - e)x^2 + (qy - f)x + (ry - g) = 0.$$

For real values of x

$$(qy - f)^2 - 4(py - e)(ry - g) \geqslant 0.$$

This is an inequality of the form $ay^2 + by + c \geqslant 0$ and is dealt with as in the examples on pages 31–4. The method is illustrated in the following example.

Example
Given that x is real, find the set of possible values of $(x^2 - 5)/(x - 3)$.

Let
$$y = \frac{x^2 - 5}{x - 3}.$$

Then
$$yx - 3y = x^2 - 5,$$

i.e.
$$x^2 - yx + (3y - 5) = 0.$$

x is real and so $b^2 - 4ac \geqslant 0$ giving

$$y^2 - 4(3y - 5) \geqslant 0.$$
$$\therefore \qquad y^2 - 12y + 20 \geqslant 0$$
and
$$(y - 2)(y - 10) \geqslant 0$$
$$\therefore \qquad y \leqslant 2 \quad \text{or} \quad y \geqslant 10.$$

Hence
$$\frac{x^2 - 5}{x - 3} \leqslant 2 \quad \text{or} \quad \frac{x^2 - 5}{x - 3} \geqslant 10.$$

Exercises 2.4(ii)

1 Given that x is real, find the set of possible values of y when

   (a) $y = \dfrac{x^2 - 3}{x - 2}$ (b) $y = \dfrac{x^2 + 3}{x + 1}$ (c) $y = \dfrac{2x^2 - 16}{x - 3}$.

2 Given that x is real, find the set of possible values of y when

   (a) $y = \dfrac{x + 1}{x^2 + 3}$ (b) $y = \dfrac{x + 2}{x^2 + 5}$ (c) $y = \dfrac{2x + 1}{2(x^2 + 2)}$.

3 Given that x is real, find the set of possible values of y when

   (a) $y = \dfrac{x^2 + 6x}{x + 2}$

   (b) $y = \dfrac{x^2 + x + 1}{x + 1}$

   (c) $y = \dfrac{x^2 + 3x + 3}{x + 2}$

   (d) $y = \dfrac{x^2 + x + 1}{x^2 + 1}$

   (e) $y = \dfrac{x^2 + 3x + 1}{x^2 + 1}$

   (f) $y = \dfrac{x^2 + x + 1}{x^2 + 4x + 1}$.

2.5 Inequalities involving the modulus sign

$|x|$, the modulus of x, is the magnitude of x and is always positive.

Thus	$\|x\| = x$	when x is positive
and	$\|x\| = -x$	when x is negative
and	$\|x\|^2 = x^2$	for all values of x.

Consider	$\|ax+b\| > c,$	where $c > 0$;
then	$\|ax+b\|^2 > c^2,$	
but	$\|ax+b\|^2 = (ax+b)^2$	for all values of x.
$\therefore$	$(ax+b)^2 > c^2.$	
i.e.	$a^2x^2 + 2abx + (b^2 - c^2) > 0.$	

The values of x which satisfy this inequality can be found by the method used in examples 1 and 2 on pages 31–3.

Example 1

Find the set of real values of x for which $|3x - 2| > 4$.

	$\|3x - 2\| > 4$
$\therefore$	$\|3x - 2\|^2 = (3x - 2)^2 > 16$
i.e.	$9x^2 - 12x - 12 > 0$
and	$9(x - 2)(x + \frac{2}{3}) > 0$

Thus the required set is the union of the sets $\{x : x < -\frac{2}{3}\}$ and $\{x : x > 2\}$, i.e. the set $\{x \in \mathbb{R} : x < -\frac{2}{3} \text{ or } x > 2\}$.

Alternatively,

for $x > \frac{2}{3}$, $3x - 2 > 4$, i.e. $x > 2$,

for $x < \frac{2}{3}$, $2 - 3x > 4$, i.e. $x < -\frac{2}{3}$.

Example 2

Find the set of real values of x for which $|3 - 2x| < |4 + x|$.

	$\|3 - 2x\|^2 < \|4 + x\|^2$
$\therefore$	$(3 - 2x)^2 < (4 + x)^2$
i.e.	$3x^2 - 20x - 7 < 0$
and	$(3x + 1)(x - 7) < 0$
	$3(x + \frac{1}{3})(x - 7) < 0$
giving	$-\frac{1}{3} < x < 7.$

Alternatively, the critical values are $x = \frac{3}{2}$ and $x = -4$.

For $\qquad x < -4$, $3 - 2x < -x - 4$, i.e. $x > 7$

which is impossible since $x < -4$.

For $\qquad -4 < x < \frac{3}{2}$, $3 - 2x < 4 + x$ giving $x > -\frac{1}{3}$;

for $\qquad \frac{3}{2} < x$, $2x - 3 < 4 + x$ giving $x < 7$.

Hence the required set is $\{x \in \mathbb{R} : -\frac{1}{3} < x < 7\}$.

Exercises 2.5(i)

1 Find the set of real values of x for which
   (a) $|x-3| > 4$ (d) $|x+2| \leqslant 1$ (g) $|2x+5| \geqslant 3$
   (b) $|x-3| < 4$ (e) $|1-x| > 3$ (h) $|3-4x| < 3$.
   (c) $|x+2| \geqslant 1$ (f) $|5-x| \leqslant 2$

2 Find the set of real values of x for which
   (a) $|2+x| > |1+2x|$ (e) $|3x+2| > |2x-3|$
   (b) $|5-x| < |1+x|$ (f) $|1-5x| < |3-4x|$
   (c) $|2-x| \geqslant |1+2x|$ (g) $|1-2x| \leqslant |3x-1|$
   (d) $|4-3x| \leqslant |2x-1|$ (h) $|4x-2| \geqslant |2-3x|$.

These inequalities can also be treated graphically.

Consider the graph of $y = |x|$;

$$y = x \quad \text{when} \quad x \text{ is positive}$$
$$y = -x \quad \text{when} \quad x \text{ is negative.}$$

Figure 2.5 shows the graph of $y = |x|$.

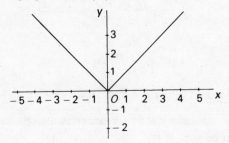

Fig. 2.5

Next consider the graph of $y = |ax + b|$

When $y = 0, ax + b = 0$, i.e. $x = -b/a$. First draw the graph of $y = ax + b$. The graph of $y = |ax + b|$ has two parts:
(i) that part of the line $y = ax + b$ which is above the x-axis,
(ii) the reflection in the x-axis of that part of the line $y = ax + b$ which is below the x-axis.
Figure 2.6 shows the graph of $y = |ax + b|$ in the case when $a > 0$ and $b > 0$.

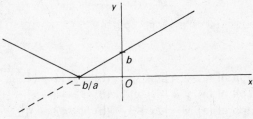

Fig. 2.6

The graphical method for dealing with inequalities of the form $|ax+b| \gtreqless c$ can be illustrated by reconsidering examples 1 and 2 on page 36.

Example 1

Find the set of real values of x for which $|3x-2| > 4$.

The graphs of $y = |3x-2|$ and $y = 4$ are shown on the same axes in Fig. 2.7. From Fig. 2.7 it can be seen that

$$|3x-2| > 4 \quad \text{when} \quad x < -\tfrac{2}{3} \text{ and when } x > 2.$$

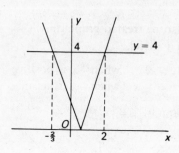

Fig. 2.7

Example 2

Find the set of real values of x for which $|3-2x| < |4+x|$.

The graphs of $y = |3-2x|$ and $y = |4+x|$ are shown on the same axes in Fig. 2.8. From Fig. 2.8 it can be seen that

$$|3-2x| < |4+x| \quad \text{when} \quad -\tfrac{1}{3} < x < 7.$$

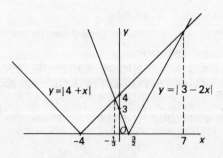

Fig. 2.8

Exercises 2.5(ii)

1   Sketch the graph of (a) $y = |x-2|$ (b) $y = |x+1|$.
2   Sketch the graph of (a) $y = |5+x|$ (b) $y = |4-x|$.
3   Sketch the graph of (a) $y = |2x+3|$ (b) $y = |3x-4|$.
4   Sketch the graph of (a) $y = |3-2x|$ (b) $y = |-1-2x|$.

5 Find graphically the set of values of x for which
   (a) $|x-2| > 4$ (c) $|3+x| \geqslant 5$
   (b) $|x+1| < 4$ (d) $|5-x| \leqslant 3$.
6 Find graphically the set of values of x for which
   (a) $|2x+3| > 6$ (c) $|3-2x| \leqslant 4$
   (b) $|3x-4| < 5$ (d) $|1-2x| \geqslant 3$.
7 Find graphically the set of values of x for which
   (a) $|1-x| > |2x-5|$ (c) $|3-4x| \geqslant |2-x|$
   (b) $|2x+1| < |x+3|$ (d) $|5-x| \leqslant |2x+3|$.

2.6 Location of the roots of an equation

Consider the graph of $y = (x-1)(x-3)(x-5)$ which is shown in Fig. 2.9.

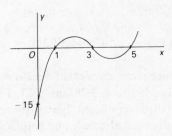

Fig. 2.9

Let $f(x) = (x-1)(x-3)(x-5)$. Then $f(x) = 0$ when $x = 1, 3, 5$.

$$\begin{array}{ll} \text{When } x = 0 & f(x) \text{ is negative,} \\ \text{when } x = 2 & f(x) \text{ is positive,} \\ \text{when } x = 4 & f(x) \text{ is negative,} \\ \text{when } x = 6 & f(x) \text{ is positive.} \end{array}$$

These values illustrate the fact that $f(x)$ changes sign as the curve $y = f(x)$ crosses the x-axis. Thus if $y = f(x)$ is a continuous curve and $f(x)$ changes sign as x goes from a to b, i.e. $f(a)$ and $f(b)$ have different signs, the curve must cross the x-axis at least once between $x = a$ and $x = b$. Thus $f(x) = 0$ for some value of x between a and b, i.e. there is a root of the equation $f(x) = 0$ between $x = a$ and $x = b$. This can be useful in locating roots of an equation.

Example 1
Locate approximately the roots of the equation $x^3 - 7x^2 + 3x + 17 = 0$.

$$\begin{array}{lll} \text{When } x = -2 & f(x) = -25, \text{ i.e. } & f(-2) = -25, \\ \text{when } x = -1 & f(x) = +6, & f(-1) = +6, \\ \text{when } x = 0 & f(x) = +17, & f(0) = +17, \\ \text{when } x = 1 & f(x) = +14, & f(1) = +14, \end{array}$$

$$\text{when } x = 2 \quad f(x) = +3, \quad f(2) = +3,$$
$$\text{when } x = 3 \quad f(x) = -10, \quad f(3) = -10,$$
$$\text{when } x = 4 \quad f(x) = -19, \quad f(4) = -19,$$
$$\text{when } x = 5 \quad f(x) = -18, \quad f(5) = -18,$$
$$\text{when } x = 6 \quad f(x) = -1, \quad f(6) = -1,$$
$$\text{when } x = 7 \quad f(x) = +38, \quad f(7) = +38.$$

Thus $f(x)$ changes sign between $x = -2$ and $x = -1$ and also between $x = 2$ and $x = 3$ and again between $x = 6$ and $x = 7$. Hence the three roots lie between -2 and -1, between 2 and 3 and between 6 and 7.

Example 2

Find the only real root of the equation $x^3 - 7x - 12 = 0$.

Let $f(x) = x^3 - 7x - 12$.
$f(0) = -12$, $f(1) = -18$, $f(2) = -18$, $f(3) = -6$, $f(4) = +24$.
$f(x)$ changes sign between $x = 3$ and $x = 4$, i.e. there is a root between 3 and 4.

Repeating the process $f(3 \cdot 2) = -1 \cdot 632$ and $f(3 \cdot 3) = +0 \cdot 837$, so the root lies between $3 \cdot 2$ and $3 \cdot 3$.

Repeating the process once more shows that $f(3 \cdot 26)$ is negative and $f(3 \cdot 27)$ is positive. Hence the root lies between $3 \cdot 26$ and $3 \cdot 27$. Furthermore $f(3 \cdot 265)$ is negative and so, correct to three significant figures, the required root is $3 \cdot 27$.

The solution to this equation can also be found graphically using the following method.

The given equation can be written in the form $x^3 = 7x + 12$. If the graphs of $y = x^3$ and $y = 7x + 12$ are drawn on the same axes, at any point of intersection of the two graphs, $y = x^3$ and also $y = 7x + 12$. Thus the x-coordinate of a point of intersection gives a root of the equation $x^3 = 7x + 12$, i.e. $x^3 - 7x - 12 = 0$.

Figure 2.10 shows the graphs of $y = x^3$ and $y = 7x + 12$ drawn on the same axes. The two graphs intersect between $x = 3 \cdot 2$ and $x = 3 \cdot 3$, hence the real root of the equation lies between $3 \cdot 2$ and $3 \cdot 3$.

In Fig. 2.11, the two graphs, $y = x^3$ and $y = 7x + 12$, have been drawn again on a much larger scale for values of x from $3 \cdot 2$ to $3 \cdot 3$. It has been assumed that the graph of $y = x^3$ approximates to a straight line over the range from $x = 3 \cdot 2$ to $x = 3 \cdot 3$. The root of the equation $x^3 - 7x - 12 = 0$ is seen to be between $3 \cdot 26$ and $3 \cdot 27$.

If the root is required to a greater degree of accuracy the graphs could be drawn again on a still larger scale for values of x between $3 \cdot 26$ and $3 \cdot 27$.

In this way the root of an equation can be found graphically to any degree of accuracy which may be required. The only limitation on the accuracy which may be achieved by this method is that imposed by the method of calculating values prior to drawing the graphs.

Instead of drawing the two graphs, $y = x^3$ and $y = 7x + 12$, and finding their point of intersection we could have drawn the graph of $y = x^3 - 7x - 12$ and found its point of intersection with $y = 0$, i.e. the x-axis.

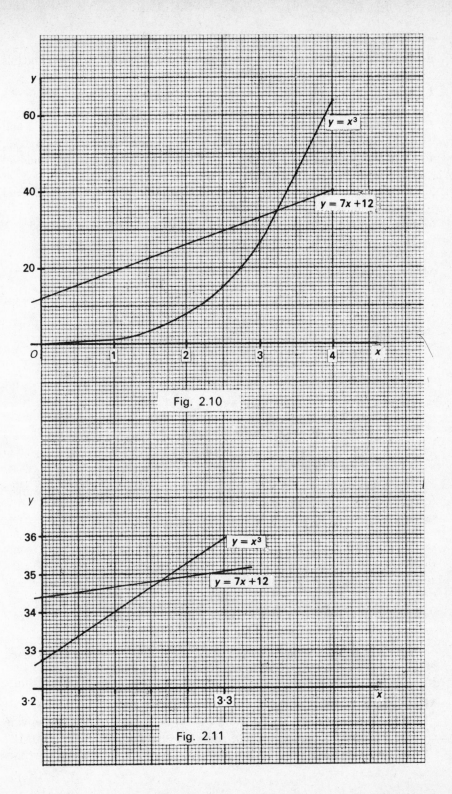

Fig. 2.10

Fig. 2.11

Exercises 2.6

1   Show that the equation $x^3 - x - 3 = 0$ has a root between 1 and 2 and find this root to three significant figures.

2   Show that the equation $x^3 + 4x + 10 = 0$ has a root between -1 and -2. Find this root correct to three significant figures by a graphical method.

3   Show that the equation $x^3 - x^2 - 1 = 0$ has a root between 1 and 2 and graphically find this root to two places of decimals.

4   Show that the equation $x^3 + x^2 - 16 = 0$ has a root between 2 and 3 and find this root to two decimal places.

5   Show that the equation $x^4 - 4x^3 + 10 = 0$ has a root between 1 and 2. Find this root to three significant figures.

6   Show that the equation $x^4 + 6x^3 + 40 = 0$ has a root between -3 and -2. Find this root to two decimal places.

Miscellaneous examples 2

1   Find the least value of

$$(2p^2 + 1)x^2 + 2(4p^2 - 1)x + 4(2p^2 + 1)$$

for real values of p and x.

$$(2p^2 + 1)x^2 + 2(4p^2 - 1)x + 4(2p^2 + 1)$$
$$= 2p^2x^2 + 8p^2x + 8p^2 + x^2 - 2x + 4$$
$$= 2p^2(x + 2)^2 + (x - 1)^2 + 3.$$

Thus the given expression has its least value 3 when $x = 1$ and $p = 0$.

2   If $f(x) = 9 + 2(k + 4)x + 2kx^2$ $(k \neq 0)$, find the set of values of k for which $f(x)$ is positive for all real values of x.

The conditions for $ax^2 + bx + c$ to be positive for all real values of x are
(i) $a > 0$ (ii) $b^2 - 4ac < 0$.
Thus for $f(x)$ to be positive for all real values of x
(i) $2k > 0$, i.e. $k > 0$,
(ii) $4(k + 4)^2 - 4(2k)9 < 0$.
Condition (ii) reduces to

$$4k^2 - 40k + 64 < 0,$$
i.e. $$4(k - 2)(k - 8) < 0.$$
$$\therefore \qquad 2 < k < 8.$$

This also satisfies condition (i) and so the required set of values of k is given by $2 < k < 8$.

Miscellaneous exercises 2

1   If p, q are the roots of the equation $2x^2 + 7x - 3 = 0$, find, without solving the equation, an equation whose roots are $2p + 1, 2q + 1$.   [JMB]

2   If α and β are the roots of the equation $ax^2 + bx + c = 0$, show that the roots

of the equation $acx^2 - (b^2 - 2ac)x + ac = 0$ are α/β and β/α.   [AEB]

3   The roots of the equation $x^3 + 3x + 2 = 0$ are α, β and γ. Find the equation whose roots are $\alpha + \dfrac{1}{\alpha}$, $\beta + \dfrac{1}{\beta}$, $\gamma + \dfrac{1}{\gamma}$.   [AEB]

4   If p, q are the roots of the equation

$$5x^2 - x - 2 = 0,$$

find, without solving the given equation, an equation whose roots are $3/p$, $3/q$.   [JMB]

5   Given that the roots of the equation $x^2 - x - 1 = 0$ are α and β, find, in its simplest form, the quadratic equation with numerical coefficients whose roots are

$$\frac{1+\alpha}{2-\alpha} \quad \text{and} \quad \frac{1+\beta}{2-\beta}.$$ [JMB]

6   If α and β are the roots of the equation $ax^2 + bx + c = 0$, show that the equation whose roots are α^3 and β^3 is

$$a^3 x^2 + b(b^2 - 3ac)x + c^3 = 0.$$ [L]

7   Form a quadratic equation with roots which exceed by 2 the roots of the quadratic equation

$$3x^2 - (p - 4)x - (2p + 1) = 0.$$

Find the values of p for which the given equation has equal roots.   [L]

8   If α, β are the roots of $x^2 - 2x + 3 = 0$, show that $\alpha^3 + \beta^3 = -10$. Hence, or otherwise, construct the quadratic equation with numerical coefficients whose roots are $1/(\alpha^4 \beta)$ and $1/(\alpha \beta^4)$.   [AEB]

9   If the equation $x^2 - qx + r = 0$ has roots $\alpha + 2$, $\beta + 1$, where α, β are the real roots of the equation $2x^2 - bx + c = 0$ and $\alpha \geqslant \beta$, find q and r in terms of b and c.

In the case $\alpha = \beta$, show that $q^2 = 4r + 1$.   [AEB]

10   The roots of the quadratic equation $x^2 - px + q = 0$ are α and β. Form, in terms of p and q, the quadratic equation whose roots are $\alpha^3 - p\alpha^2$ and $\beta^3 - p\beta^2$.   [AEB]

11   If α, β, γ are the roots of the equation

$$x^3 - x^2 - 4x + 5 = 0,$$

find cubic equations whose roots are
(a) 2α, 2β and 2γ,
(b) $1/\alpha$, $1/\beta$ and $1/\gamma$,
(c) $\alpha + \beta$, $\beta + \gamma$ and $\gamma + \alpha$.
Evaluate $\Sigma(\alpha + \beta)^2$.   [L]

12   Given that α, β, γ are the three roots of the equation $x^3 - x^2 + 3x + 10 = 0$ show that $\alpha^2 + \beta^2 + \gamma^2 = -5$.

Hence, or otherwise, find the value of $\alpha^3 + \beta^3 + \gamma^3$.

Find also a cubic equation whose roots are α^{-1}, β^{-1}, γ^{-1}.   [L]

13 Given that α, β, γ are the roots of the equation $x^3 + px^2 + qx + r = 0$, where the coefficients p, q, r are real, show that $\alpha^2 + \beta^2 + \gamma^2 = p^2 - 2q$.

Find $(\alpha - \beta)^2 + (\beta - \gamma)^2 + (\gamma - \alpha)^2$ in terms of p and q and show that α, β, γ cannot all be real if $p^2 < 3q$.   [L]

14 If $\alpha + \beta + \gamma = \alpha^2 + \beta^2 + \gamma^2 = \alpha^3 + \beta^3 + \gamma^3 = 2$, find by considering values of $(\alpha + \beta + \gamma)^2$ and $(\alpha + \beta + \gamma)^3$, or otherwise, the values of (i) $\alpha\beta + \beta\alpha + \gamma\alpha$, (ii) $\alpha\beta\gamma$.

Hence find the equation whose roots are α, β and γ.   [AEB]

15 Three numbers p, q, and r are such that $p + q + r = 6$ and $p^2 + q^2 + r^2 = 38$. Show that

$$pq + qr + rp = -1.$$

If also $pqr = -30$, form a cubic equation whose roots are p, q, and r. Hence, or otherwise, find the three numbers.   [L]

16 If α and β are the roots of the equation $x^2 + 2ax + b = 0$, show that $(\alpha - \beta)^2 = 4(a^2 - b)$. Express in their simplest form in terms of α and β the roots of the equation

$$bx^2 + 2a(b + 1)x + (b + 1)^2 = 0. \qquad \text{[AEB]}$$

17 If a, b and c are rational, prove that the roots of the following equation are rational

$$(2b - c)x^2 + 2(a + b)x + (2a + c) = 0.$$

Establish the condition on a, b and c for the roots to be equal.   [AEB]

18 (i) If α, β, γ are the roots of the equation

$$x^3 + 2x^2 + 2x + k = 0,$$

find the equation whose roots are α^2, β^2, γ^2.

Show that

$$(\beta^2 + \gamma^2)^3 + (\gamma^2 + \alpha^2)^3 + (\alpha^2 + \beta^2)^3 + 3k^2 = 0.$$

(ii) In the equation $x^3 - 4x^2 + ax + 4 = 0$ one root is equal to the sum of the other two. Find a and solve the equation.   [JMB]

19 If a is a positive constant, find the set of values of x for which $a(x^2 + 2x - 8)$ is negative. Find the value of a if this function has a minimum value of -27.   [L]

20 Prove that for any real values of a and x the value of the expression

$$(a^2 + 1)x^2 - 2a^2x + (a^2 - 1)$$

can never be less than -1.   [L]

21 Find the value of $k(\neq 1)$ such that the quadratic function of x

$$k(x + 2)^2 - (x - 1)(x - 2)$$

is equal to zero for only one value of x.

Find also (a) the range of values of k for which the function possesses a minimum value, (b) the range of values of k for which the value of the function never exceeds 12·5.

Sketch the graph of the function for $k = \frac{1}{2}$ and for $k = 2\frac{1}{2}$.　　　　[L]

22　If $ax^2 + bx + c = a(x + p)^2 + q, (a \neq 0)$, express p and q in terms of a, b and c. Deduce that, if $b^2 < 4ac$, the expression has the same sign as a for all real values of x.

If $g(x) = (k - 6) + (k - 3)x - x^2$, find the set of values of k for which $g(x)$ is always negative.　　　　[L]

23　Prove that $ax^2 + bx + c$ is positive for all real values of x if $a > 0$ and $b^2 < 4ac$.

Find the range of values of k for which

$$x^2 + kx + 3 + k$$

is positive for all real values of x. Deduce the range of values of k for which

$$k(x^2 + kx + 3 + k)$$

is positive for all real values of x.　　　　[JMB]

24　(a) Find the range, or ranges, of values of the constant k for which the equation

$$x^2 + kx + 2k = 3$$

has real distinct roots.

(b) Sketch a graph of the function

$$y = px^2 + 4x + p + 3$$

in each of the cases $p = -3$ and $p = 2$.

Find the ranges of values of the constant p for which the function y is (i) positive for all values of x, (ii) negative for all values of x.　　　　[JMB]

25　Find the set of values of k for which

$$f(x) = 3x^2 - 5x - k$$

is greater than unity for all real values of x.

Show that, for all k, the minimum value of $f(x)$ occurs when $x = 5/6$. Find k if this minimum value is zero.　　　　[L]

26　Prove that for real values of x the values of the function

$$\frac{6x + 5}{3x^2 + 4x + 2}$$

cannot lie outside the interval $-3/2$ to 3.　　　　[L]

27　(i) If x is real, find the set of possible values of

$$\frac{(2x + 1)}{(x^2 + 2)}.$$

(ii) The roots of the equation $x^2 + px + q = 0$ are α and β. Without solving this equation form the quadratic equation whose roots are $(\alpha + \beta)^2$ and $(\alpha - \beta)^2$. [L]

28 Show that the range of values taken by the function

$$\frac{x^2 + x + 7}{x^2 - 4x + 5}$$

for real values of x is from $\frac{1}{2}$ to $13\frac{1}{2}$ inclusive. Sketch the graph of the function, indicating its asymptote. [JMB]

29 If x is real, find the range of possible values of

$$\frac{4(x - 2)}{4x^2 + 9}.$$ [AEB]

30 If x is real, show that the expression

$$y = \frac{x^2 + x + 1}{x + 1}$$

can have no real value between -3 and 1. [L]

31 Find the range of values within which the function

$$\frac{(x - 1)\,(x + 3)}{(x - 2)}$$

cannot lie for all real values of x. [AEB]

32 Find the range of values of k for which the function

$$\frac{x^2 - 1}{(x - 2)\,(x + k)},$$

where x is real, takes all real values. [JMB]

33 Given that x is real and that

$$y = \frac{x^2 + 2}{2x + 1},$$

show that y cannot lie between -2 and 1. [L]

34 If x is real and

$$y = \frac{x + \lambda}{(x + 2)\,(x + 3)},$$

show that y cannot lie between $3 - 2\sqrt{2}$ and $3 + 2\sqrt{2}$ when $\lambda = 1$.

With this value of λ, find the set of values of x for which y is positive. Find also the set of values λ for which y can take all real values. [L]

35 Prove that, as x varies through all real values, the expression

$$\frac{x^2 + 3x + 1}{x^2 + 5x + 2}$$

takes every real value twice, except the value unity. Sketch, without numerical details, the shape of the graph of the above function.   [JMB]

36   Find the solution set of the inequality

$$\frac{4-x}{x-2} > 3.$$ [C]

37   (i) For what range of values of x is $(2+3x)(4-5x)$ positive?
(ii) Factorise completely $x^4 - 5x^2 + 4$, and determine the ranges of values of x for which this expression is negative.   [JMB]

38   Find the solution sets of the inequalities

(i) $\dfrac{3}{2} < \dfrac{2x-3}{x-5}$ (ii) $3x(x-5) < 2(2x-3)$.   [C]

39   (i) Prove that $x^4 - 7x^2 - 18 < 0$ only when $-3 < x < 3$.
(ii) If the roots of $x^2 + px + q = 0$ are α and β, where α and β are non-zero, form the equation whose roots are $2/\alpha$, $2/\beta$.   [L]

40   Find the set of real values of x for which

$$\frac{x^2+12}{x} > 7.$$ [L]

41   Find the complete set of real values of x for which

$$\frac{x^2+6}{x} > 5$$ [AEB]

42   Find the solution set of the inequality

$$\frac{12}{x-3} < x+1.$$ [C]

43   (a) State the range of values of x for which the following expressions are negative:

(i) $2x^2 + 5x - 12$,

(ii) $\dfrac{x-2}{(x-1)(x-3)}$.

(b) The value of the constant a is such that the quadratic function $f(x) = x^2 + 4x + a + 3$ is never negative. Determine the nature of the roots of the equation $af(x) = (x^2+2)(a-1)$. Deduce the value of a for which this equation has equal roots.   [AEB]

44   Find the sets of values of x ($x \in \mathbb{R}$) for which

(a) $\left|\dfrac{2x+3}{x-1}\right| < 1$,   (b) $\dfrac{2x+3}{x-1} < 1$.   [L]

45 Show that, if x is real, $2x^2 + 6x + 9$ is always positive. Hence, or otherwise, solve the inequality

$$\frac{(x+1)(x+3)}{x} > \frac{(x+6)}{3}.$$

[JMB]

46 Find the set of values of x for which

$$-2 \leqslant \frac{3x-6}{(x-1)(x-3)} \leqslant 2.$$

[L]

47 Find the set of values of x for which

$$-3 \leqslant \frac{(x-1)(x-5)}{(x-3)} \leqslant 3.$$

[L]

48 Find the set of values of x for which

$$\frac{5x-4}{x^2+2} > \frac{1}{2}\left[\frac{1}{x-2} + \frac{5}{x+2}\right].$$

[L]

3 Algebra 2

3.1 Sequences

$$1, 3, 5, 7, 9$$
$$1, 4, 9, 16, 25, \ldots$$
$$1, -1, 1, -1, 1, \ldots$$
$$5, 4, 3, 2, 1$$

These are examples of sequences of numbers. In a *sequence* the numbers appear in a defined order and there is usually a definite law relating each number to the other numbers. Each number is called a *term* of the sequence.

The sequence 1, 3, 5, 7, 9 has five terms and is an example of a *finite* sequence.

The sequence $1^2, 2^2, 3^2, \ldots, n^2, \ldots$ does not stop and so does not have a finite number of terms. This is an example of an *infinite* sequence.

To define a sequence it is not sufficient to give the first few terms. For instance the sequence 1, 2, 3, 2, 3, 4, 3, 4, 5, ... would not be defined by the first three terms. The definition of a sequence requires

(a) the first term,

(b) the number of terms,

(c) the law by which successive terms can be found.

In the sequence $u_1, u_2, u_3, \ldots, u_r, \ldots, u_n$, the first term is u_1, the sequence has n terms and the law relating the terms will be given by the general term u_r.

Consider the sequence of ten terms 1, 3, 5, 7, 9, 11, 13, 15, 17, 19. In this sequence each term is 2 more than the preceding term.

$$u_1 = 1, \; u_2 = 1 + 2 = 3, \; u_3 = 1 + 2(2) = 5, \; u_4 = 1 + 3(2) = 7,$$

and
$$u_r = 1 + (r-1)(2) = 2r - 1.$$

When the difference between consecutive numbers in a sequence is constant (as it is in the example just considered), an expression for the general term u_r can be obtained in the form $ar + b$.

With the infinite sequence 4, 7, 10, 13, ... there is a constant difference of 3 between consecutive terms.

Let $u_r = ar + b$, then $u_2 = 2a + b = 7$
and $u_3 = 3a + b = 10$.

From these simultaneous equations in a and b we get $a = 3$, $b = 1$. Thus $u_r = 3r + 1$.

Example

Given that the general term u_r of the infinite sequence $9, 4, -1, -6, \ldots$ is of the form $ar + b$, find a and b.

Let $u_r = ar + b$, then $u_2 = 2a + b = 4$
and $u_3 = 3a + b = -1$

Solving we get $a = -5$ and $b = 14$, and so $u_r = -5r + 14$ or $14 - 5r$. We can easily check that this result is true for the other given terms of the series.

Exercises 3.1

1   Write down the first four terms of the sequences in which the general term u_r is

(a) $r + 1$

(b) $2r - 3$

(c) $4r + 2$

(d) $3r - 5$

(e) $(r + 1)^2$

(f) $1/r$

(g) $r^2 - 1$

(h) $r!$

(i) $(r + 1)(r + 2)$

(j) $(2r - 1)(2r + 1)(2r + 3)$

(k) $\dfrac{1}{r + 2}$

(l) $\dfrac{r}{r + 1}$

(m) $\dfrac{1}{(3r + 1)(3r + 4)}$

(n) 2^r

(o) $(\tfrac{1}{3})^r$

(p) $(-1)^r r^2$

(q) $(-1)^r (2r - 3)$.

2   Write out the finite sequences of n terms with the general term u_r when

(a) $u_r = r - 1$, $n = 5$

(b) $u_r = 3r - 2$, $n = 4$

(c) $u_r = 2r + 3$, $n = 4$

(d) $u_r = \dfrac{r}{r + 2}$, $n = 5$

(e) $u_r = r^2 + 3$, $n = 4$

(f) $u_r = \dfrac{1}{r(r + 1)}$, $n = 6$

(g) $u_r = 3^{r-1}$, $n = 5$

(h) $u_r = (0.1)^r$, $n = 4$

(i) $u_r = \dfrac{r^2}{(3r - 2)(3r + 1)}$, $n = 4$

(j) $u_r = \dfrac{1}{2^r}$, $n = 6$

(k) $u_r = (-1)^r \dfrac{r}{r + 1}$, $n = 4$.

3   The general term u_r of a sequence is of the form $ar + b$.
Given that $u_3 = 0$ and $u_5 = 8$, show that $u_8 = 20$.

4   The general term u_n of a sequence is of the form $an^2 + bn + c$.
Given that $u_1 = u_8$ and that $u_2 = u_7$, show that $u_3 = u_6$.
Given also that $u_1 = 1$ and $u_2 = -2$, show that $u_9 = 5$.

3.2   Series

If the terms of a sequence are added together we get a series. $1 + 2 + 3 + 4$ is a series of four terms and is an example of a *finite* series.
$1^2 + 2^2 + 3^2 + \ldots + r^2 + \ldots$ is an example of an *infinite* series.

In general the sum of the terms of the sequence

$$u_1, u_2, u_3, \ldots, u_n$$

gives the series $u_1 + u_2 + u_3 + \ldots + u_n$.

This can be expressed more briefly by using the sigma (Σ) notation.

$$\sum_{r=1}^{n} u_r \text{ is defined as } u_1 + u_2 + u_3 + \ldots + u_n.$$

$$\sum_{r=1}^{n} u_r, \text{ which is often written as } \sum_{1}^{n} u_r,$$

is read as the sum of all the terms u_r as r takes all positive integral values from 1 to n.

Exercises 3.2

1 Write out and evaluate

(a) $\displaystyle\sum_{r=1}^{4} r$

(g) $\displaystyle\sum_{r=1}^{5} r!$

(b) $\displaystyle\sum_{r=1}^{5} r^2$

(h) $\displaystyle\sum_{r=1}^{4} (-r/2)^2$

(c) $\displaystyle\sum_{r=2}^{5} (2r-1)$

(i) $\displaystyle\sum_{1}^{3} 3^r$

(d) $\displaystyle\sum_{r=1}^{4} (-1)^r(2r+1)$

(j) $\displaystyle\sum_{2}^{4} (-1)^r \frac{2}{r}$

(e) $\displaystyle\sum_{r=2}^{4} \frac{1}{r(r+1)}$

(k) $\displaystyle\sum_{0}^{3} (2)^r(r-1)$

(f) $\displaystyle\sum_{r=3}^{5} (-1)^r(r+1)(r+3)$

(l) $\displaystyle\sum_{1}^{4} (-1)^{r+1} r(r+1).$

2 Express in Σ notation

(a) $1+2+3+4+5$

(g) $1.2+2.3+3.4+4.5+5.6$

(b) $7+11+15+19$

(c) $3+1-1-3-5$

(h) $1.4+4.7+7.10+10.13$

(d) $\dfrac{1}{3}+\dfrac{1}{5}+\dfrac{1}{7}+\dfrac{1}{9}$

(i) $2+1+\dfrac{1}{2}+\dfrac{1}{4}+\dfrac{1}{8}$

(e) $-1+4-9+16-25$

(j) $x+2x^2+3x^3+4x^4$

(f) $1-\dfrac{1}{4}+\dfrac{1}{9}-\dfrac{1}{16}+\dfrac{1}{25}$

(k) $1-a+a^2-a^3+a^4.$

3.3 Series of powers of the natural numbers

Consider S_n, the sum of the first n natural numbers.

Then $\quad S_n = \quad 1 \quad + \quad 2 \quad + \quad 3 \quad + \ldots \quad +(n-1)+ \quad n$

also $\quad S_n = \quad n \quad +(n-1)+(n-2)+ \ldots \quad + \quad 2 \quad + \quad 1$

Adding $\quad 2S_n = (n+1) +(n+1)+(n+1)+ \ldots \quad +(n+1)+(n+1)$

$= n(n+1)$

Thus $\quad S_n = \tfrac{1}{2}n(n+1) = \displaystyle\sum_{r=1}^{n} r$, the sum of the first n natural numbers.

The series $1 + 2 + 3 + \ldots + n$ is an example of an *arithmetic progression*, the general form of which is

$$a + (a+d) + (a+2d) + \ldots + (l-d) + l.$$

where a is the first term, l is the last term and d is the common difference between successive terms.

Again $\quad S_n = \quad a \quad + (a+d) + (a+2d) + \ldots + (l-d) + \quad l$
and $\quad\;\; S_n = \quad l \quad + (l-d) + (l-2d) + \ldots + (a+d) + \quad a$

Adding $\quad 2S_n = (a+l) + (a+l) + (a+l) + \ldots + (a+l) + (a+l)$
$$= n(a+l)$$

Thus for an arithmetic progression, $S_n = \tfrac{1}{2}n(a+l)$.

$$l = \text{the } n\text{th term} = a + (n-1)d$$

giving
$$S_n = \tfrac{1}{2}n[a + a + (n-1)d]$$
$$= \tfrac{1}{2}n[2a + (n-1)d].$$

Next consider the series $1^2 + 2^2 + 3^2 + \ldots + n^2$.

$$(r+1)^3 \;-\; r^3 \;\equiv\; 3r^2 \;+\; 3r \;+\; 1$$

Putting in turn $r = 1, 2, 3, \ldots, (n-2), (n-1), n$ in this identity we get

$$
\begin{array}{rclcrcrcr}
2^3 & - & 1^3 & \equiv & 3(1)^2 & + & 3(1) & + & 1 \\
3^3 & - & 2^3 & \equiv & 3(2)^2 & + & 3(2) & + & 1 \\
4^3 & - & 3^3 & \equiv & 3(3)^2 & + & 3(3) & + & 1
\end{array}
$$

$$
\begin{array}{rclcrcrcr}
(n-1)^3 & - & (n-2)^3 & \equiv & 3(n-2)^2 & + & 3(n-2) & + & 1 \\
n^3 & - & (n-1)^3 & \equiv & 3(n-1)^2 & + & 3(n-1) & + & 1 \\
(n+1)^3 & - & n^3 & \equiv & 3n^2 & + & 3n & + & 1
\end{array}
$$

Adding $\quad (n+1)^3 \;-\; 1^3 \;=\; 3\sum\limits_{r=1}^{n} r^2 + 3\sum\limits_{r=1}^{n} r \;+\; n$

i.e. $\quad 3\sum\limits_{r=1}^{n} r^2 = (n+1)^3 - 1^3 - 3\sum\limits_{r=1}^{n} r - n$
$$= n^3 + 3n^2 + 3n - 3 \times \tfrac{1}{2}n(n+1) - n$$

giving $\quad \sum\limits_{r=1}^{n} r^2 = \tfrac{1}{6}n(n+1)(2n+1).$

Example

Evaluate $\sum\limits_{r=1}^{n} r(r+1)$.

The series can be written as $1(2) + 2(3) + 3(4) + \ldots + n(n+1)$.
Let the sum of the series be S_n.

Then
$$S_n = \sum_{r=1}^{n} r(r+1) = \sum_{r=1}^{n} (r^2 + r)$$

$$= \sum_{r=1}^{n} r^2 + \sum_{r=1}^{n} r$$

$$= \frac{1}{6}n(n+1)(2n+1) + \tfrac{1}{2}n(n+1)$$

$$= \frac{1}{6}n(n+1)[(2n+1) + 3]$$

$$= \frac{1}{3}n(n+1)(n+2).$$

Exercises 3.3
1 Evaluate

(a) $\displaystyle\sum_{r=1}^{n} (2r-1)$ (e) $\displaystyle\sum_{r=1}^{n} r(r-1)$

(b) $\displaystyle\sum_{r=1}^{n} (3r+2)$ (f) $\displaystyle\sum_{r=1}^{n} (r+1)(r+2)$

(c) $\displaystyle\sum_{r=1}^{2n} (2r-3)$ (g) $\displaystyle\sum_{r=0}^{n} (r+1)^2$

(d) $\displaystyle\sum_{r=1}^{n} r(r+2)$ (h) $\displaystyle\sum_{r=1}^{2n} (-1)^r r.$

2 (a) Find the sum of the squares of the first $2n$ natural numbers.
(b) Find the sum of the squares of the first n even natural numbers.
(c) Find the sum of the squares of the first n odd natural numbers.

3 Use the identity $(r+1)^4 - r^4 \equiv 4r^3 + 6r^2 + 4r + 1$ to show that

$$\sum_{r=1}^{n} r^3 = \tfrac{1}{4}n^2(n+1)^2 = \left[\sum_{r=1}^{n} r\right]^2.$$

4 Evaluate

(a) $\displaystyle\sum_{1}^{n} r^2(r+1)$ (b) $\displaystyle\sum_{1}^{n} r(r+1)(r+2)$ (c) $\displaystyle\sum_{1}^{n} r(r+2)(r+4).$

3.4 Mathematical induction
The results of the previous section could also be proved by the method of
mathematical induction. This method is used to prove a given result which is in
terms of a positive integer, usually represented by the letter n. The method will be
illustrated in the following proof of the result

$$\sum_{r=1}^{n} r = \tfrac{1}{2}n(n+1).$$

Let T be the set of integers for which $\sum_{r=1}^{n} r = \frac{1}{2}n(n+1)$, i.e. T is the truth set.

(i) Is the result true for $n = 1$?

If $n = 1$, the series consists of the first term only. Also $\frac{1}{2}n(n+1) = 1$ when $n = 1$ and so the result is true for $n = 1$, i.e. $1 \in T$.

(ii) If the result is true for a particular value of n, say k, is it inevitably true also for $n = k+1$, i.e. does $k \in T \Rightarrow (k+1) \in T$?

Assume that $k \in T$, i.e. $1 + 2 + 3 + \ldots + k = \frac{1}{2}k(k+1)$. By adding $(k+1)$ to each side we get

$$1 + 2 + 3 + \ldots + k + (k+1) = \frac{1}{2}k(k+1) + (k+1)$$
$$= \frac{1}{2}(k+1)(k+2),$$

which is $\frac{1}{2}n(n+1)$ when $n = (k+1)$.

Thus if the result is true for $n = k$, then it is also true for $n = k+1$, i.e. $k \in T \Rightarrow k+1 \in T$.

From (i) $1 \in T$ and so by (ii) $2 \in T$ and then by (ii) again $3 \in T$ and so on indefinitely, showing that all values of $n \in T$,

$$\text{i.e. the result} \quad \sum_{r=1}^{n} r = \frac{1}{2}n(n+1)$$

is true for all values of the positive integer n.

In the proof that a result is true for all positive integral values of n there are two essential steps:

(i) to show that the result is true for $n = 1$, i.e. that $1 \in T$,

(ii) to show that, if the result is true for $n = k$, it must also be true for $n = k+1$, i.e. that $k \in T \Rightarrow k+1 \in T$.

Although step (i) is often trivial, it must *never* be omitted or assumed.

Example

Evaluate $\sum_{r=1}^{n} \dfrac{1}{r(r+1)}$.

$$S_n = \frac{1}{1(2)} + \frac{1}{2(3)} + \frac{1}{3(4)} + \ldots + \frac{1}{n(n+1)}.$$

By simple addition it can be seen that $S_1 = 1/2$, $S_2 = 2/3$, $S_3 = 3/4$, $S_4 = 4/5$, suggesting that $S_n = n/(n+1)$. This result can be proved by mathematical induction.

(i) If $n = 1$, $S_1 = \dfrac{1}{2}$ and $\dfrac{n}{n+1} = \dfrac{1}{2}$ and so $1 \in T$.

(ii) Assume that $S_k = \dfrac{k}{k+1}$, i.e. $k \in T$,

so that $\quad \dfrac{1}{1(2)} + \dfrac{1}{2(3)} + \dfrac{1}{3(4)} + \ldots + \dfrac{1}{k(k+1)} = \dfrac{k}{k+1}$.

Then
$$\frac{1}{1(2)}+\frac{1}{2(3)}+\frac{1}{3(4)}+\ldots+\frac{1}{k(k+1)}+\frac{1}{(k+1)(k+2)}$$

$$=\frac{k}{k+1}+\frac{1}{(k+1)(k+2)}$$

$$=\frac{k(k+2)+1}{(k+1)(k+2)}$$

$$=\frac{(k+1)^2}{(k+1)(k+2)}$$

$$=\frac{k+1}{k+2}, \text{ which is } \frac{n}{n+1} \text{ when } n=k+1.$$

Thus on the assumption that $S_n = n/(n+1)$ is true for $n=k$, this result is also true for $n=k+1$, i.e. $k \in T \Rightarrow k+1 \in T$.

Thus having shown that (i) $1 \in T$ and (ii) $k \in T \Rightarrow k+1 \in T$, we know, by mathematical induction, that $S_n = n/(n+1)$ is true for all values of the positive integer n.

It should be noted that the method of mathematical induction is not restricted in its use to results concerned with the summation of series.

Exercises 3.4

1 Use the method of mathematical induction to show that

(a) $\displaystyle\sum_{r=1}^{n} r^2 = \frac{1}{6}n(n+1)(2n+1)$

(b) $\displaystyle\sum_{r=1}^{n} r^3 = \frac{1}{4}n^2(n+1)^2$

(c) $\displaystyle\sum_{r=1}^{n} r(r+1) = \frac{1}{3}n(n+1)(n+2)$

(d) $\displaystyle\sum_{r=1}^{n} r(r+1)(r+2) = \frac{1}{4}n(n+1)(n+2)(n+3)$

(e) $\dfrac{1}{1(3)}+\dfrac{1}{3(5)}+\dfrac{1}{5(7)}+\ldots+\dfrac{1}{(2n-1)(2n+1)}=\dfrac{n}{2n+1}.$

3.5 Geometric progressions

The series
$$1+\frac{1}{2}+\frac{1}{4}+\frac{1}{8}+\ldots+\frac{1}{2^{n-1}} \qquad \ldots (1)$$

and
$$1+2+4+8+\ldots+2^{n-1} \qquad \ldots (2)$$

are both *geometric progressions*, i.e. series of the form

$$a+ar+ar^2+\ldots+ar^{n-1},$$

where a is the first term and r is the common ratio of successive terms. In series

(1), $a = 1$ and $r = \frac{1}{2}$ and in series (2) $a = 1$ and $r = 2$.

Let $$S_n = a + ar + ar^2 + \ldots + ar^{n-1}$$
Multiplying by r, $$rS_n = ar + ar^2 + \ldots + ar^{n-1} + ar^n$$
Subtracting, $$(1-r)S_n = a \qquad\qquad\qquad\qquad - ar^n$$

giving $$S_n = \frac{a(1-r^n)}{1-r}, \ (r \neq 1).$$

If l, m and n are three consecutive terms of a geometric progression, then m is said to be a *geometric mean* of l and n.

The common ratio $r = \dfrac{m}{l} = \dfrac{n}{m}$ gives $m^2 = ln$,

i.e. there are two geometric means of l and n, $+\sqrt{(ln)}$ and $-\sqrt{(ln)}$. (cf. the arithmetic mean of l and n which is $\frac{1}{2}(l+n)$.)

Example 1
If the second term of a geometric progression is 6 and the fifth term is -162, find the first term and the common ratio.

Let the progression be $a + ar + ar^2 + \ldots$. Then $ar = 6$ and $ar^4 = -162$.
Dividing, we get $r^3 = (-162)/6 = -27$ giving $r = -3$ and $a = -2$.

Example 2
Find the least number of terms for which the sum of the geometric progression
$\dfrac{1}{2} + \dfrac{3}{4} + \dfrac{9}{8} + \ldots$ exceeds 100.

$a = \frac{1}{2}$ and $r = 3/2$ and so

$$S_n = \frac{\frac{1}{2}[1-(3/2)^n]}{1-(3/2)} \quad \text{or} \quad \frac{\frac{1}{2}[(3/2)^n - 1]}{(3/2) - 1}$$

$$S_n = \left(\frac{3}{2}\right)^n - 1 > 100$$

$\therefore$ $$1\cdot5^n > 101$$
$\therefore$ $$n\log_{10}1\cdot5 > \log_{10}101$$
i.e. $$0\cdot176 ln > 2\cdot0043$$
$$n > \frac{2\cdot0043}{0\cdot1761} \quad (\approx 11\cdot38).$$

Thus the least number of terms required is 12.

Example 3
Find the least number of terms for which the sum of the series $1 + \dfrac{1}{2} + \dfrac{1}{4} + \dfrac{1}{8} + \ldots$

differs from 2 by less than 0·0001.

$a = 1$ and $r = \frac{1}{2}$ and so

$$S_n = \frac{1[1 - (\frac{1}{2})^n]}{1 - \frac{1}{2}} = 2[1 - (\frac{1}{2})^n].$$

We require the least value of n for which

$$2 - 2[1 - (\frac{1}{2})^n] < 0 \cdot 0001$$

i.e. $\qquad\qquad 2\,(\frac{1}{2})^n < 0 \cdot 0001 \quad \text{or} \quad \dfrac{2}{0 \cdot 0001} < 2^n$

$\therefore \qquad\qquad\qquad 20\,000 < 2^n$

$\therefore \qquad\qquad \log_{10} 20\,000 < n \log_{10} 2$

$\therefore \qquad\qquad\qquad\qquad n > \dfrac{4 \cdot 3010}{0 \cdot 3010} \quad (\approx 14 \cdot 29).$

Thus the least number of terms required is 15.

Example 4

A man borrows £10000 on 1st January 1979. Compound interest at the rate of 12 per cent is added on the 31st December each year. The man is to repay the loan in twenty equal annual instalments of £x on 1st January of each year, the first repayment being made on 1st January 1980. Calculate the value of x to the nearest integer.

On 31 Dec '79 the man owes £10 000 + 12% of £10 000, i.e. £10 000 × 1·12
On 1 Jan '80 the man owes £10 000 × 1·12 − £x (after one repayment)
On 31 Dec '80 the man owes £(10 000 × 1·12 − x) × 1·12
On 1 Jan '81 the man owes £(10 000 × 1·12 − x) × 1·12 − £x
i.e. $\qquad\qquad$ £10 000 × 1·12² − £x(1 + 1·12)
$\qquad\qquad\qquad\qquad\qquad\qquad$ (after two repayments)
On 31 Dec '81 the man owes £[10 000 × 1·12² − x(1 + 1·12)] × 1·12
On 1 Jan '82 the man owes £[10 000 × 1·12² − x(1 + 1·12)] × 1·12 − £x
i.e. $\qquad\qquad$ £10 000 × 1·12³ − £x(1 + 1·12 + 1·12²)
$\qquad\qquad\qquad\qquad\qquad\qquad$ (after three repayments).

After 20 repayments therefore he owes

$$£[10\,000 \times 1 \cdot 12^{20} - x(1 + 1 \cdot 12 + 1 \cdot 12^2 + \ldots + 1 \cdot 12^{19})]$$

and with the 20th repayment the debt is cleared.

$\therefore \qquad 10\,000 \times 1 \cdot 12^{20} - x(1 + 1 \cdot 12 + 1 \cdot 12^2 + \ldots + 1 \cdot 12^{19}) = 0.$

$1 + 1 \cdot 12 + 1 \cdot 12^2 + \ldots + 1 \cdot 12^{19}$ is a geometric progression in which $a = 1$, $r = 1 \cdot 12$ and $n = 20$.

$$\therefore \qquad 10\,000 \times 1 \cdot 12^{20} - x\left[\frac{1(1 \cdot 12^{20} - 1)}{1 \cdot 12 - 1}\right] = 0$$

$$x = \frac{(10\,000 \times 1 \cdot 12^{20})(1 \cdot 12 - 1)}{1 \cdot 12^{20} - 1} \approx 1339.$$

Exercises 3.5

1 Find the sum of the first n terms of the following geometric series

    (a) $3 + 6 + 12 + 24 + \ldots$ when $n = 6$

    (b) $2 + 1 + \frac{1}{2} + \frac{1}{4} + \ldots$ when $n = 10$

    (c) $-1 + 4 - 16 + 64 - \ldots$ when $n = 12$

    (d) $3 - \frac{9}{4} + \frac{27}{16} - \frac{81}{64} + \ldots$ when $n = 15$

    (e) $\frac{1}{x^2} + \frac{1}{x} + 1 + x + \ldots$ when $n = 20$

    (f) $a - a^2 + a^3 - a^4 + \ldots$ when $n = 12$

2 Find the common ratio and the rth term of the geometric series S when

    (a) $S = \frac{1}{2} + \frac{3}{4} + \frac{9}{8} + \ldots$ and $r = 6$

    (b) $S = 2 - 6 + 18 - \ldots$ and $r = 8$

    (c) $S = 2 + \frac{1}{2} + \frac{1}{8} + \ldots$ and $r = 6$

    (d) $S = 4 - 6 + 9 - \ldots$ and $r = 5$

    (e) $S = x - 2x^2 + 4x^3 - \ldots$ and $r = 7$

    (f) $S = \frac{1}{a} + \frac{b}{a^2} + \frac{b^2}{a^3} + \ldots$ and $r = 10$.

3 Evaluate $\displaystyle\sum_{r=1}^{n} \left(\frac{1}{x}\right)^{2r-1}$.

4 Evaluate $\displaystyle\sum_{r=1}^{n} x^{2r-3}$.

5 Find the first five terms of the geometric progression whose third term is $0 \cdot 09$ and whose 6th term is $2 \cdot 43$.

6 Find the geometric means of (a) $1 \cdot 44$ and 1, (b) $\frac{3}{4}$ and $\frac{1}{3}$.

7 If the value of the pound decreases at the rate of 20 per cent per year, find the value of the pound after 10 years and show that the values of the pound at the end of successive years form a geometric progression.

8 A man borrows £1000 to buy a car on 1st January 1978. On the last day of each month, compound interest at the rate of 1 per cent per month is added. The man repays the debt in twelve equal instalments made on the first day of each month, the first instalment being paid on 1st February 1978. Find the value of each instalment.

3.6 Infinite geometric progressions

If $S_n = 1 + 2 + 4 + \ldots + 2^{n-1}$,

then $S_1 = 1, S_2 = 3, S_3 = 7, S_4 = 15, S_5 = 31, \ldots, S_n = 2^n - 1$. This is a sequence of ever increasing numbers. Furthermore S_n can be made to exceed any chosen number, however large, by taking a large enough value for n, i.e. S_n increases indefinitely. The series

$$1 + 2 + 4 + \ldots + 2^{n-1} + \ldots$$

is said to diverge or to be divergent.

Consider now the series

$$S_n = 1 + \frac{1}{2} + \frac{1}{4} + \ldots + \left(\frac{1}{2}\right)^{n-1}.$$

Here $S_1 = 1$, $S_2 = 3/2$, $S_3 = 7/4$, $S_4 = 15/8$, $S_5 = 31/16$, $\ldots$, $S_n = 2 - (\frac{1}{2})^{n-1}$. Again we have a sequence of ever increasing numbers but, although each number is larger than the preceding number, every number in the sequence is less than 2. Indeed the difference between 2 and S_n can be made as small as we choose by taking a large enough value for n. (See example 3 in section 3.5.) We say that this sequence tends to a limit 2 as n tends to infinity. The infinite series

$$1 + \frac{1}{2} + \frac{1}{4} + \ldots + \left(\frac{1}{2}\right)^{n-1} + \ldots$$

is said to converge or to be convergent.

The limiting value of the sum of n terms of this series as n tends to infinity is 2. This limiting value is sometimes called the 'sum to infinity' of the series. The difference in the meaning of the word 'sum' in 'sum to infinity' and in the 'sum to n terms' should be noted. The sum to n terms is an actual arithmetical sum and can be obtained by simply adding together the first n terms of the series. On the other hand the 'sum to infinity' will never equal the sum of any number of terms no matter how large the number of terms taken may be. The 'sum to infinity' is the limit to which the sum to n terms converges as n gets larger and larger.

Consider the geometric progression

$$a + ar + ar^2 + \ldots + ar^{n-1} + \ldots.$$

$$S_n = \frac{a(1 - r^n)}{1 - r} = \frac{a}{1 - r} - \frac{ar^n}{1 - r}.$$

If $|r| < 1$, then $|r^n|$ and also $\left|\dfrac{r^n}{1-r}\right|$ can be made as small as we choose by taking n sufficiently large, i.e. the limit of S_n as n tends to infinity is $a/(1-r)$. In other words this geometric progression converges and the limiting value of its sum to n terms as n tends to infinity, or the 'sum to infinity', is $a/(1-r)$.

If $|r| \geqslant 1$, then S_n does not tend to a limit and the series does not converge. The series is then said to be divergent.

Example

Express the recurring decimal $0 \cdot \dot{2} \dot{3}$ in the form a/b, where a and b are integers without a common factor.

$$0 \cdot \dot{2} \dot{3} = 0 \cdot 232\,323\,23 \ldots = 0 \cdot 23 + 0 \cdot 0023 + 0 \cdot 000\,023 + \ldots.$$

This is a geometric progression with $a = 0 \cdot 23$ and $r = 0 \cdot 01$. Since $|r| < 1$, the series is convergent and the limiting value of its sum to n terms as n tends to infinity, or its 'sum to infinity', is

$$\frac{a}{1-r} = \frac{0 \cdot 23}{1-0 \cdot 01} = \frac{23}{99}.$$

An alternative method is to let

$$n = 0 \cdot 232\,323\,23 \ldots$$

then
$$100n = 23 \cdot 232\,323\,23 \ldots$$

subtracting
$$99n = 23, \text{ i.e. } n = \frac{23}{99}.$$

Exercises 3.6

1 Does the infinite geometric progression $u_1 + u_2 + u_3 + \ldots u_r + \ldots$ converge, and if it does find the sum to infinity when

(a) $u_r = 2(\frac{2}{3})^{r-1}$

(d) $u_r = (-1)^{r-1}\left(\frac{1}{20}\right)\left(\frac{3}{2}\right)^{r-1}$

(b) $u_r = \frac{1}{20}\left(\frac{3}{2}\right)^{r-1}$

(e) $u_r = a^r$ if $|a| < 1$

(c) $u_r = 100(\frac{1}{2})^{r-1}$

(f) $u_r = \left(\frac{1}{a}\right)^r$ if $|a| > 1$.

2 Express the recurring decimal $0 \cdot \dot{1}\dot{2}$ in the form a/b, where a and b are integers without a common factor.

3 If the second term of a geometric progression is 12 and the fifth term is $0 \cdot 0405$, show that the series converges and find its sum to infinity.

4 The first term of an infinite geometric progression is 10 and its sum to infinity is 25. Find the first four terms of the series.

5 The second term of a geometric progression is $1 \cdot 8$ and its sum to infinity is 20. Find the first four terms of each of the possible progressions.

3.7 The binomial expansion

$(1+x)^0 = 1$
$(1+x)^1 = 1+x$
$(1+x)^2 = 1+2x+x^2$
$(1+x)^3 = 1+3x+3x^2+x^3$
$(1+x)^4 = 1+4x+6x^2+4x^3+x^4$
$(1+x)^5 = 1+5x+10x^2+10x^3+5x^4+x^5$

These results can be obtained by continued multiplication but this method is clearly a tedious one. The coefficients can be obtained readily from Pascal's triangle given in the following diagram.

$$
\begin{array}{ccccccccccccc}
&&&&&& 1 &&&&&& \\
&&&&& 1 && 1 &&&&& \\
&&&& 1 && 2 && 1 &&&& \\
&&& 1 && 3 && 3 && 1 &&& \\
&& 1 && 4 && 6 && 4 && 1 && \\
& 1 && 5 && 10 && 10 && 5 && 1 & \\
&&& . & . & . & . & . & . & . & . &
\end{array}
$$

From the above expansions and from Pascal's triangle, it is clear that the coefficients of the powers of x do follow a pattern and from this we can investigate the expansion of $(1+x)^n$, where n is any positive integer.

It appears that the first two terms of the expansion of $(1+x)^n$ will be $1+nx$. It can be checked that, in the above expansions, the coefficient of x^2 is given by

$$
\frac{n(n-1)}{(1)(2)} \text{ for } n = 2, 3, 4, 5.
$$

This suggests
$$
\frac{n(n-1)(n-2)}{(1)(2)(3)}
$$

for the coefficient of x^3 and checking shows that this is true for $n = 3, 4, 5$. If this pattern is maintained

$$
(1+x)^n = 1 + nx + \frac{n(n-1)}{2!}x^2 + \frac{n(n-1)(n-2)}{3!}x^3 + \ldots
$$

$$
+ \frac{n(n-1)\ldots(n-r+1)}{r!}x^r + \ldots + \frac{n(n-1)\ldots(2)(1)}{n!}x^n,
$$

where $n! = 1.2 \ldots (n-1)n$.

The method of mathematical induction will now be used to show that this is true for any positive integer n.

(i) If $n = 1$, LHS $= 1 + x$, RHS $= 1 + x + 0 + 0 + \ldots$, i.e. $1 \in T$.

(ii) Assume that the result is true for $n = k$, i.e. assume $k \in T$. Then

$$
(1+x)^k = 1 + kx + \frac{k(k-1)}{2!}x^2 + \ldots + \frac{k(k-1)\ldots(k-r+1)}{r!}x^r + \ldots
$$

and
$$
x(1+x)^k = \quad x + \quad kx^2 + \ldots + \frac{k(k-1)\ldots(k-r+2)}{(r-1)!}x^r + \ldots
$$

$$
(1+x)(1+x)^k = 1 + (k+1)x + \left[\frac{k(k-1)}{2} + k\right]x^2 + \ldots
$$

$$
+ \left[\frac{k(k-1)\ldots(k-r+1)}{r!} + \frac{k(k-1)\ldots(k-r+2)}{(r-1)!}\right]x^r + \ldots
$$

$$
\therefore \quad (1+x)^{k+1} = 1 + (k+1)x + \frac{(k+1)k}{2!}x^2 + \ldots + Ax^r + \ldots,
$$

where
$$A = \left[\frac{k(k-1)\ldots(k-r+2)}{(r-1)!} \right] \left[\frac{k-r+1}{r} + 1 \right]$$
$$= \frac{(k+1)k(k-1)\ldots(k-r+2)}{r!},$$

showing that the result is true for $n = k+1$, i.e. $k \in T \Rightarrow k+1 \in T$. Hence by mathematical induction the result is true for all values of the positive integer n.

The coefficient of x^r in this expansion of $(1+x)^n$ can also be expressed as

$$\frac{n!}{(n-r)!r!}$$

When n is a positive integer it is clear that the expansion, which is known as the binomial expansion, gives a finite series. For instance, if $n = 3$, the expansion becomes

$$(1+x)^3 = 1 + 3x + \frac{(3)(2)}{(1)(2)}x^2 + \frac{(3)(2)(1)}{(1)(2)(3)}x^3 + \frac{(3)(2)(1)(0)}{(1)(2)(3)(4)}x^4 + \ldots,$$

where the coefficients of x^4 and higher powers of x contain a factor of zero and so the series terminates with the term in x^3 and therefore has just four terms. When n is a positive integer the series given is a finite one with $(n+1)$ terms.

However if n is negative, the numerators of the coefficients will never contain a factor of zero since, in that case, $n-r+1$ will never equal zero. Then the expansion does not terminate, and the result is an infinite series. This is also the case when n is fractional.

Let us consider a particular case when n is not a positive integer. When $x = 1$ and $n = -1$, the expansion would give

$$(1+1)^{-1} = 1 + (-1)(1) + \frac{(-1)(-2)}{(1)(2)}(1)^2 + \frac{(-1)(-2)(-3)}{(1)(2)(3)}(1)^3 + \ldots$$
$$= 1 - 1 + 1 - 1 + \ldots,$$

whereas $(1+1)^{-1} = 2^{-1} = \frac{1}{2}$. Clearly the expansion is not valid in this particular case when $x = 1$ and $n = -1$.

Nevertheless it can be shown, although not at this stage, that the binomial expansion is valid for all real values of n (and not just for positive integral values of n) when $|x| < 1$, i.e. when $-1 < x < 1$.

The expansion of $(1+x)^n$ is a special case of the more general binomial expansion of $(a+x)^n$.

$$(a+x)^n = \left[a\left(1 + \frac{x}{a}\right) \right]^n = a^n \left(1 + \frac{x}{a}\right)^n$$

$$= a^n \left[1 + n\left(\frac{x}{a}\right) + \frac{n(n-1)}{2!}\left(\frac{x}{a}\right)^2 + \ldots + \frac{n(n-1)\ldots(n-r+1)}{r!}\left(\frac{x}{a}\right)^r + \ldots \right]$$

$$(a+x)^n = a^n + na^{n-1}x + \frac{n(n-1)}{2!}a^{n-2}x^2 + \ldots$$

$$+ \frac{n(n-1)\ldots(n-r+1)}{r!}a^{n-r}x^r + \ldots$$

The expression $\dfrac{n(n-1)\ldots(n-r+1)}{r!}$ is usually written as $\dbinom{n}{r}$.

Then $\quad \dbinom{n}{1} = n, \quad \dbinom{n}{2} = \dfrac{n(n-1)}{2!}, \quad \dbinom{n}{3} = \dfrac{n(n-1)(n-2)}{3!}$, etc.

which is obtained by multiplying the numerator and denominator by $(n-r)!$. The reader may well find it easier to remember the $(1+x)^n$ form of the binomial expansion and adapt it for use with expressions of the $(a+x)^n$ type as shown in part (b) of the following example.

Example 1
Expand (a) $(1-2x)^4$ (b) $(2+x)^{-1}$.

(a) Here $n = 4$ and replacing x by $(-2x)$ we get

$$(1-2x)^4 = 1 + 4(-2x) + \frac{(4)(3)}{(1)(2)}(-2x)^2 + \frac{(4)(3)(2)}{(1)(2)(3)}(-2x)^3$$

$$+ \frac{(4)(3)(2)(1)}{(1)(2)(3)(4)}(-2x)^4$$

$$= 1 - 8x + 24x^2 - 32x^3 + 16x^4.$$

(b) $(2+x)^{-1} = \left[2\left(1+\dfrac{x}{2}\right)\right]^{-1} = 2^{-1}\left(1+\dfrac{x}{2}\right)^{-1}$.

Then using the $(1+x)^n$ expansion with $n = -1$ and replacing x by $\dfrac{x}{2}$ we get

$$(2+x)^{-1} = 2^{-1}\left[1+\frac{x}{2}\right]^{-1}$$

$$= \tfrac{1}{2}\left[1 - \frac{x}{2} + \frac{(-1)(-2)}{(1)(2)}\left(\frac{x}{2}\right)^2 + \frac{(-1)(-2)(-3)}{(1)(2)(3)}\left(\frac{x}{2}\right)^3 + \ldots\right]$$

$$= \tfrac{1}{2}\left(1 - \frac{x}{2} + \frac{x^2}{4} - \frac{x^3}{8} + \ldots\right).$$

This is a geometric progression with common ratio $(-x/2)$. It will converge provided that $-2 < x < 2$.

Example 2
Find approximations to (a) $1 \cdot 03^5$ (b) $\sqrt{1 \cdot 02}$ (c) $\sqrt{3 \cdot 96}$.

(a) $1.03^5 = (1+0.03)^5$

$$= 1 + 5(0.03) + \frac{(5)(4)}{(1)(2)}(0.03)^2 + \frac{(5)(4)(3)}{(1)(2)(3)}(0.03)^3 + \frac{(5)(4)(3)(2)}{(1)(2)(3)(4)}(0.03)^4$$

$$+ (0.03)^5$$

$$= 1 + 0.15 + 0.009 + 0.000\,27 + 0.000\,004\,05 + 0.000\,000\,024\,3$$

$$= 1.1593 \quad \text{to 5 significant figures.}$$

(b) $\sqrt{1.02} = (1+0.02)^{\frac{1}{2}}$.

Here the binomial expansion is valid because, although $n = \frac{1}{2}$, we have $x = 0.02$ so that $-1 < x < 1$.

$$\therefore \quad \sqrt{1.02} = (1+0.02)^{\frac{1}{2}}$$

$$= 1 + (\tfrac{1}{2})(0.02) + \frac{(\frac{1}{2})(-\frac{1}{2})}{(1)(2)}(0.02)^2 + \frac{(\frac{1}{2})(-\frac{1}{2})(-\frac{3}{2})}{(1)(2)(3)}(0.02)^3 + \; \ldots$$

$$= 1 + 0.01 - 0.000\,05 + 0.000\,000\,5 - \; \ldots$$

$$= 1.009\,95 \quad \text{to 5 decimal places.}$$

(c) $\sqrt{3.96} = (4 - 0.04)^{\frac{1}{2}} = [4(1 - 0.01)]^{\frac{1}{2}} = 4^{\frac{1}{2}}(1-0.01)^{\frac{1}{2}} = 2(1-0.01)^{\frac{1}{2}}$.

Here again the binomial expansion is valid because, although $n = \frac{1}{2}$, we have $x = -0.01$ and so $-1 < x < 1$.

$$\sqrt{3.96} = 2\left[1 + (\tfrac{1}{2})(-0.01) + \frac{(\frac{1}{2})(-\frac{1}{2})}{(1)(2)}(-0.01)^2 + \frac{(\frac{1}{2})(-\frac{1}{2})(-\frac{3}{2})}{(1)(2)(3)}(-0.01)^3 + \; \ldots\right]$$

$$= 2[1 - 0.005 - 0.000\,125 \ldots]$$

$$\approx 1.990$$

Example 3

The value of g, the acceleration due to gravity, is to be found experimentally by measuring the length l and the period of oscillation T of a simple pendulum. If there is an error of $+1$ per cent in the measured length, an error of -1 per cent in the measured period T and g is calculated from the formula

$$g = \frac{4\pi^2 l}{T^2},$$

find the percentage error in the resulting value of g.

Let T_0, l_0 and g_0 be the true values and T_1, l_1 and g_1 be the experimental values. Then

$$T_1 = T_0 - (0.01)T_0 = T_0(1-0.01) \quad \text{and} \quad l_1 = l_0(1+0.01)$$

and

$$g_1 = \frac{4\pi^2 l_0(1+0.01)}{T_0{}^2(1-0.01)^2} = g_0(1+0.01)(1-0.01)^{-2}$$

$$= g_0(1+0.01)\left[1 + (-2)(-0.01) + \frac{(-2)(-3)}{(1)(2)}(-0.01)^2 + \; \ldots\right]$$

$$= g_0(1+0.01)(1+0.02+0.0003+ \; \ldots)$$

$$= g_0(1+0.03+0.0005+ \; \ldots)$$

$$\approx 1.0305 g_0$$

i.e. g_1 is approximately $3\cdot05$ per cent too large.

Example 4
Expand $\sqrt{(1+x-2x^2)}$ in terms of ascending powers of x up to and including the term in x^2.

$$\sqrt{(1+x-2x^2)} = \sqrt{(1-x)(1+2x)} = (1-x)^{\frac{1}{2}}(1+2x)^{\frac{1}{2}}$$

$$(1-x)^{\frac{1}{2}} = 1 + (\tfrac{1}{2})(-x) + \frac{(\tfrac{1}{2})(-\tfrac{1}{2})}{(1)(2)}(-x)^2 + \ldots$$

$$= 1 - \tfrac{1}{2}x - \tfrac{1}{8}x^2 + \ldots$$

$$(1+2x)^{\frac{1}{2}} = 1 + (\tfrac{1}{2})(2x) + \frac{(\tfrac{1}{2})(-\tfrac{1}{2})}{(1)(2)}(2x)^2 + \ldots$$

$$= 1 + x - \tfrac{1}{2}x^2 + \ldots$$

$$\sqrt{(1+x-2x^2)} = (1 - \tfrac{1}{2}x - \tfrac{1}{8}x^2 + \ldots)(1 + x - \tfrac{1}{2}x^2 + \ldots)$$

$$= 1 + \frac{1}{2}x - \frac{9}{8}x^2 + \ldots$$

Another method is to put $y = x - 2x^2$ in the expansion of $(1+y)^{1/2}$.

Exercises 3.7
1 Expand (a) $(1-x)^5$ (b) $(1+2x)^4$ (c) $(3+2x)^3$ (d) $(4-x)^5$.
2 Expand in terms of ascending powers of x up to and including the term in x^3

(a) $(1-x)^{\frac{1}{2}}$ (c) $\sqrt[3]{(1+3x)}$ (e) $\dfrac{1}{3-x}$

(b) $\dfrac{1}{(1+x)^2}$ (d) $\sqrt{(4+x)}$ (f) $\dfrac{1}{\sqrt[3]{(8+x)}}$.

3 Expand in terms of ascending powers of x up to and including the term in x^3
(a) $(1-x-2x^2)^5$ (b) $(2-x-x^2)^{-2}$.
4 Without using tables or calculator, find approximately

(a) $\sqrt{0\cdot98}$ (b) $\sqrt{9\cdot27}$ (c) $\dfrac{1}{\sqrt[3]{8\cdot08}}$.

5 If $x = ab^2/c$ and $y = a/(bc)$, find the approximate percentage errors in x and y given that there are errors of $+1\%$, -1% and -2% in the values of a, b and c respectively.

3.8 The remainder and factor theorems
Let
$$P(x) = a_n x^n + a_{n-1}x^{n-1} + \ldots + a_1 x + a_0,$$

where $a_n, a_{n-1}, \ldots, a_1, a_0$ are constants, n is a positive integer and $a_n \neq 0$. Then $P(x)$ is called a polynomial of degree n in x.

When $P(x)$ is divided by $(x - \alpha)$, where α is a constant, let the remainder be R. Then

$$P(x) \equiv (x - \alpha)\,Q(x) + R,$$

where $Q(x)$ is a polynomial of degree $(n-1)$. Putting $x = \alpha$ we get $P(\alpha) = R$. Then when the polynomial $P(x)$ is divided by $(x - \alpha)$ the remainder is $P(\alpha)$, i.e. the value of the polynomial when $x = \alpha$. This result is known as the *remainder theorem*.

Let the remainder be R_1 when the polynomial $P(x)$ is divided by $(px + q)$, where p and q are constants. Then

$$P(x) \equiv (px + q)\phi(x) + R_1,$$

where $\phi(x)$ is a polynomial of degree $(n-1)$. Putting $x = -q/p$ so that $px + q = 0$, we get $P(-q/p) = R_1$, i.e. when the polynomial $P(x)$ is divided by $(px + q)$ the remainder is $P(-q/p)$.

An important special case is when $(x - \alpha)$ is a factor of $P(x)$. Then $R = 0$ and so $P(\alpha) = 0$. This is known as the *factor theorem* and can be stated as: 'If $(x - \alpha)$ is a factor of the polynomial $P(x)$, then $P(\alpha) = 0$'. Also if $(px + q)$ is a factor of a polynomial $P(x)$, then $P(-q/p) = 0$.

The factor theorem is particularly useful in factorising polynomials and hence in solving polynomial equations.

Example 1

If $P(x) \equiv x^3 + 2x^2 - x - 2$, factorise $P(x)$ and find the remainders when $P(x)$ is divided by $(x + 3)$ and $(2x - 1)$.

Since the constant term of $P(x)$ is -2 this suggests that $x \pm 1$ or $x \pm 2$ may be factors.

When $P(x)$ is divided by

(a) $(x - 2)$, the remainder $P(2) = 8 + 8 - 2 - 2 \neq 0$, i.e. $x - 2$ is not a factor,

(b) $(x + 2)$, the remainder $P(-2) = -8 + 8 + 2 - 2 = 0$, i.e. $x + 2$ is a factor.

$\therefore \qquad P(x) = (x + 2)\,Q(x) = (x + 2)\,(x^2 - 1) = (x + 2)\,(x - 1)\,(x + 1)$.

That $(x - 1)$ and $(x + 1)$ are factors can be checked by showing that $P(1) = 0$ and $P(-1) = 0$.

When $P(x)$ is divided by

(a) $(x + 3)$, the remainder $P(-3) = -27 + 18 + 3 - 2 = -8$,

(b) $(2x - 1)$, the remainder $P(\tfrac{1}{2}) = \tfrac{1}{8} + \tfrac{1}{2} - \tfrac{1}{2} - 2 = -15/8$.

Example 2

$P(x) = x^3 + ax^2 + bx + c$ and when $P(x)$ is divided by $(x - 1)$, the remainder is -8. If $x^2 - x - 2$ is a factor of $P(x)$, find the values of a, b and c and solve the equation $P(x) = 0$.

When $P(x)$ is divided by $(x - 1)$ the remainder is -8.

$\therefore \qquad P(1) = 1 + a + b + c = -8 \qquad \qquad \ldots (1)$

$x^2 - x - 2 = (x - 2)(x + 1)$ and since $x^2 - x - 2$ is a factor of $P(x)$, then both $(x - 2)$ and $(x + 1)$ are factors of $P(x)$.

$$\therefore \qquad P(2) = 8 + 4a + 2b + c = 0 \qquad\qquad \ldots (2)$$
$$\text{and } P(-1) = -1 + a - b + c = 0 \qquad\qquad \ldots (3)$$

Solving equations 1, 2 and 3 we get $a = 2$, $b = -5$, $c = -6$.

$$\therefore \qquad\qquad P(x) = x^3 + 2x^2 - 5x - 6 = (x - 2)(x + 1)(x + 3).$$

Hence if $P(x) = 0$, then $x = 2$, -1 or -3.

Exercises 3.8

1 Find the remainder when $x^3 + 2x^2 + 3x + 4$ is divided by
   (a) $x - 2$ (b) $x + 1$ (c) $2x - 1$ (d) $3x + 2$ (e) x.
2 Find the remainder when $2x^3 - x^2 + 3x - 5$ is divided by
   (a) $x - 1$ (b) $2x + 1$ (c) $3x - 2$ (d) $2x$.
3 Factorise
   (a) $x^3 - 6x^2 + 11x - 6$ (d) $x^3 - a^3$
   (b) $4x^3 + 8x^2 - x - 2$ (e) $x^3 + a^3$
   (c) $x^3 + 3x^2 + 3x + 2$ (f) $x^3 - 7x^2y + 7xy^2 + 15y^3$.
4 Solve the equations
   (a) $x^3 - 7x - 6 = 0$
   (b) $x^4 + 5x^3 + 5x^2 - 5x - 6 = 0$
   (c) $3x^3 + x^2 - 5x + 2 = 0$.

3.9 Partial fractions

In arithmetic $\dfrac{5}{6}$ can be expressed as $\frac{1}{2} + \frac{1}{3}$. Similarly in algebra we can express

$\dfrac{5x + 7}{(x + 1)(x + 2)}$ as the sum of $\dfrac{2}{x + 1}$ and $\dfrac{3}{x + 2}$, which are then known as *partial fractions*.

Example 1

If $f(x) \equiv \dfrac{4x + 1}{(2x + 3)(x - 1)}$, express $f(x)$ in partial fractions.

Assuming that $f(x)$ can be expressed as

$$\frac{A}{2x + 3} + \frac{B}{x - 1},$$

where A and B are constants,

i.e. $\qquad\qquad \dfrac{4x + 1}{(2x + 3)(x - 1)} \equiv \dfrac{A}{2x + 3} + \dfrac{B}{x - 1},$

then
$$\frac{4x+1}{(2x+3)(x-1)} \equiv \frac{A(x-1)+B(2x+3)}{(2x+3)(x-1)}$$

$$\therefore \qquad 4x+1 \equiv A(x-1)+B(2x+3).$$

If two polynomials are identically equal, then
 (i) the corresponding coefficients of powers of x are equal,
(ii) the polynomials are equal for all values of x.
Using (i) gives $4 = A+2B$ and $1 = -A+3B$.
Solving gives $A = 2$ and $B = 1$.
Alternatively using (ii) and putting $x = 1$ gives $5 = 5B$, i.e. $B = 1$, and putting $x = -\frac{3}{2}$ we get $-5 = A(-\frac{5}{2})$, i.e. $A = 2$.

$$\therefore \qquad \frac{4x+1}{(2x+3)(x-1)} \equiv \frac{2}{2x+3} + \frac{1}{x-1}$$

Example 2

If $f(x) \equiv \dfrac{4x-5}{(x-2)^2}$, express $f(x)$ in partial fractions.

Assuming that $f(x)$ can be expressed as $\dfrac{A}{x-2} + \dfrac{B}{(x-2)^2}$, where A and B are constants, (cf. $\frac{3}{4} = \frac{1}{2} + \frac{1}{4}$ in arithmetic)

i.e.
$$\frac{4x-5}{(x-2)^2} \equiv \frac{A}{x-2} + \frac{B}{(x-2)^2},$$

then
$$4x-5 \equiv A(x-2)+B.$$

Putting $x = 2$, $3 = B$.
Comparing coefficients of x, $4 = A$.

$$\therefore \qquad \frac{4x-5}{(x-2)^2} \equiv \frac{4}{x-2} + \frac{3}{(x-2)^2}.$$

Example 3

Express $\dfrac{8x^2+4x+1}{(x^2+1)(2x-1)}$ in partial fractions.

Assume that $\dfrac{8x^2+4x+1}{(x^2+1)(2x-1)} \equiv \dfrac{Ax+B}{x^2+1} + \dfrac{C}{2x-1}$, where A, B and C are constants.

Then
$$8x^2+4x+1 \equiv (Ax+B)(2x-1)+C(x^2+1).$$

Putting $x = \frac{1}{2}$, $5 = C(\frac{5}{4})$, i.e. $C = 4$.

Comparing coefficients of x^2, $8 = 2A+C$, giving $A = 2$.
Comparing coefficients of x, $4 = -A+2B$, giving $B = 3$.

$$\frac{8x^2 + 4x + 1}{(x^2 + 1)(2x - 1)} \equiv \frac{2x + 3}{x^2 + 1} + \frac{4}{2x - 1}.$$

Example 4

If $f(x) \equiv \dfrac{4x^2 + x - 3}{(x - 1)(2x + 1)}$, express $f(x)$ in partial fractions.

Here the numerator is the same degree as the denominator and so we start by dividing the numerator by the denominator to give

$$f(x) \equiv 2 + \frac{3x - 1}{(x - 1)(2x + 1)}.$$

Proceeding as in example 1 we have

$$\frac{3x - 1}{(x - 1)(2x + 1)} \equiv \frac{A}{x - 1} + \frac{B}{2x + 1}.$$

Then $\qquad\qquad\qquad 3x - 1 \equiv A(2x + 1) + B(x - 1).$

Putting $x = 1$, $\qquad\qquad 2 = 3A$ i.e. $A = \frac{2}{3}$.

Comparing coefficients of x, $\quad 3 = 2A + B$, i.e. $B = 5/3$.

$$\frac{4x^2 + x - 3}{(x - 1)(2x + 1)} \equiv 2 + \frac{2}{3(x - 1)} + \frac{5}{3(2x + 1)}.$$

From these examples it will be seen that the procedure for expressing an algebraic fraction in partial fractions is as follows:

(1) If the degree of the numerator is equal to or higher than the degree of the denominator, the first step is to divide the numerator by the denominator.

(2) For each linear factor $ax + b$ in the denominator, assume there to be a partial fraction of the form

$$\frac{A}{ax + b},$$

where A is a constant.

(3) For each repeated linear factor $(ax + b)^r$ in the denominator, assume there to be r partial fractions of the form

$$\frac{A_1}{ax + b}, \quad \frac{A_2}{(ax + b)^2}, \quad \cdots, \quad \frac{A_r}{(ax + b)^r},$$

where $A_1, A_2, \ldots, A_r$ are constants.

(4) For each quadratic factor $ax^2 + bx + c$ in the denominator, assume there to be a partial fraction of the form

$$\frac{Ax + B}{ax^2 + bx + c},$$

where A and B are constants.

Exercises 3.9

Express in partial fractions

1   $\dfrac{x}{(x+1)(x+2)}$

9   $\dfrac{5x^2+3}{(x-1)(x^2+1)}$

2   $\dfrac{7x-12}{(x-1)(x-2)}$

10   $\dfrac{(x-1)(x-3)}{(4x-1)(x^2+2)}$

3   $\dfrac{5x+4}{(2x+1)(3x+2)}$

11   $\dfrac{x^2-x+2}{x(x^2-x+1)}$

4   $\dfrac{2(x+2)}{(2x-3)(1+4x)}$

12   $\dfrac{2x^3+7x^2+6x+5}{(x+1)^2(x^2+1)}$

5   $\dfrac{4x+6}{(x+1)(x+2)(x+3)}$

13   $\dfrac{x^3-6x-4}{(x-1)^2(x^2+x+1)}$

6   $\dfrac{2(x^2+x+1)}{(x-1)(x+1)(x+2)}$

14   $\dfrac{x^2+8x+9}{(x+1)(x+2)}$

7   $\dfrac{2x+5}{(x+2)^2}$

15   $\dfrac{-x^3+6x^2-13x+11}{(x-2)^2}$

8   $\dfrac{7-8x}{(2x-1)^2}$

16   $\dfrac{x^3+x^2+6x+3}{(x+1)(x^2+1)}.$

3.10 Use of partial fractions with expansions

Partial fractions may be used in expressing algebraic fractions in the form of power series.

Example

Express $\dfrac{4x+3}{(x+2)(2x-1)}$ as a series of terms of ascending powers of x up to and including the term in x^2.

Using the methods of the previous section we get

$$\frac{4x+3}{(x+2)(2x-1)} = \frac{1}{x+2} + \frac{2}{2x-1}$$

$$\frac{1}{x+2} = (2+x)^{-1} = \left[2\left(1+\frac{x}{2}\right)\right]^{-1} = 2^{-1}\left[1+\frac{x}{2}\right]^{-1} = \tfrac{1}{2}\left(1+\frac{x}{2}\right)^{-1}$$

$$= \frac{1}{2}\left[1+(-1)\left(\frac{x}{2}\right)+\frac{(-1)(-2)}{(1)(2)}\left(\frac{x}{2}\right)^2 + \ldots\right] \quad \text{(using the binomial}$$

expansion)

$$= \frac{1}{2} - \frac{1}{4}x + \frac{1}{8}x^2 + \ldots.$$

$$\frac{2}{2x-1} = \frac{-2}{1-2x} = -2(1-2x)^{-1}$$

$$= -2\left[1+(-1)(-2x)+\frac{(-1)(-2)}{(1)(2)}(-2x)^2+\ \ldots\right]$$

$$= -2-4x-8x^2+\ \ldots$$

$$\therefore \quad \frac{4x+3}{(x+2)(2x-1)} = \left[\frac{1}{2}-\frac{1}{4}x+\frac{1}{8}x^2\ \ldots\right]+\left[-2-4x-8x^2\ \ldots\right]$$

$$= -\frac{3}{2}-\frac{17}{4}x-\frac{63}{8}x^2+\ \ldots\ .$$

The expansion for $\left(1+\dfrac{x}{2}\right)^{-1}$ is only valid for $\left|\dfrac{x}{2}\right| < 1$, i.e. $|x| < 2$.

The expansion for $(1-2x)^{-1}$ is only valid for $|2x| < 1$, i.e. $|x| < \frac{1}{2}$.

Thus the expansion for $\dfrac{4x+3}{(x+2)(2x-1)}$ is only valid for the intersection of the sets
$\{x \in \mathbb{R}: |x| < 2\}$ and $\{x \in \mathbb{R}: |x| < \frac{1}{2}\}$,
i.e. for the set $\{x \in \mathbb{R}: |x| < \frac{1}{2}\}$ or $\{x \in \mathbb{R}: -\frac{1}{2} < x < \frac{1}{2}\}$.

Exercises 3.10
Find the first three non-zero terms in the expansion in ascending powers of x of $f(x)$ when $f(x)$ is equal to

1 $\dfrac{x}{1+x}$ 4 $\dfrac{x}{2x+1}$ 7 $\dfrac{x}{(x+1)(x+2)}$

2 $\dfrac{1}{x+3}$ 5 $\dfrac{3}{3x-1}$ 8 $\dfrac{2x+5}{(x+2)^2}$

3 $\dfrac{2+x}{1-x}$ 6 $\dfrac{2x+1}{2x-1}$ 9 $\dfrac{5x^2+3}{(x-1)(x^2+1)}$.

In each case state the set of values of x for which the expansion is valid.

3.11 Use of partial fractions in the summation of series

Partial fractions may also be useful in the summation of certain series.

Example

Evaluate $\displaystyle\sum_{r=1}^{n} \frac{2}{(2r-1)(2r+1)}$.

By partial fractions $\dfrac{2}{(2r-1)(2r+1)} = \dfrac{1}{2r-1} - \dfrac{1}{2r+1}$.

Thus $\displaystyle\sum_{r=1}^{n} \frac{2}{(2r-1)(2r+1)} = \sum_{r=1}^{n}\left[\frac{1}{2r-1} - \frac{1}{2r+1}\right]$.

Putting $r = 1$ gives the 1st term of the series $= \dfrac{1}{1} - \dfrac{1}{3}$

$\qquad r = 2 \qquad$ 2nd term $\qquad = \dfrac{1}{3} - \dfrac{1}{5}$

$\qquad r = 3 \qquad$ 3rd term $\qquad = \dfrac{1}{5} - \dfrac{1}{7}$

. . . .
. . . .

$\qquad r = n - 1 \qquad (n-1)$th term $\qquad = \dfrac{1}{2n-3} - \dfrac{1}{2n-1}$

$\qquad r = n \qquad\quad$ nth term $\qquad = \dfrac{1}{2n-1} - \dfrac{1}{2n+1}$

Adding gives the sum of the series $\qquad = 1 - \dfrac{1}{2n+1}$ $\qquad$ ...(1)

$$= \dfrac{2n}{2n+1} \qquad \text{...(2)}$$

From form (1) of the sum we can see that, as n gets larger and larger, $\dfrac{1}{2n+1}$ gets

smaller and smaller and so the sum of the series tends to 1. Thus the sum to infinity of this series is 1.

Exercises 3.11

Evaluate

1 (a) $\displaystyle\sum_{r=1}^{n} \dfrac{1}{r(r+1)}$ $\qquad$ (b) $\displaystyle\sum_{r=1}^{\infty} \dfrac{1}{r(r+1)}$.

2 (a) $\displaystyle\sum_{r=1}^{n} \dfrac{1}{(3r-2)(3r+1)}$ $\qquad$ (b) $\displaystyle\sum_{r=1}^{\infty} \dfrac{1}{(3r-2)(3r+1)}$.

3 (a) $\displaystyle\sum_{r=1}^{n} \dfrac{1}{r(r+1)(r+3)}$ $\qquad$ (b) $\displaystyle\sum_{r=1}^{\infty} \dfrac{1}{r(r+1)(r+3)}$.

4 (a) $\displaystyle\sum_{r=1}^{n} \dfrac{1}{(2r-1)(2r+1)(2r+3)}$ $\qquad$ (b) $\displaystyle\sum_{r=1}^{\infty} \dfrac{1}{(2r-1)(2r+1)(2r+3)}$.

Miscellaneous exercises 3

1 Find the sum of the first n terms of the arithmetic progression
   $2 + 5 + 8 + \ldots$.
   Find the value of n for which the sum of the first $2n$ terms will exceed the sum of the first n terms by 224. $\qquad$ [AEB]

2 The sum of the first twenty terms of an arithmetic progression is 45, and the sum of the first forty terms is 290. Find the first term and the common difference.
   Find the number of terms in the progression which are less than 100.
   $\qquad$ [JMB]

3 The first and last terms of an arithmetic progression are a and l, respectively. If the progression has n terms, prove from first principles that its sum is $\frac{1}{2}n(a+l)$.

An array consists of 100 rows of numbers, in which the rth row contains $r+1$ numbers. In each row, the numbers form an arithmetic progression whose first and last terms are 1 and 99, respectively. Calculate the sum of the numbers in the rth row, and hence determine the sum of all the numbers in the array. [JMB]

4 Prove that the sum of the first n terms of the arithmetical progression

$$a + (a+d) + (a+2d) + \ldots,$$

is $\qquad\qquad \frac{1}{2}n[2a+(n-1)d].$

Hence, or otherwise, show that

$$1^2 - 2^2 + 3^2 - 4^2 + \ldots + (2n-1)^2 - (2n)^2 = -n(2n+1).$$

Deduce the sums of the series
(a) $1^2 - 2^2 + 3^2 - 4^2 + \ldots + (2n-1)^2 - (2n)^2 + (2n+1)^2$, and
(b) $21^2 - 22^2 + 23^2 - 24^2 + \ldots + 39^2 - 40^2$. [C]

5 (a) An arithmetic progression whose first term is 2 and whose nth term is 32 has the sum of its first n terms equal to 357. Find n.

If the smallest term in this series which has a value exceeding 100 is the rth term, find r.

(b) A geometric progression has first term a and common ratio r, and the sum of the first n terms is denoted by S_n. State the set of values of r for which the progression has a sum to infinity, and give a formula for this sum.

Given that $a > 0$ and that $r = 0 \cdot 8$, find the least value of n for which S_n exceeds 99% of the sum to infinity. [C]

6 By using $\sum_{r=1}^{n} r^2 = \frac{1}{6}n(n+1)(2n+1)$, or otherwise, find the value of

(a) $\sum_{r=1}^{n} 3(r+1)(r+2)$ (b) $\sum_{r=n}^{2n} r^2/n^2$. [L]

7 (i) Simplify $r(r+1)(r+2) - (r-1)r(r+1)$ and use your result to prove that

$$\sum_{r=1}^{n} r(r+1) = \frac{1}{3}n(n+1)(n+2).$$

Deduce that $\sum_{r=1}^{n} r^2 = \frac{n}{6}(n+1)(2n+1).$

(ii) Find the sum of the series

$$(1)(2)(3) + (3)(4)(5) + (5)(6)(7) + \ldots + (2n-1)(2n)(2n+1). \quad [L]$$

8 Prove by induction, or otherwise, that the sum of the cubes of the first n natural numbers is $\frac{1}{4}n^2(n+1)^2$.

Deduce the sum of the cubes of the first n even numbers.
Find the sum of the series

$$(1)(3)(5) + (2)(4)(6) + (3)(5)(7) + \ldots + (n-2)\,n\,(n+2).$$

Show that $\displaystyle\sum_{n=3}^{\infty} \frac{1}{(n-2)\,n\,(n+2)} = \frac{11}{96}.$ [AEB]

9 Prove that the sum of the cubes of the first n even numbers is $2n^2(n+1)^2$ and hence, or otherwise, find the sum of the cubes of the first n odd numbers.

[The formula, $\displaystyle\sum_{r=1}^{n} r^3 = \tfrac{1}{4}n^2(n+1)^2$, may be quoted without proof.]

[JMB]

10 Find the sum of the series

$$1^2 - 2^2 + 3^2 - 4^2 + \ldots + (-1)^{n+1} n^2$$

(i) when n is even, (ii) when n is odd. In each case express your answer as a product of factors linear in n. [JMB]

11 Prove that

$$1^2 + 2^2 + 3^2 + \ldots + n^2 = \tfrac{1}{6}n(n+1)(2n+1).$$

Show that

$$a^2 + (a+d)^2 + (a+2d)^2 + \ldots + (a+nd)^2 =$$

$$\tfrac{1}{6}(n+1)\big[6a(a+nd) + d^2 n(2n+1)\big].$$

Hence, or otherwise, prove that

$$2^2 + 4^2 + 6^2 + \ldots + l^2 = \tfrac{1}{6}l(l+1)(l+2), \quad l \text{ even}$$

and

$$1^2 + 3^2 + 5^2 + \ldots + l^2 = \tfrac{1}{6}l(l+1)(l+2), \quad l \text{ odd}. \quad \text{[JMB]}$$

12 Prove by mathematical induction that

$$\sum_{1}^{n} r(r+3) = \frac{n}{3}(n+1)(n+5). \quad \text{[AEB]}$$

13 Prove by induction, or otherwise, that

$$\frac{1}{(1)(2)} + \frac{1}{(2)(3)} + \frac{1}{(3)(4)} + \ldots + \frac{1}{n(n+1)} = \frac{n}{n+1}. \quad \text{[L]}$$

14 Prove by induction or otherwise that

$$(2)1! + (5)2! + (10)3! + \ldots + (n^2+1)n! = n(n+1)! \quad \text{[L]}$$

15 (a) Find the sum of the first n terms of the series

$$(1)(3)(5) + (3)(5)(7) + (5)(7)(9) + \ldots + (2r-1)(2r+1)(2r+3) + \ldots.$$

(b) Prove by induction that the sum of the first n terms of the series

$$\frac{1}{2} + \frac{(1)(3)}{(2)(4)} + \frac{(1)(3)(5)}{(2)(4)(6)} + \frac{(1)(3)(5)(7)}{(2)(4)(6)(8)} + \ldots + \frac{(1)(3)(5) \ldots (2r-1)}{(2)(4)(6) \ldots 2r} + \ldots$$

is

$$\frac{(1)(3)(5) \ldots (2n+1)}{(2)(4) \ldots 2n} - 1. \qquad \text{[AEB]}$$

16 (a) Show, by induction or otherwise, that

$$\sum_{r=1}^{n} (r-1)(r+1) = \tfrac{1}{6}n(n-1)(2n+5).$$

(b) Evaluate $\displaystyle\sum_{r=1}^{n} \frac{1}{(2r-1)(2r+1)(2r+3)}$

and show that the sum to infinity is $\tfrac{1}{12}$. [AEB]

17 Prove by induction that, for all positive integers n,

$$1^3 + 2^3 + \ldots + n^3 = \tfrac{1}{4}n^2(n+1)^2.$$

Deduce that

$$(n+1)^3 + (n+2)^3 + \ldots + (2n)^3 = \tfrac{1}{4}n^2(3n+1)(5n+3). \quad \text{[JMB]}$$

18 Show that the sum of the first n terms of the series

$$1^2 - 2^2 + 3^2 - 4^2 + \ldots + (-1)^{r+1}r^2 \ldots$$

is equal to $\tfrac{1}{2}n(n+1)$ if n is odd, and is equal to $-\tfrac{1}{2}n(n+1)$ if n is even.
[JMB]

19 Show that the sum of the series

$$(1)(1) + (2)(3) + (3)(5) + \ldots + n(2n-1) \text{ is } \tfrac{1}{6}n(n+1)(4n-1).$$

Hence determine the sum of the series

$$1(2n-1) + 2(2n-3) + 3(2n-5) + \ldots + n[2n - (2n-1)]. \qquad \text{[JMB]}$$

20 Find the set of values of x for which the geometric series with first term 10 and common ratio 10^x is convergent.

Calculate to one decimal place the sum to infinity of the series when $x = -0.1$, and find the smallest integer n such that the sum of the first n terms exceeds 99% of the sum to infinity [L]

21 Starting from first principles, prove that the sum of the first n terms of a geometric progression whose first term is a and whose common ratio is r (where $r \neq 1$) is

$$S_n = \frac{a(1-r^n)}{1-r}.$$

Show that

$$\frac{S_{3n} - S_{2n}}{S_n} = r^{2n}.$$

Given that $r = \frac{1}{2}$, find

$$\sum_{n=1}^{\infty} \frac{S_{3n} - S_{2n}}{S_n}.$$

[JMB]

22 Prove that the sum of the geometric series

$$\sum_{k=1}^{n} ar^{k-1}$$

is $\dfrac{a(1-r^n)}{1-r}$ for $r \neq 1$.

Let $S = \displaystyle\sum_{k=1}^{n} kx^{k-1}$ for $x \neq 1$. By considering $S - xS$, or otherwise, find the sum of the series S. [JMB]

23 State the nth term and the sum of n terms of the geometric series

$$(b-1) + \frac{(b-1)^2}{b} + \frac{(b-1)^3}{b^2} + \ldots$$

and show that the series has a finite sum to infinity for all values of b greater than $\frac{1}{2}$.

If b_1, b_2 are two such values of $b(b_1 > b_2)$ and if S_1, S_2 are the corresponding sums to infinity, prove that $S_1 > S_2$. [AEB]

24 A man borrowed £1000 on 1st January 1973 at an interest rate of 1% per calendar month, the interest being added on the last day of each month. The man repaid the loan in 30 equal monthly instalments each of £x, made on the first day of each succeeding month, the first repayment being made on 1st February 1973. Show that the sum the man owed immediately after he had made the third repayment was

$$£[1000(1 \cdot 01)^3 - x(1 + 1 \cdot 01 + 1 \cdot 01^2)].$$

Calculate the value of x to 2 significant figures. [AEB]

25 (a) By using an infinite geometric progression show that $0 \cdot 43\dot{2}\dot{1}$ i.e. $0 \cdot 4321212121 \ldots$ is equal to $713/1650$.

(b) Write down the term containing p^r in the binomial expansion of $(q + p)^n$ where n is a positive integer. If $p = \frac{1}{6}$, $q = \frac{5}{6}$ and $n = 30$, find the value of r for which this term has the numerically greatest value. [AEB]

26 Write down the first three terms in the expansion, in ascending powers of x, of $(1 - ax)^{-n}$.

If the coefficients of x and of x^2 in the expansions of $(1 - x^2)^{15}$ and of $(1 + 2x)^p (1 - 3x)^{-q}$ in ascending powers of x are equal, find p and q. [AEB]

27 Show that the first three terms in the expansion in ascending powers of x of

$$(1 + 8x)^{\frac{1}{4}}$$

are the same as the first three terms in the expansion of

$$\frac{1+5x}{1+3x}.$$

Use the corresponding approximation

$$(1+8x)^{\frac{1}{4}} \approx \frac{1+5x}{1+3x}$$

to obtain an approximation to $(1 \cdot 16)^{\frac{1}{4}}$ as a rational fraction in its lowest terms. [JMB]

28 Write down the expansion in ascending powers of x up to the term in x^2 of (i) $(1+x)^{\frac{1}{2}}$, (ii) $(1-x)^{-\frac{1}{2}}$, and simplify the coefficients. Hence, or otherwise, expand

$$\sqrt{\left(\frac{1+x}{1-x}\right)}$$

in ascending powers of x up to the term in x^2. By using $x = 1/10$ obtain an estimate, to three decimal places, for $\sqrt{11}$. [JMB]

29 Write down the expansion of $(1+y)^n$ in ascending powers of y, giving the first four terms.

Given that x is so small that its cube and higher powers are negligible compared with unity, find the constants a, b and c in the approximate formula

$$\left(\frac{1+3x}{8-3x}\right)^{\frac{1}{3}} \approx a + bx + cx^2.$$
[JMB]

30 Express $(1+x)^{-1}$ and $(1-2x)^{-2}$ as series in ascending powers of x in each case up to and including the term in x^2. Hence show that

$$\frac{1}{(1+x)(1-2x)^2} = 1 + 3x + 9x^2,$$

when x is such that terms in x^n, $n \geqslant 3$, can be neglected.

The value of g is calculated from the period T of a pendulum, where $T = \dfrac{c}{\sqrt{(gh)}}$. If errors of -2% in T and $+1\%$ in h are made, use the expansion derived for

$$\frac{1}{(1+x)(1-2x)^2}$$

to find the percentage error in the calculated value of g, giving your answer to 2 significant figures. [AEB]

31 (a) Find the value of the constant a if the coefficient of x^2 in the expansion of

$$(1 - x + ax^2)^4$$

is zero, and obtain the coefficient of x^3 when a has this value.

(b) Show that, if terms in x^3 and higher powers of x in the expansion in ascending powers of x are neglected, then

$$\left(\frac{1-x}{1+x}\right)^n = 1 - 2nx + 2n^2x^2.$$

Hence find an approximation to the fourth root of 19/21, in the form p/q where p and q are positive integers. [C]

32 (a) Show that, if x is small compared with y,

$$\sqrt{(x+y)} - \sqrt{y} \approx \frac{x}{2\sqrt{y}}.$$

(b) The binomial expansion of $(1+x)^n$, n a positive integer, may be written in the form

$$(1+x)^n = 1 + c_1 x + c_2 x^2 + c_3 x^3 + \ldots + c_r x^r + \ldots$$

Show that, if c_{s-1}, c_s and c_{s+1} are in arithmetic progression, then $(n-2s)^2 = n+2$. Find possible values of s when $n = 62$. [AEB]

33 Write down expressions for the roots of the quadratic equation

$$x^2 - x + a = 0,$$

where a is a constant. If a is so small that a^4 and higher powers of a may be neglected, show that the roots are approximately

$$1 - a - a^2 - 2a^3 \quad \text{and} \quad a + a^2 + 2a^3.$$

Using these approximations, estimate to three decimal places the roots of the equation

$$100x^2 - 100x + 3 = 0.$$ [JMB]

34 Expand $(1+x-3x^2)^4$ in ascending powers of x as far as the term in x^3, showing that the coefficients of x and x^2 are 4 and -6 respectively.

If $a > b$ and the first three terms in the expansion, in ascending powers of x, of

$$\frac{1+ax}{\sqrt{(1+bx)}}$$

are the same as those in the previous expansion, find a and b.

State the set of values of x for which the second expansion is valid, and show that the coefficients of x^3 in the two expansions differ by 48. [L]

35 Prove that the remainder on dividing a polynomial $f(x)$ by $(x-a)$ is $f(a)$.

The remainder when a polynomial $f(x)$ is divided by $(x-2)$ is 3 and the remainder when it is divided by $(x+1)$ is 6. If the remainder when $f(x)$ is divided by $(x-2)(x+1)$ is $px + q$, find the numerical values of p and q. [JMB]

36 When a polynomial is divided by $(x-2)$ the remainder is 2, and when it is divided by $(x-3)$ the remainder is 5. When it is divided by $(x-2)(x-3)$ the remainder is $ax+b$. Find the values of a and b.　　　　　[AEB]

37 The polynomial x^3+ax^2+bx+c leaves the remainders 2, 2 and 6 when divided by $x+1$, x and $x-1$ respectively. Calculate a, b and c. Find the range, or ranges, of values of x for which the polynomial is positive.　　　[JMB]

38 Determine the quadratic function $f(x)$ which is exactly divisible by $(2x+1)$ and has remainders -6 and -5 when divided by $(x-1)$ and $(x-2)$ respectively. Determine $g(x) \equiv (px+q)^2 f(x)$, where p, q are constants, given that, on division by $(x-2)^2$, the remainder is $-39-3x$.　　　[AEB]

39 When the polynomial $P(x)$ is divided by $(x-2)$ the remainder is 4, and when $P(x)$ is divided by $(x-3)$ the remainder is 7. Find, by writing

$$P(x) \equiv (x-2)(x-3)Q(x) + ax + b,$$

the remainder when $P(x)$ is divided by $(x-2)(x-3)$. If also $P(x)$ is a cubic in which the coefficient of x^3 is unity, and $P(1)=1$, determine $Q(x)$.[JMB]

40 Prove that when a polynomial $f(x)$ is divided by $ax+b$, where $a \neq 0$, the remainder is $f(-b/a)$.

Find the polynomial in x of the third degree, which vanishes when $x = -1$ and $x = 2$, has the value 8 when $x = 0$ and leaves the remainder $16/3$ when divided by $3x+2$.　　　　　[JMB]

41 A polynomial $f(x)$ is divided by x^2-a^2, where $a \neq 0$, and the remainder is $px+q$. Prove that

$$p = \frac{1}{2a}[f(a)-f(-a)],$$
$$q = \tfrac{1}{2}[f(a)+f(-a)].$$

Find the remainder when $x^n - a^n$ is divided by $x^2 - a^2$ for the cases when (i) n is even, (ii) n is odd.　　　　　[JMB]

42 Express in partial fractions

(a) $\dfrac{2}{x(x^2+2)}$　　(b) $\dfrac{4}{x(x+2)^2}$　　　　　　　　　[L]

43 Express $\dfrac{1}{(1+x)(1-2x)}$ in partial fractions.

Hence find the first three terms of the expansion of

$$\frac{1}{(1+x)(1-2x)}$$

in ascending powers of x.

Find the coefficient of x^n and state the set of values of x for which the expansion is valid.　　　　　[AEB]

44 Express the following function in partial fractions:

$$f(x) = \frac{1}{(1+x)(1-3x)}$$

Find the first four terms in the expansion of $f(x)$ in ascending powers of x, and obtain the coefficient of x^n. [JMB]

45 Express in partial fractions

(a) $\dfrac{3x^2 + x - 2}{(x-2)^2(1-2x)}$ (b) $\dfrac{7+x}{1+x+x^2+x^3}$.

Assuming that $-1 < x < 1$ obtain an expansion for the function

$$\frac{7+x}{1+x+x^2+x^3}$$

in the form

$$a_0 + a_1 x + a_2 x^2 + a_3 x^3 + \dots$$

finding the values of the coefficients as far as a_5 inclusive. [L]

46 Express

$$f(x) \equiv \frac{2}{2 - 3x + x^2}$$

in partial fractions and hence, or otherwise, obtain $f(x)$ as a series of ascending powers of x, giving the first four non-zero terms of this expansion.

State the set of values of x for which this expansion is valid, and find the coefficient of x^n in the expansion. [C]

47 Express $\dfrac{1-7x}{(1+x)(1-3x)}$ in partial fractions and obtain the coefficient of x^r in its expansion in ascending powers of x. If this coefficient is denoted by a_r, show that, if s is any integer greater than zero,

$$\frac{a_{s+2} - a_s}{a_{s+1} - a_{s-1}} = 3.$$ [L]

48 Express

$$f(x) = \frac{2x + 4}{(x-1)(x+3)}$$

in partial fractions.

(i) If x is small, obtain an expansion of $f(x)$ in ascending powers of x as far as the term in x^2.

(ii) If x is large, obtain an expansion of $f(x)$ in ascending powers of $\dfrac{1}{x}$ as far as the term in $\dfrac{1}{x^3}$. [C]

49 Find the sum of the first n terms of each of the series whose rth terms are

(i) $\dfrac{1}{2r(2r+2)}$ (ii) $\dfrac{1}{(2r-1)(2r+1)}$.

Hence, or otherwise, find as a single fraction the sum of the first $2n$ terms of the series whose rth term is

$$\frac{1}{r(r+2)}.$$

[JMB]

50 Express in partial fractions

$$\frac{1}{x(x+1)(x+2)}.$$

Hence, or otherwise, find the sum of the first n terms of the series

$$\frac{1}{(1)(2)(3)} + \frac{1}{(2)(3)(4)} + \frac{1}{(3)(4)(5)} + \cdots$$

State the sum to infinity, S, of the series and find how many terms of the series must be taken so that their sum differs from S by less than $\dfrac{1}{1000}$.

[AEB]

4   Trigonometry 1

4.1   Angles of any magnitude

The reader will already have met the sine, cosine and tangent of acute angles in the solution of right-angled triangles.

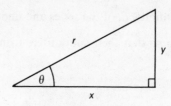

Fig. 4.1

If x, y, r and θ are as shown in Fig. 4.1, we have

$$\text{sine } \theta = \frac{y}{r}, \quad \text{cosine } \theta = \frac{x}{r}, \quad \text{tangent } \theta = \frac{y}{x}.$$

When $\theta = 0°$, $y = 0$ and $x = r$,

$$\text{sine } 0° = 0, \quad \text{cosine } 0° = 1, \quad \text{tangent } 0° = 0.$$

When $\theta = 90°$, $x = 0$ and $y = r$,

$$\text{sine } 90° = 1, \quad \text{cosine } 90° = 0 \quad \text{but tangent } 90° \text{ is undefined.}$$

However as θ tends to $90°$, tangent θ tends to infinity.

These definitions need to be generalised to cover angles of all magnitudes, including positive and negative angles. The definitions for angles of all sizes must incorporate the definitions for acute angles.

Measuring from OA (Fig. 4.2), anti-clockwise rotation is taken to be positive and clockwise rotation is taken to be negative. Thus the angle $AOP = +\theta$ and the angle $AOQ = -\theta$.

As OP rotates in an anti-clockwise sense so that P moves through B, C and D and then back to A, the angle AOP increases through $+90°$, $+180°$, $+270°$ to $+360°$. OP may continue with its anti-clockwise rotation turning through angles greater than $+360°$.

If P moves anti-clockwise right round the circle back to A and then on to B

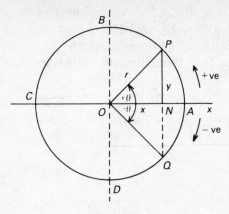

Fig. 4.2

again, the angle AOP will now be $+450°$. OP can continue to rotate and so produce angles of any magnitude.

If OP rotates in a clockwise sense, then negative angles will be produced and these can similarly be of any magnitude.

Taking OA along the x-axis Ox and OB along the y-axis Oy, the point P will have cartesian coordinates (x, y) and polar coordinates (r, θ). Thus x and y can be positive or negative according to the usual convention, θ will be positive for anti-clockwise rotation and negative for clockwise rotation with r taken always to be positive.

Then sine, cosine and tangent (abbreviated to sin, cos and tan) are defined for angles of all magnitudes as follows

$$\sin \theta = \frac{y}{r}, \quad \cos \theta = \frac{x}{r}, \quad \tan \theta = \frac{y}{x} \left(= \frac{\sin \theta}{\cos \theta} \right).$$

As x and y may be positive or negative, then clearly these ratios may also be positive or negative for different values of θ.

Again when $x = 0$, and $\cos \theta = 0$, $\tan \theta$ is undefined but as θ tends to $90°$, $\cos \theta$ tends to 0 and $\tan \theta$ tends to $\pm$ infinity.

Some calculators find the sine, cosine and tangent only for angles between $0°$ and $90°$. In finding the sine, cosine or tangent of an angle outside this range the problem is (i) how to find its magnitude, (ii) how to find its sign.

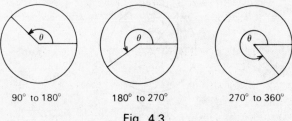

90° to 180°   180° to 270°   270° to 360°

Fig. 4.3

To deal with (ii) first, Fig. 4.3 shows examples of the angle θ in the three intervals 90° to 180°, 180° to 270° and 270° to 360°.

The following table shows, for the different intervals of values of θ, the signs of x and y and, deduced from these, the signs of sin θ, cos θ and tan θ.

θ	x	y	sin θ	cos θ	tan θ
0°– 90°	+	+	+	+	+
90°–180°	−	+	+	−	−
180°–270°	−	−	−	−	+
270°–360°	+	−	−	+	−

These results can be summarised in the following diagram (Fig. 4.4) which shows which of the three trigonometrical ratios, sine, cosine and tangent, are positive in each of the four quadrants.

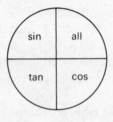

Fig. 4.4

Next to deal with the magnitudes.

90°–180°

$$|\sin \theta| = \frac{PN}{r} = \sin \alpha,$$

where $\alpha = 180° - \theta$ (Fig. 4.5). Thus the magnitude of sin θ is the same as the magnitude of sin $(180° - \theta)$.

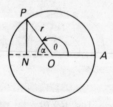

Fig. 4.5

Similarly for cos θ and tan θ.

$$|\sin \theta| = \sin (180° - \theta)$$
$$|\cos \theta| = \cos (180° - \theta)$$
$$|\tan \theta| = \tan (180° - \theta)$$

To find the required magnitude, enter the acute angle $(180° - \theta)$ in the calculator.

180°–270°

$$|\sin \theta| = \frac{PN}{r} = \sin \beta,$$

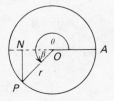

Fig. 4.6

where $\beta = \theta - 180°$ (Fig. 4.6). Thus $|\sin \theta| = \sin (\theta - 180°)$.
Similarly for cos θ and tan θ. Enter the acute angle $(\theta - 180°)$ in the calculator.

270°–360°

$$|\sin \theta| = \frac{PN}{r} = \sin \gamma,$$

where $\gamma = 360° - \theta$ (Fig. 4.7). Thus $|\sin \theta| = \sin (360° - \theta)$.

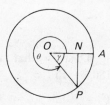

Fig. 4.7

Similarly for cos θ and tan θ. Enter the acute angle $(360° - \theta)$ in the calculator.
The finding of the magnitudes and signs of sines, cosines and tangents of angles of any size is summarised in the one diagram shown in Fig. 4.8. In each quadrant the positive ratios and the related acute angle are shown. This angle is always obtained by a combination of θ, 180° and 360°–*never* 90° or 270°.

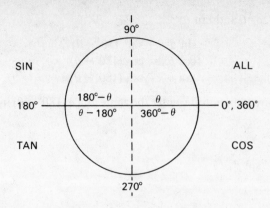

Fig. 4.8

The special cases of 0°, 90°, 180°, 270° and 360° are important. We have seen that

$$\sin 0° \;=\; 0, \quad \cos 0° \;=\; 1, \quad \tan 0° = 0,$$
and $\quad \sin 90° \;=\; 1, \quad \cos 90° \;=\; 0, \quad \tan 90° \text{ is undefined.}$

Also $\quad \sin 180° = \;\;\; 0, \quad \cos 180° = -1, \quad \tan 180° = 0,$
and $\quad \sin 270° = -1, \quad \cos 270° = \;\;\; 0, \quad \tan 270° \text{ is undefined.}$

$$\sin 360° = \sin 0° = 0, \quad \cos 360° = \cos 0° = 1, \quad \tan 360° = \tan 0° = 0.$$

Example 1
Find cos 140°.

Since 90° < 140° < 180°, Fig. 4.8 shows that cos 140° is negative and that the related acute angle is 180° − 140° = 40°.

$$\therefore \qquad\qquad\qquad \cos 140° \;=\; -\cos 40° \;=\; -0\cdot7660.$$

Example 2
Find tan 230°.

Since 180° < 230° < 270°, Fig. 4.8 shows that tan 230° is positive and that the related acute angle is 230° − 180° = 50°.

$$\therefore \qquad\qquad\qquad \tan 230° \;=\; +\tan 50° \;=\; +1\cdot1918.$$

Example 3
Find sin 330°.

Since 270° < 330° < 360°, Fig. 4.8 shows that sin 330° is negative and that the related acute angle is 360° − 330° = 30°.

$$\sin 330° \;=\; -\sin 30° \;=\; -0\cdot5$$

Example 4
Find cos 850°.

The angle 850° amounts to two complete revolutions (720°) + 130°.

$$\therefore \qquad \cos 850° = \cos 130° = -\cos 50° = -0\cdot6428.$$

Exercises 4.1(i)
1 Find the sine, cosine and tangent of
   (a) 100°   (f) 320°   (k) 206·4°
   (b) 200°   (g) 137°   (l) 308·7°
   (c) 300°   (h) 253°   (m) 161·1°
   (d) 110°   (i) 298°   (n) 237·8°
   (e) 240°   (j) 141·5°   (o) 316·5°.
2 Find the sine, cosine and tangent of
   (a) 400°   (g) 430°   (m) 503·8°
   (b) 800°   (h) 520°   (n) 614·2°
   (c) 1100°   (i) 940°   (o) 908·6°
   (d) 500°   (j) 475°   (p) 529·2°
   (e) 600°   (k) 583°   (q) 700·6°
   (f) 1200°   (l) 1311°   (r) 1287·7°.
3 Without using a calculator write down the values of the sine, cosine and
   tangent of
   (a) 450°   (d) 720°   (g) 990°
   (b) 540°   (e) 810°   (h) 1080°
   (c) 630°   (f) 900°   (i) 1260°.

Next we consider the trigonometrical ratios for negative angles.

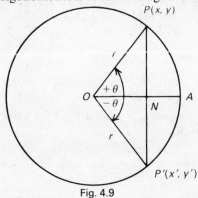

Fig. 4.9

In Fig. 4.9 the angle $AOP = +\theta$ and the angle $AOP' = -\theta$. From the
triangles OPN and $OP'N$ it is clear that the magnitudes of $\sin(-\theta)$, $\cos(-\theta)$
and $\tan(-\theta)$ are the same as the magnitudes of $\sin\theta$, $\cos\theta$ and $\tan\theta$
respectively.

If the coordinates of P' are (x', y'), then $x' = x$ and $y' = -y$.

Thus
$$\sin(-\theta) = \frac{y'}{r} = \frac{-y}{r} = -\sin\theta,$$

$$\cos(-\theta) = \frac{x'}{r} = \frac{x}{r} = \cos\theta,$$

$$\tan(-\theta) = \frac{y'}{x'} = \frac{-y}{x} = -\tan\theta.$$

Here we have considered only the case of $0° \leqslant \theta \leqslant 90°$. The reader should verify that these results hold for all values of θ.

Example 1
Find $\cos(-80°)$.
$$\cos(-80°) = \cos 80° = 0.1736.$$

Example 2
Find $\sin(-200°)$.
$$\sin(-200°) = -\sin 200° = -(-\sin 20°) = +0.3420.$$

Example 3
Find $\tan(-500°)$.
$$\tan(-500°) = -\tan 500° = -\tan 140° = -(-\tan 40°)$$
$$= +0.8391.$$

Exercises 4.1(ii)

1 Find the sine, cosine and tangent of
   (a) $-50°$ (h) $-261°$ (o) $-700°$
   (b) $-72°$ (i) $-234.9°$ (p) $-473°$
   (c) $-61.5°$ (j) $-300°$ (q) $-871.5°$
   (d) $-140°$ (k) $-314°$ (r) $-903.7°$
   (e) $-103°$ (l) $-299.3°$ (s) $-817.6°$
   (f) $-116.7°$ (m) $-400°$ (t) $-1001.3°$
   (g) $-210°$ (n) $-600°$ (u) $-2331.9°$

2 Tabulate the values of $\sin x$, $\cos x$ and $\tan x$ for values of x at $30°$ intervals from $-360°$ to $+360°$.

x	$-360°$	$-330°$	...	$-30°$	$0°$	$30°$	...	$330°$	$360°$
$\sin x$	0	0.500	...	-0.500	0	0.500	...	-0.500	0
$\cos x$	1	0.866	...	0.866	1	0.866	...	0.866	1
$\tan x$	0	0.577	...	-0.577	0	0.577	...	-0.577	0

4.2 Graphs of $y = \sin x$, $y = \cos x$ and $y = \tan x$

From the table obtained in exercises 4.1(ii) question 2, graphs of $y = \sin x$, $y = \cos x$ and $y = \tan x$ can be drawn. These are shown below in Fig. 4.10.

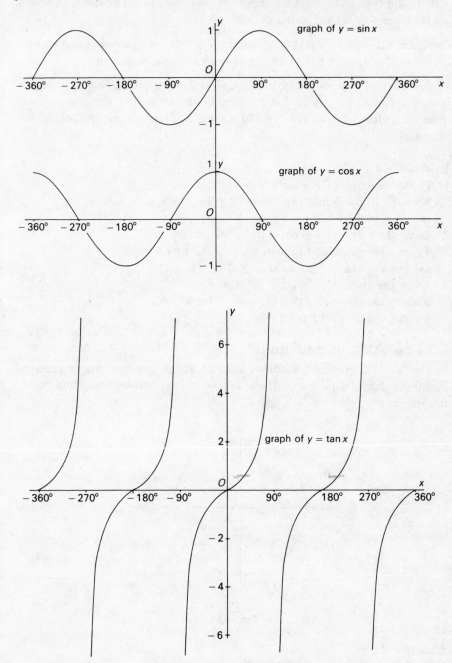

Fig. 4.10

It can be seen from these graphs and from the preceding paragraphs that sin x and cos x are the same as sin $(x + $ a multiple of $360°)$ and cos $(x + $ a multiple of $360°)$ respectively. The graphs of $y = \sin x$ and $y = \cos x$ repeat themselves after $360°$, i.e. sin $(x + 360°) = \sin x$ and cos $(x + 360°) = \cos x$. Thus sin x and cos x are both periodic with a period of $360°$.

Definition   If $f(x + p) = f(x)$ for all values of x, then f is a periodic function with period p. The smallest positive value of p is called the period of f.

From the table of values and from the graph of $y = \tan x$ it can be seen that tan x is also periodic. In this case the values in the table and the graph repeat themselves after $180°$, i.e. tan $(x + 180°) = \tan x$. Thus tan x is periodic with a period of $180°$.

Exercises 4.2

1   Draw the graph of $y = \sin x$ for $-180° \leqslant x \leqslant 180°$.
    From your graph find two values of x for which
    (a) $\sin x = 0.4$ (b) $\sin x = -0.6$.

2   Draw the graph of $y = \cos x$ for $-360° \leqslant x \leqslant 360°$.
    From your graph find four values of x for which
    (a) $\cos x = 0.6$ (b) $\cos x = -0.2$.

3   Draw the graph of $y = \tan x$ for $0° \leqslant x \leqslant 360°$.
    From your graph find two values of x for which
    (a) $\tan x = 2$ (b) $\tan x = -3$.

4.3   Solution of equations

In exercises 4.2 graphical solutions were obtained to simple trigonometrical equations. Next, similar equations will be solved without resorting to the drawing of a graph.

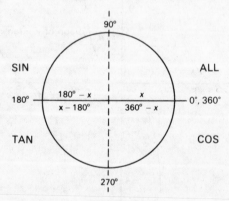

Fig. 4.11

Example 1
Solve the equation $\cos x = 0.4$ for $0° \leqslant x \leqslant 360°$.

Cos x is positive and so from Fig. 4.11 the angle x is in the first quadrant, i.e. between 0° and 90°, or in the fourth quadrant, i.e. between 270° and 360°. The acute angle with a cosine of 0·4 is 66·4°.

1st quadrant: $x = 66·4°$

4th quadrant: $360° - x = 66·4°$ giving $x = 293·6°$.

∴ Required solution is $x = 66·4°$ or 293·6°.

Example 2

Solve the equation $\sin x = -0·7$ for $0° \leqslant x \leqslant 360°$.

Sin x is negative and so from Fig. 4.11 the angle x is in the third quadrant, i.e. between 180° and 270°, or in the fourth quadrant, i.e. between 270° and 360°. The related acute angle is 44·4°.

3rd quadrant: $x - 180° = 44·4°$ giving $x = 224·4°$

4th quadrant: $360° - x = 44·4°$ giving $x = 315·6°$.

∴ Required solution is $x = 224·4°$ or 315·6°.

Example 3

Solve the equation $\tan x = 1·5$ for $0° \leqslant x \leqslant 360°$.

Tan x is positive and so from Fig. 4.11 the angle x is in the first quadrant, i.e. between 0° and 90°, or in the third quadrant, i.e. between 180° and 270°. The related acute angle is 56·3°.

1st quadrant: $x = 56·3°$

3rd quadrant: $x - 180° = 56·3°$ giving $x = 236·3°$

∴ Required solution is $x = 56·3°$ or 236·3°.

The procedure for solving equations of this kind for $0° \leqslant x \leqslant 360°$ can be summarised as follows:

(1) Use the diagram of Fig. 4.11 to find the quadrants in which the two values of x lie.

(2) Find the related acute angle whose corresponding trigonometrical ratio has the same magnitude as the given value.

(3) Put 'related angle' $= x,$ $180° - x,$ $x - 180°,$ $360° - x$

in the   1st,   2nd,   3rd,   4th   quadrants,

choosing the quadrants found in (1), thus getting the required values of x.

Exercises 4.3

1   Solve, for $0° \leqslant x < 360°$, the following equations

   (a) $\sin x = 0·5$ (f) $\sin x = 1$

   (b) $\sin x = 0·373$ (g) $\sin x = -1$

   (c) $\sin x = -0·35$ (h) $\sin x = \frac{2}{3}$

   (d) $\sin x = -0·3456$ (i) $\sin x = -\frac{1}{2}$

   (e) $\sin x = 0$ (j) $\sin x = -\frac{3}{4}$.

2   Solve, for $0° \leqslant x \leqslant 360°$, the following equations
 (a) $\cos x = 0.5$ (f) $\cos x = 1$
 (b) $\cos x = 0.745$ (g) $\cos x = -1$
 (c) $\cos x = -0.866$ (h) $\cos x = \frac{1}{4}$
 (d) $\cos x = -0.1234$ (i) $\cos x = -\frac{1}{2}$
 (e) $\cos x = 0$ (j) $\cos x = -\frac{2}{3}$.

3   Solve, for $0° \leqslant x < 360°$, the following equations
 (a) $\tan x = 1$ (f) $\tan x = -1$
 (b) $\tan x = 2.5$ (g) $\tan x = \frac{3}{4}$
 (c) $\tan x = -1.5$ (h) $\tan x = -\frac{2}{3}$
 (d) $\tan x = -0.2345$ (i) $\tan x = -3\frac{1}{4}$
 (e) $\tan x = 0$ (j) $\tan x = -\sqrt{3}$.

4.4   General solution of equations

From the periodic nature of the functions sine, cosine and tangent illustrated in the graphs of $y = \sin x$, $y = \cos x$ and $y = \tan x$ shown in Fig. 4.10, it is clear that there is an infinite set of angles with a given sine, cosine or tangent. For instance, if $\sin x = 0.5$, then $x = 30°$ or $150°$, or $30° \pm$ any multiple of $360°$, or $150° \pm$ any multiple of $360°$.

Consider the general solution of the equation $\tan x = a$. Let α be the angle between $0°$ and $180°$ for which $\tan \alpha = a$. The period of $\tan x$ is $180°$ and so the solutions of the equation $\tan x = a$ are α and $\alpha \pm$ any multiple of $180°$. Thus the general solution of the equation $\tan x = a$, i.e. the complete set of values of x for which $\tan x = a$, is

$$x = \alpha + 180n°,$$

where n is any integer, positive or negative, or is zero, i.e. $n \in \mathbb{Z}$.

Consider the general solution of the equation $\cos x = b$. Let α be the angle between $0°$ and $180°$ for which $\cos \alpha = b$. Since $\cos(-\alpha) = \cos \alpha$, $-\alpha$ also has a cosine equal to b. Thus between $-180°$ and $+180°$ there are two solutions, $x = \pm\alpha$. The period of $\cos x$ is $360°$. Hence the solutions of $\cos x = b$ are $x = \pm\alpha \pm$ any multiple of $360°$. Thus the general solution is

$$x = 360n° \pm \alpha,$$

where n is any integer, positive or negative, or is zero, i.e. $n \in \mathbb{Z}$.

Consider the general solution of the equation $\sin x = c$.
$c > 0$: Let α be the acute angle whose sine is c. Then the sine of the angle $(180° - \alpha)$ is also equal to c. In the range $0°$ to $360°$ we have two solutions

$$x = \alpha \quad \text{and} \quad x = 180° - \alpha.$$

Hence the general solution of the equation $\sin x = c$, where $c > 0$, is given by $x = \alpha + 360n°$ and $x = (180° - \alpha) + 360n°$, where $n \in \mathbb{Z}$, which can be written in the form

$$x = 180n° + (-1)^n \alpha, \quad \text{where } n \in \mathbb{Z}.$$

$c < 0$: Let α be the angle between $0°$ and $-90°$ whose sine is c. Then the sine of the angle $(180° - \alpha)$ is also equal to c. Hence the general solution is the same as in the case when $c > 0$.

General solution of equations – summary

(1) If $\tan x = \tan \alpha$, then $x = 180n° + \alpha$.
(2) If $\cos x = \cos \alpha$, then $x = 360n° \pm \alpha$.
(3) If $\sin x = \sin \alpha$, then $x = 360n° + \alpha$ or $360n° + (180° - \alpha)$
$$\text{i.e.} \quad x = 180n° + (-1)^n \alpha$$
(where n is any positive or negative integer or is zero, i.e. $n \in \mathbb{Z}$).

Example 1
Find the general solution of the equation $\tan x = 2$.

From calculator $\alpha = 63 \cdot 4°$.
$\therefore$ General solution is $x = 180n° + 63 \cdot 4°$

Example 2
Find the general solution of the equation $\cos x = -0 \cdot 5$.

$\cos 120° = -0 \cdot 5$, $\therefore \alpha = 120°$.
$\therefore$ General solution is $x = 360n° \pm 120°$.

Example 3
Find the general solution of the equation $\sin x = -0 \cdot 3$.

$$\sin(-17 \cdot 5°) = -0 \cdot 3 \quad \text{or} \quad \sin 197 \cdot 5° = -0 \cdot 3,$$
$$\therefore \qquad \alpha = -17 \cdot 5° \quad \text{or} \quad \alpha = 197 \cdot 5°.$$
$\therefore$ General solution is

$$x = 180n° + (-1)^n(-17 \cdot 5°) \quad \text{or} \quad x = 180n° + (-1)^n(197 \cdot 5°).$$
By giving particular values to n it will be seen that both these forms of the general solution give the same set of values of x.

Exercises 4.4(i)
1   Find the general solutions of the following equations
   (a) $\tan x = 2 \cdot 3$ (f) $\cos x = -1$ (k) $\sin x = 1$
   (b) $\tan x = -0 \cdot 85$ (g) $\cos x = \frac{1}{2}$ (l) $\sin x = -1$
   (c) $\tan x = 0$ (h) $\cos x = -\frac{1}{2}$ (m) $\sin x = -\frac{1}{2}$
   (d) $\cos x = 0$ (i) $\cos x = -0 \cdot 754$ (n) $\sin x = 0 \cdot 25$
   (e) $\cos x = 1$ (j) $\sin x = 0$ (o) $\sin x = -\frac{2}{3}$.
2   Find the general solution of the following equations and in each case list the solutions in the range $0° \leqslant x \leqslant 360°$.
   (a) $(\sin x - 0 \cdot 1)(\sin x + 0 \cdot 2) = 0$ (d) $\tan^2 x - \tan x - 2 = 0$
   (b) $(3 \cos x - 1)(2 \cos x + 1) = 0$ (e) $6 \cos^2 x - \cos x - 1 = 0$.
   (c) $4 \sin^2 x - 1 = 0$

Example 1
Find the general solution of the equation $\tan x = \tan 30°$.

$\tan x = \tan 30°. \therefore x = 180n° + 30°$.

Example 2
Find the general solution of the equation $\cos 2x = \cos x$.

$\cos 2x = \cos x. \therefore 2x = 360n° \pm x$
$\therefore x = 360n°$ or $120n°$
$\therefore$ General solution is $x = 120n°$ since this includes all values given by $360n°$.

Example 3
Find the general solution of the equation $\cos 2x = \sin x$.

$\cos 2x = \sin x = \cos(90° - x)$
$\therefore 2x = 360n° \pm (90° - x)$
$\therefore 3x = 360n° + 90°$ or $x = 360n° - 90°$
$\therefore$ General solution is $x = 120n° + 30°$ or $360n° - 90°$.

Exercises 4.4(ii)
1 Find the general solutions of the following equations
   (a) $\tan 2x = \tan x$ (e) $\sin 3x = \sin 2x$
   (b) $\tan 2x = \tan(90° - x)$ (f) $\sin 4x - \sin x = 0$
   (c) $\cos 3x = \cos x$ (g) $\cos x = \sin 3x$
   (d) $\cos 4x - \cos 2x = 0$ (h) $\cos(x + 30°) = \cos(60° - x)$.

4.5 Relations between trigonometrical ratios
In section 4.1 sine, cosine and tangent were defined. Cotangent, secant and cosecant (abbreviated to cot, sec and cosec) are defined as follows:

$$\cot\theta = \frac{1}{\tan\theta}, \quad \sec\theta = \frac{1}{\cos\theta}, \quad \cosec\theta = \frac{1}{\sin\theta}.$$

The Pythagorean identities
From the definitions $\sin\theta = y/r$ and $\cos\theta = x/r$ and from Pythagoras' theorem

$$x^2 + y^2 = r^2,$$

we get
$$\frac{x^2}{r^2} + \frac{y^2}{r^2} = 1,$$

i.e. $\cos^2\theta + \sin^2\theta = 1$ ... (A)

Dividing this equation in turn by $\cos^2\theta$ and $\sin^2\theta$ we get

$$1 + \tan^2\theta = \sec^2\theta \qquad \qquad \text{... (B)}$$

and
$$1 + \cot^2\theta = \cosec^2\theta \qquad \qquad \text{... (C)}$$

The identities (A), (B) and (C) are illustrated in Fig. 4.12. These identities are useful in solving some trigonometrical equations.

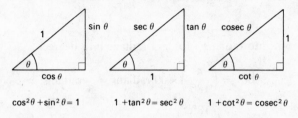

$$\cos^2\theta + \sin^2\theta = 1 \qquad 1 + \tan^2\theta = \sec^2\theta \qquad 1 + \cot^2\theta = \operatorname{cosec}^2\theta$$

Fig. 4.12

Example
Find the general solution of the equation

$$\sec^2 x - 5\tan x + 5 = 0,$$

and list the solutions between $0°$ and $360°$.

Replacing $\sec^2 x$ by $(1 + \tan^2 x)$ we get

$$\tan^2 x - 5\tan x + 6 = 0,$$
$$(\tan x - 2)(\tan x - 3) = 0,$$
$$\therefore \qquad \tan x = 2 \quad \text{or} \quad \tan x = 3,$$

$\therefore$ General solution is

$$x = 180n° + 63\!\cdot\!4° \quad \text{or} \quad 180n° + 71\!\cdot\!6°.$$

Putting $n = 0$ and 1 gives the solutions between $0°$ and $360°$ as $x = 63\!\cdot\!4°$, $71\!\cdot\!6°$, $243\!\cdot\!4°$, $251\!\cdot\!6°$.

Exercises 4.5
1   Find the general solutions of the following equations and in each case list the solutions in the range $0° \leqslant x \leqslant 360°$.
   (a) $\sec^2 x - 3\tan x + 1 = 0$
   (b) $\sec^2 x - \tan x - 7 = 0$
   (c) $\sec^2 x - \tan x - 1 = 0$
   (d) $\sin^2 x + \cos x - 1 = 0$
   (e) $6\sin^2 x - \cos x - 5 = 0$
   (f) $2\operatorname{cosec}^2 x - 3\cot x - 1 = 0$
   (g) $2\operatorname{cosec}^2 x - 5\cot x - 5 = 0.$

4.6   Circular measure

So far, angles have been expressed in degrees. For these units a complete rotation is divided into 360 parts to give the degree. The degree is divided into 60 parts to give the minute and the minute is divided into 60 parts to give the second.

The unit angle, the radian, is defined as the angle subtended at the centre of a circle of radius r by an arc of length equal to the radius r. Thus

$$2\pi \text{ radians} = 360°, \text{ i.e. } \pi \text{ radians} = 180°, \text{ and } 1° = \frac{\pi}{180} \text{ radians.}$$

The abbreviation for radian is rad.

Example
Convert (a) 60° into radians   (b) $\pi/4$ radians into degrees.

$$60° = \left(\frac{\pi}{180}\right) \times 60 \text{ radians} = \frac{\pi}{3} \text{ radians;}$$

$$\frac{\pi}{4} \text{ radians} = \left(\frac{180°}{\pi}\right) \times \frac{\pi}{4} = 45°.$$

Exercises 4.6(i)
1   Convert into radians
    30°,   45°,   90°,   120°,   135°,   150°,   240°,   270°,   300°.
2   Convert into degrees
    $\pi/4$, $\pi/3$, $7\pi/6$, $\pi/5$, $5\pi/3$.
3   Convert into radians (to 4 decimal places)
    25°,   110°,   200°,   40·5°,   54·3°,   144·7°,   283·2°.
4   Convert into degrees (to one decimal place)
    2 rad,   14 rad,   1·4 rad,   2·5 rad,   1·76 rad,   1·234 rad.
5   Show that
    (a) $\sin\left(\frac{\pi}{2} - x\right) = \cos x, \quad \sin\left(\frac{\pi}{2} + x\right) = \cos x,$

    (b) $\cos\left(\frac{\pi}{2} - x\right) = \sin x, \quad \cos\left(\frac{\pi}{2} + x\right) = -\sin x,$

    (c) $\tan\left(\frac{\pi}{2} - x\right) = \cot x, \quad \tan\left(\frac{\pi}{2} + x\right) = -\cot x.$

6   Show that
    (a) $\sin(\pi - x) = \sin x, \quad \sin(\pi + x) = -\sin x,$
    (b) $\cos(\pi - x) = -\cos x, \quad \cos(\pi + x) = -\cos x,$
    (c) $\tan(\pi - x) = -\tan x, \quad \tan(\pi + x) = \tan x.$
7   Show that, if n is an integer,
    (a) $\sin(n\pi) = 0, \cos(n\pi) = (-1)^n,$

    (b) when n is even, $\sin\left((2n+1)\frac{\pi}{2}\right) = 1,$

    (c) when n is odd, $\sin\left((2n+1)\frac{\pi}{2}\right) = -1.$

In section 4.4 there was a summary giving the general solutions to simple trigonometrical equations in terms of degrees. In terms of radians this summary becomes:

(1) If $\tan x = \tan \alpha$, then $x = n\pi + \alpha$.

(2) If $\cos x = \cos \alpha$, then $x = 2n\pi \pm \alpha$.

(3) If $\sin x = \sin \alpha$, then $x = 2n\pi + \alpha$ or $2n\pi + (\pi - \alpha)$,

$$\text{i.e. } x = n\pi + (-1)^n \alpha.$$

Exercises 4.6(ii)

1 Find, in radians, the general solutions of the following equations

   (a) $\tan x = 1\cdot 2$ (c) $\sin x = -1$ (e) $\cos x = -\frac{1}{4}$

   (b) $\cos x = \frac{1}{2}$ (d) $\tan x = -2$ (f) $\sin x = -0\cdot 2345$.

2 Find, in radians, the general solutions of the following equations

   (a) $(\tan x + 2)(\tan x - 3) = 0$ (d) $\tan 3x = \tan x$

   (b) $(3 \sin x + 1)(2 \sin x - 1) = 0$ (e) $\cos 5x - \cos 2x = 0$.

   (c) $12 \cos^2 x + \cos x - 6 = 0$

4.7 Graphs of $y = a \sin (bx + c)$ and $y = a \cos (bx + c)$

In chapter 2 the effects of the values of a, b and c on the nature of the curve $y = ax^2 + bx + c$ were considered. The effects of the values of a, b and c on the curves $y = a \sin (bx + c)$ and $y = a \cos (bx + c)$ will now be considered.

Figure 4.13 shows, on the same axes, the graphs of

$$y = \tfrac{1}{2} \sin x, \quad y = \sin x, \quad y = 2 \sin x, \quad y = 3 \sin x,$$

for $-360° \leqslant x \leqslant 360°$.

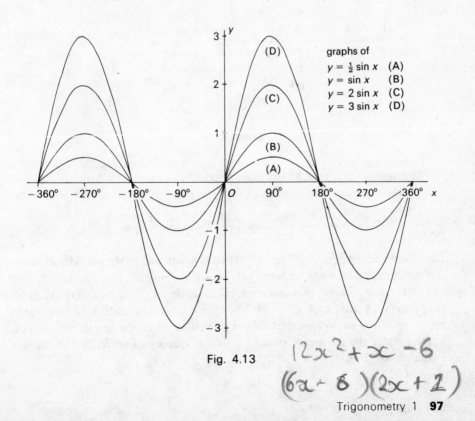

graphs of
$y = \frac{1}{2} \sin x$ (A)
$y = \sin x$ (B)
$y = 2 \sin x$ (C)
$y = 3 \sin x$ (D)

Fig. 4.13 $12x^2 + x - 6$

$(6x - 8)(2x + 2)$

The graphs clearly all have the same period. Indeed the only difference between the graphs is in the range of values taken by y.

Since $-1 \leqslant \sin x \leqslant 1$, it follows that

$$-\tfrac{1}{2} \leqslant \tfrac{1}{2}\sin x \leqslant \tfrac{1}{2}, \quad -2 \leqslant 2\sin x \leqslant 2, \quad -3 \leqslant 3\sin x \leqslant 3.$$

Thus for $y = a\sin x$ the values of y range from $-a$ to $+a$.

Figure 4.14 shows, on the same axes, the graphs of

$$y = \sin \tfrac{1}{2}x, \quad y = \sin x, \quad y = \sin 2x, \quad y = \sin 3x$$

for $-2\pi \leqslant x \leqslant 2\pi$.

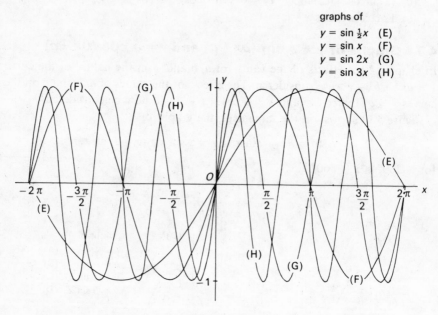

graphs of
$y = \sin \tfrac{1}{2}x$ (E)
$y = \sin x$ (F)
$y = \sin 2x$ (G)
$y = \sin 3x$ (H)

Fig. 4.14

The graphs clearly all take the same range of values of y, i.e. from -1 to $+1$. The periods however are different. From the graphs the periods are seen to be

y	$\sin \tfrac{1}{2}x$	$\sin x$	$\sin 2x$	$\sin 3x$
period	4π	2π	π	$2\pi/3$

For $y = \sin bx$ the period will be $2\pi/b$. Thus as b increases so the period decreases.

From the graphs it can be seen that, between $x = 0$ and $x = 2\pi$, there is one complete cycle of $\sin x$, two complete cycles of $\sin 2x$, three complete cycles of $\sin 3x$ and only half a cycle of $\sin \tfrac{1}{2}x$. Thus the value of b indicates how frequently the graph of $y = \sin bx$ repeats itself in a given interval, i.e. the greater the value of b, the greater the frequency of $\sin bx$. The frequency of $\sin bx$ is b times the frequency of $\sin x$.

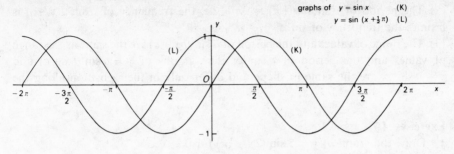

graphs of $y = \sin x$ (K)
           $y = \sin (x + \frac{1}{2}\pi)$ (L)

Fig. 4.15

Figure 4.15 shows, on the same axes, the graphs of

$$y = \sin x \text{ and } y = \sin (x + \tfrac{1}{2}\pi) \quad \text{for } -2\pi \leqslant x \leqslant 2\pi.$$

Figure 4.16 shows, on the same axes, the graphs of

$$y = \cos x \text{ and } y = \cos (x - \tfrac{1}{3}\pi) \quad \text{for } -2\pi \leqslant x \leqslant 2\pi.$$

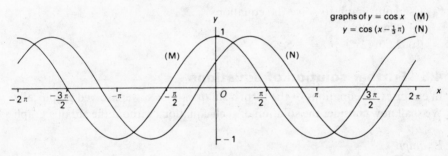

graphs of $y = \cos x$ (M)
          $y = \cos (x - \frac{1}{3}\pi)$ (N)

Fig. 4.16

From Fig. 4.15 it can be seen that the graph of $y = \sin (x + \frac{1}{2}\pi)$ is the same as the graph of $y = \sin x$ but its position is different. If the graph of $y = \sin x$ is moved $\frac{1}{2}\pi$ to the left it coincides with the graph of $y = \sin (x + \frac{1}{2}\pi)$.

Similarly it can be seen from Fig. 4.16 that when the graph of $y = \cos x$ is moved $\frac{1}{3}\pi$ to the right it coincides with the graph of $y = \cos (x - \frac{1}{3}\pi)$.

Thus changes in the value of c have no effect on the range of values taken by $\sin (x + c)$ nor do they affect its period. They simply have the effect of moving the graph along the x-axis.

Clearly the same applies to $\cos (x + c)$. Indeed it will have been noticed that the graph of $y = \sin (x + \frac{1}{2}\pi)$ is exactly the same as the graph of $y = \cos x$, i.e. $\sin (x + \frac{1}{2}\pi) = \cos x$.

Summary

(1) The range of values taken by $a \sin (bx + c)$ is from $-a$ to $+a$.

(2) The period of $a \sin (bx + c)$ is $2\pi/b$, i.e. the frequency of $a \sin (bx + c)$ is b times the frequency of $\sin x$.

(3) The range of values and the period of $a \sin (bx + c)$ are the same as the range of values and the period of $a \sin bx$. The graphs of $y = a \sin (bx + c)$ and $y = a \sin bx$ are the same in shape and differ only in their position along the x-axis.

Exercises 4.7

1 Draw the graph of $y = 5 \sin (2x + \frac{1}{3}\pi)$ for $0 \leqslant x \leqslant \pi$.
   Use your graph to solve the equations
   (a) $5 \sin (2x + \frac{1}{3}\pi) = 0$
   (b) $5 \sin (2x + \frac{1}{3}\pi) = 5$
   (c) $5 \sin (2x + \frac{1}{3}\pi) = -3$.

2 Draw the graph of $y = 4 \cos (\frac{1}{2}x - \frac{1}{4}\pi)$ for $-2\pi \leqslant x \leqslant 2\pi$.
   Use your graph to solve the equations
   (a) $4 \cos (\frac{1}{2}x - \frac{1}{4}\pi) = -4$
   (b) $4 \cos (\frac{1}{2}x - \frac{1}{4}\pi) = 2$.

3 Draw the graph of $y = 2 \tan (2x + \frac{3}{4}\pi)$ for $-\frac{1}{2}\pi \leqslant x \leqslant \frac{1}{2}\pi$.
   Use your graph to solve the equations
   (a) $2 \tan (2x + \frac{3}{4}\pi) = 2$
   (b) $2 \tan (2x + \frac{3}{4}\pi) = -5$.

4.8 Further solution of equations

In exercises 4.7 equations of the form $a \sin (bx + c) = d$ were solved graphically. We shall now consider the solution of such equations without the aid of a graph.

Example 1
Find the general solution of the equation $\cos \frac{1}{2}x = 0.5$.

Since $\cos 60° = \cos \pi/3 = 0.5$ the general solution is

$$\tfrac{1}{2}x = 2n\pi \pm \tfrac{1}{3}\pi \quad \text{or} \quad 360n° \pm 60°,$$

i.e.
$$x = 4n\pi \pm \tfrac{2}{3}\pi \quad \text{or} \quad 720n° \pm 120°.$$

Example 2
Find the general solution of the equation $\sin 2x = \sqrt{3}/2$. List all the solutions between 0 and 2π radians.

Consider the equilateral triangle ABC (Fig. 4.17) with altitude AD and with $AB = BC = CA = 2$. Then $BD = 1$ and $AD = \sqrt{3}$.
   From $\triangle ABC$ we get

$$\sin 60° = \frac{\sqrt{3}}{2}, \quad \cos 60° = \frac{1}{2}, \quad \tan 60° = \frac{\sqrt{3}}{1},$$

$$\sin 30° = \frac{1}{2}, \quad \cos 30° = \frac{\sqrt{3}}{2}, \quad \tan 30° = \frac{1}{\sqrt{3}}.$$

Fig. 4.17

(Similarly from an isosceles right-angled triangle whose equal sides are of length l the ratios for $45°$ can also be found.)

Since $\sin 60° = \sin \frac{1}{3}\pi = \dfrac{\sqrt{3}}{2}$, the general solution of $\sin 2x = \dfrac{\sqrt{3}}{2}$ is

$$2x = n\pi + (-1)^n(\tfrac{1}{3}\pi) \quad \text{or} \quad 180n° + (-1)^n(60°),$$

i.e. $\qquad x = \tfrac{1}{2}n\pi + (-1)^n(\pi/6) \quad \text{or} \quad 90n° + (-1)^n(30°).$

Note that to get all the values of x between 0 and 2π it is necessary to find all the values of $2x$ between 0 and 4π.

$2x = \pi/3,\ 2\pi/3,\ 7\pi/3,\ 8\pi/3$ or $60°,\ 120°,\ 420°,\ 480°,$
$\ x = \pi/6,\ \pi/3,\ 7\pi/6,\ 4\pi/3$ or $30°,\ 60°,\ 210°,\ 240°.$

Exercises 4.8(i)

1 Find, in degrees, the values of x between $0°$ and $360°$ for which
(a) $\cos 2x = -0{\cdot}5$ (b) $\sin \frac{1}{2}x = 0{\cdot}5$.
2 Find, in radians, the values of x between 0 and 2π for which
(a) $\cos 3x = 0{\cdot}4$ (b) $\sin \frac{1}{3}x = \sqrt{3}/2$.
3 Find, to the nearest tenth of a degree, the general solutions of the equations
(a) $\sin 4x = 0{\cdot}5$ (b) $\cos \frac{1}{4}x = 0{\cdot}6$.
4 Find, in radians, the general solutions of the equations
(a) $\sin 3x = \sqrt{3}/2$ (b) $\cos \frac{1}{3}x = 0{\cdot}7$.

Example
Find the general solution of the equation

$$2 \sin(2x + \tfrac{1}{3}\pi) = \sqrt{2}.$$

$$\sin(2x + \tfrac{1}{3}\pi) = 1/\sqrt{2} = \sin \tfrac{1}{4}\pi.$$

$\therefore$ The general solution is

$$2x + \tfrac{1}{3}\pi = n\pi + (-1)^n(\tfrac{1}{4}\pi) \quad \text{or} \quad 180n° + (-1)^n(45°)$$

$\therefore \qquad 2x = n\pi + (-1)^n(\tfrac{1}{4}\pi) - \tfrac{1}{3}\pi \quad \text{or} \quad 180n° + (-1)^n(45°) - 60°$

and $\qquad x = \tfrac{1}{2}n\pi - (\pi/6) + (-1)^n(\pi/8) \quad \text{or} \quad 90n° - 30° + (-1)^n(22\tfrac{1}{2}°).$

Exercises 4.8(ii)

1 Find the general solutions, in degrees, of the equations
(a) $3 \cos(x - 30°) = 1$ (b) $4 \sin(x + 60°) = 3$.

2 Find the general solutions, in radians, of the equations
  (a) $2\cos(x+\pi/6) = 1$ (b) $2\sin(x-\tfrac{1}{4}\pi) = \sqrt{3}$.

Graphical methods can be used to solve other trigonometrical equations. The method is that used in section 2.6 to solve the equation $x^3 - 7x - 12 = 0$. The following example will illustrate this.

Example
Find graphically an approximation to the positive root of the equation $x - 2\sin x = 0$.

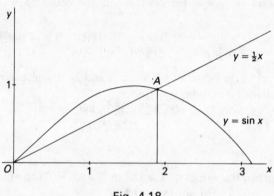

Fig. 4.18

The equation can be rearranged as $\tfrac{1}{2}x = \sin x$. Figure 4.18 shows, on the same axes, the graphs of $y = \tfrac{1}{2}x$ and $y = \sin x$ for values of x between 0 and π. The two graphs intersect at the origin and at the point A. Clearly the equation is satisfied by $x = 0$. Also at the point A the values of y are the same for both graphs. At A the value of x is approximately $1\cdot9$.

   Thus when $x \approx 1\cdot9$ we have $\tfrac{1}{2}x = \sin x$, i.e. $x - 2\sin x = 0$. From the graph it is clear that, for positive values of x, the only point of intersection of the two graphs is the point A. Hence the only positive root of the given equation is approximately $1\cdot9$, i.e. $x \approx 1\cdot9$ is the required solution.

Exercises 4.8(iii)
1 By drawing suitable graphs find an approximation to the root of the equation $x^2 - \cos x = 0$ between 0 and $\tfrac{1}{2}\pi$.
2 Find graphically a root of the equation $x + 2 = \tan x$ between 0 and $\tfrac{1}{2}\pi$.
3 Find graphically a root of the equation $2\cos x - 2\sin x = 1$ between 0 and $\tfrac{1}{2}\pi$.

Miscellaneous exercises 4
 1 Solve the equation $\cot x - 2\tan x = 0$ for $-90° < x < 90°$.
 2 Solve the equation $\sec^2 x - 2\tan x = 0$ for $0 \leqslant x \leqslant 2\pi$.

3 Find, in degrees, the general solution of the equation

$$\operatorname{cosec}^2 x - 2\cot x - 4 = 0.$$

4 Find, in radians, the general solution of the equation

$$\operatorname{cosec} x - 4\sin x = 0.$$

5 Find the solutions of the equation $\tan x + 3\cot x = 5\sec x$ for which $0 < x < 2\pi$. [L]

6 (a) Find two values of θ between $0°$ and $180°$ satisfying the equation

$$6\sin^2\theta = 5 + \cos\theta.$$

(b) Find a value of x between 0 and π satisfying the equation

$$\sin(x + \tfrac{1}{3}\pi) = \cos(x - \tfrac{1}{3}\pi).$$ [JMB]

7 Find, in radians, the general solution of the equation

$$2\sin\theta = \sqrt{3}\tan\theta.$$ [L]

8 Solve the equation $\cos(\tfrac{3}{4}x) = \tan 163°$, giving all the solutions between $0°$ and $360°$. [L]

9 Find all values of x, between 0 and 2π, for which

$$\cos(x - \alpha) = \cos 2x,$$

where α is a given positive acute angle. [C]

10 Find the values of θ between 0 and 2π for which

$$\sin 2\theta = \sin\frac{\pi}{6}.$$ [L]

11 (a) Find a general formula for all the angles θ that satisfy the equation $\tan\theta = \sqrt{3}$.
Find all the angles θ, between $0°$ and $360°$, that satisfy
  (i) $\tan 2\theta = \sqrt{3}$
  (ii) $\tan\tfrac{1}{2}\theta = \sqrt{3}$.
(b) Find all the angles θ, between $-180°$ and $+180°$, that satisfy the equation $\operatorname{cosec}^2\theta - 2\sec^2\theta = 1$. [JMB]

12 Find, in radians, the general solution of the equation

$$\cos x = \sin 3x.$$ [L]

13 If x and y are acute angles, solve the simultaneous equations $\sin x = 2\sin y$, $\tan x = 3\tan y$, giving the values of x and y in degrees.

14 Find graphically a root of the equation $x\tan x = 1$ between 0 and $\tfrac{1}{2}\pi$.

15 Draw the graph of $y = \sin x + \cos x$ for $0° \leqslant x \leqslant 360°$. Hence solve the equation $\sin x + \cos x = 1$ for $0° \leqslant x \leqslant 360°$. From your graph find the greatest and least values of $\sin x + \cos x$.

16 Draw the graph of $\sin 3x - \sin x$ for values of x between 0 and $\frac{1}{2}\pi$ radians. Hence solve the equation $\sin 3x = \sin x - x$ for values of x in this range.
[L]

17 Draw the graphs of $\sin x + \frac{1}{2}\cos x$ and $\frac{1}{4} + \frac{1}{2}x^2$ from $x = 0$ to $x = \frac{1}{2}\pi$. Hence find an approximate solution of the equation

$$4 \sin x + 2 \cos x - 2x^2 = 1.$$
[L]

18 Plot the points on the curve given by the equations $x = \cos t°$, $y = \cos 2t°$, for the values 0, 30, 60, ..., 180 of t and sketch the curve.
[L]

5 Trigonometry 2

5.1 The sine formula

In Fig. 5.1 O is the centre of the circumcircle of the triangle ABC.
$OA = OB = OC = R$ and $BD = DC = \frac{1}{2}a$, where a, b and c are the lengths of
the sides BC, CA and AB respectively.

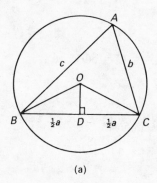 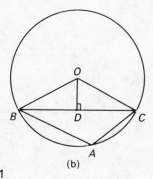

(a)   (b)

Fig. 5.1

When the angle A is acute, as in Fig. 5.1(a), we have

$$\angle BOC = 2A \quad \text{and} \quad \angle COD = A.$$

When the angle A is obtuse, as in Fig. 5.1(b), we have

$$\angle BOC = 360° - 2A \quad \text{and} \quad \angle COD = 180° - A.$$

In either case $\sin C\hat{O}D = \sin A = \dfrac{CD}{OC} = \dfrac{\frac{1}{2}a}{R} = \dfrac{a}{2R}$.

$\therefore \qquad \dfrac{a}{\sin A} = 2R$ and similarly $\dfrac{b}{\sin B} = 2R$ and $\dfrac{c}{\sin C} = 2R$.

$\therefore \qquad \dfrac{a}{\sin A} = \dfrac{b}{\sin B} = \dfrac{c}{\sin C} = 2R$.

This is known as the *sine formula* and can be used to solve triangles
when   (i) 2 angles and 1 side are known,
or   (ii) 2 sides and a non-included angle are known.

Example 1

Solve the triangle ABC in which $\angle A = 51 \cdot 2°$, $\angle B = 72 \cdot 7°$, $c = 12 \cdot 5$, and find the radius of the circumcircle of the triangle.

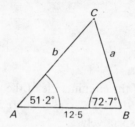

Fig. 5.2

$$51 \cdot 2° + 72 \cdot 7° + C = 180°, \therefore \angle C = 56 \cdot 1°.$$

By the sine formula

$$\frac{a}{\sin 51 \cdot 2°} = \frac{b}{\sin 72 \cdot 7°} = \frac{12 \cdot 5}{\sin 56 \cdot 1°} = 2R$$

$$a = \frac{12 \cdot 5 \sin 51 \cdot 2°}{\sin 56 \cdot 1°} = 11 \cdot 74$$

$$b = \frac{12 \cdot 5 \sin 72 \cdot 7°}{\sin 56 \cdot 1°} = 14 \cdot 38$$

$$R = \frac{12 \cdot 5}{2 \sin 56 \cdot 1°} = 7 \cdot 53$$

As a check on the calculation, we can use the identity

$$a \cos B + b \cos A = c.$$

$$\begin{aligned} a \cos B + b \cos A &= 11 \cdot 74 \cos 72 \cdot 7° + 14 \cdot 38 \cos 51 \cdot 2° \\ &= 12 \cdot 50 \text{ to 2 decimal places.} \end{aligned}$$

(If N is the foot of the perpendicular from C to AB, this identity is equivalent to $AN + NB = AB$.)

Two sides and a non-included angle are not always sufficient to define a triangle unambiguously. This is apparent if we try to construct the triangle ABC in which $a = 8$ cm, $b = 10$ cm, $\angle A = 50°$.

$AC (= 10 \text{ cm})$ and $\angle CAD (= 50°)$ are constructed (Fig. 5.3). Then with centre C and radius 8 cm, an arc of a circle is drawn. This cuts AD at two points B_1 and B_2. Thus two triangles CAB_1 and CAB_2 can be drawn to fit the data.

Since $\triangle B_1 B_2 C$ is isosceles $\alpha = \beta$,

so that $\qquad \angle AB_2 C = 180° - \beta = 180° - \alpha = 180° - \angle AB_1 C.$

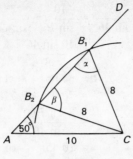

Fig. 5.3

However if we now try to construct the triangle ABC in which $a = 10$ cm, $b = 8$ cm and $\angle A = 50°$, now $CB > CA$ and the arc of radius 10 cm cuts the line segment AD in one point only, the point B (Fig. 5.4). In this case only one triangle fits the data.

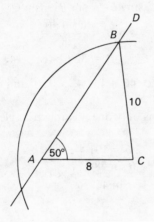

Fig. 5.4

When two sides and a non-included angle are given, two triangles are possible when the given angle is opposite to the smaller of the two given sides. However the reader may find it a sound policy always to consider the possibility of two triangles when dealing with this case of two sides and a non-included angle. This is illustrated in the following example.

Example 2
Solve the triangle ABC when
 (i) $a = 8$, $b = 10$, $\angle A = 46°$,
 (ii) $a = 8$, $b = 6$, $\angle A = 46°$.

(i) By the sine formula

$$\frac{8}{\sin 46°} = \frac{10}{\sin B} = \frac{c}{\sin C}$$

$$\therefore \qquad \sin B = \frac{10 \sin 46°}{8}$$

$$\therefore \qquad \angle B = 64\cdot0° \quad \text{or} \quad 116\cdot0°$$

and
$$\angle C = 70\cdot0° \quad \text{or} \quad 18\cdot0°.$$

$$c = \frac{8 \sin 70\cdot0°}{\sin 46°} \quad \text{or} \quad \frac{8 \sin 18\cdot0°}{\sin 46°}$$

$$= 10\cdot45 \quad \text{or} \quad 3\cdot44$$

i.e. there are two triangles which fit the data
(a) in which $\angle B = 64\cdot0°$, $\quad \angle C = 70\cdot0°$, $\quad c = 10\cdot45$,
(b) in which $\angle B = 116\cdot0°$, $\quad \angle C = 18\cdot0°$, $\quad c = 3\cdot44$.

(ii) By the sine formula

$$\frac{8}{\sin 46°} = \frac{6}{\sin B} = \frac{c}{\sin C}$$

$$\sin B = \frac{6 \sin 46°}{8}$$

$$\therefore \qquad \angle B = 32\cdot6° \quad \text{or} \quad 147\cdot4°$$

but $147\cdot4°$ is impossible since $46° + 147\cdot4° > 180°$ and the sum of the angles of a triangle cannot exceed $180°$.

Thus, in this case, there is only one possible triangle with
$$\angle B = 32\cdot6° \text{ and}$$
$$\angle C = 180° - 46° - 32\cdot6° = 101\cdot4°.$$

Then $\quad c = \dfrac{6 \sin 101\cdot4°}{\sin 46°} = \dfrac{6 \sin 78\cdot6°}{\sin 46°} = 8\cdot18.$

Exercises 5.1

1 Solve the triangle ABC given that
(a) $\angle A = 54°$, $\quad \angle B = 61°$, $\quad c = 12$
(b) $\angle A = 43\cdot5°$, $\quad \angle B = 70\cdot2°$, $\quad c = 20$
(c) $\angle B = 61\cdot5°$, $\quad \angle C = 47\cdot8°$, $\quad a = 15$
(d) $\angle B = 22\cdot1°$, $\quad \angle C = 41\cdot2°$, $\quad a = 12$
(e) $\angle A = 19\cdot3°$, $\quad \angle C = 35\cdot2°$, $\quad b = 6\cdot78$.

2 Solve the triangle ABC given that
(a) $\angle A = 37°$, $\quad a = 14$, $\quad b = 12$
(b) $\angle B = 61\cdot2°$, $\quad b = 7\cdot6$, $\quad c = 5\cdot7$
(c) $\angle C = 73\cdot7°$, $\quad a = 47\cdot5$, $\quad c = 61\cdot9$
(d) $\angle A = 37\cdot3°$, $\quad a = 11$, $\quad b = 14$
(e) $\angle B = 16\cdot8°$, $\quad b = 15$, $\quad c = 25$
(f) $\angle C = 53\cdot2°$, $\quad b = 18\cdot6$, $\quad c = 16\cdot1$
(g) $\angle A = 103\cdot2°$, $\quad a = 234$, $\quad b = 156$.

5.2 The cosine formula

A triangle may be defined by the lengths of its three sides or by the lengths of two of its sides and the magnitude of the angle between these sides. The sine formula cannot be used to solve such triangles because, of the three fractions in this formula, we shall have either the three numerators only or two of the numerators and the other denominator.

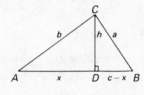

Fig. 5.5

Figure 5.5 shows an acute-angled triangle ABC in which CD is perpendicular to AB, $CD = h$ and $AD = x$.

In the triangle BCD

$$a^2 = h^2 + (c - x)^2$$
$$= h^2 + c^2 - 2cx + x^2$$
$$= (h^2 + x^2) + c^2 - 2cx.$$

In the triangle ACD $b^2 = h^2 + x^2$ and $x = b \cos A$.

$\therefore$ $\qquad\qquad a^2 = b^2 + c^2 - 2bc \cos A$

Similarly $\qquad b^2 = c^2 + a^2 - 2ca \cos B$

and $\qquad\qquad c^2 = a^2 + b^2 - 2ab \cos C.$

The reader is left to show that these results, which are known as the *cosine formula*, are also true when the triangle ABC has an obtuse angle.

The cosine formula in this form can be used to solve triangles when two sides and the included angle are given. When three sides of a triangle are given, the cosine formula is rearranged in the form

$$\cos A = \frac{b^2 + c^2 - a^2}{2bc}$$

$$\cos B = \frac{c^2 + a^2 - b^2}{2ca}$$

$$\cos C = \frac{a^2 + b^2 - c^2}{2ab}.$$

Example 1

Solve the triangle ABC in which $a = 7$, $b = 6$, $c = 5$.

$$\cos A = \frac{6^2 + 5^2 - 7^2}{2(6)(5)} = \frac{12}{60} = 0 \cdot 2$$

$$\therefore \qquad \angle A = 78 \cdot 5°$$

$$\cos B = \frac{5^2 + 7^2 - 6^2}{2(5)(7)} = \frac{38}{70} = 0 \cdot 5429$$

$$\therefore \qquad \angle B = 57 \cdot 1°.$$

$$\cos C = \frac{7^2 + 6^2 - 5^2}{2(7)(6)} = \frac{60}{84} = 0 \cdot 7143$$

$$\therefore \qquad \angle C = 44 \cdot 4°.$$

The sum $A + B + C$ provides a check on the working. There is an alternative method. After finding cos A, we can evaluate sin A, and then use the sine formula to find the angles B and C.

Example 2

Solve the triangle ABC in which $\angle A = 100°$, $b = 7 \cdot 2$, $c = 10 \cdot 6$.

$$\begin{aligned} a^2 &= b^2 + c^2 - 2bc \cos A \\ &= 7 \cdot 2^2 + 10 \cdot 6^2 - 2(7 \cdot 2)(10 \cdot 6) \cos 100° \\ &= 51 \cdot 84 + 112 \cdot 36 - 2(7 \cdot 2)(10 \cdot 6)(-\cos 80°) \\ &= 190 \cdot 7 \end{aligned}$$

$$\therefore \qquad a = 13 \cdot 81.$$

Now using the cosine formula again or the sine formula we have

$$\cos B = \frac{13 \cdot 81^2 + 10 \cdot 6^2 - 7 \cdot 2^2}{2(13 \cdot 81)(10 \cdot 6)} \quad \text{or} \quad \frac{7 \cdot 2}{\sin B} = \frac{13 \cdot 81}{\sin 100°}$$

giving $\angle B = 30 \cdot 9°$ and hence $\angle C = 49 \cdot 1°$.

Exercises 5.2

1 Solve the triangle ABC given that
   (a) $a = 7$, $\quad b = 8$, $\quad c = 9$
   (b) $a = 10 \cdot 4$, $\quad b = 6 \cdot 4$, $\quad c = 4 \cdot 8$
   (c) $a = 123$, $\quad b = 98$, $\quad c = 195$.
2 Solve the triangle ABC given that
   (a) $\quad a = 8$, $\qquad b = 6$, $\qquad \angle C = 72°$
   (b) $\angle A = 49 \cdot 5°$, $\quad b = 8 \cdot 3$, $\qquad c = 6 \cdot 9$
   (c) $\quad a = 10 \cdot 34$, $\angle B = 61 \cdot 2°$, $\quad c = 8 \cdot 321$
   (d) $\quad a = 6 \cdot 32$, $\qquad b = 7 \cdot 92$, $\angle C = 103 \cdot 8°$.

5.3 Addition theorems

In Fig. 5.6 the angles A and B are both acute and the angle $(A + B)$ is also acute.

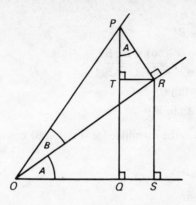

Fig. 5.6

$$\sin (A + B) = \frac{PQ}{OP} = \frac{PT + TQ}{OP} = \frac{PT + RS}{OP}$$

$$= \frac{PR \cos A + OR \sin A}{OP}$$

$$= \frac{PR}{OP} \cos A + \frac{OR}{OP} \sin A$$

$$= \sin B \cos A + \cos B \sin A$$

$$= \sin A \cos B + \cos A \sin B.$$

Similarly it can be shown that

$$\cos (A + B) = \cos A \cos B - \sin A \sin B.$$

Putting $-B$ for B we get also

$$\sin (A - B) = \sin A \cos B - \cos A \sin B$$

and
$$\cos (A - B) = \cos A \cos B + \sin A \sin B.$$

It can be shown that these results hold for all values of A and B and not just for acute angles. However, these general results are better proved by vector and matrix methods. We shall assume that these results apply to angles of all sizes.

$$\tan (A + B) = \frac{\sin (A + B)}{\cos (A + B)}$$

$$= \frac{\sin A \cos B + \cos A \sin B}{\cos A \cos B - \sin A \sin B}$$

$$= \frac{\tan A + \tan B}{1 - \tan A \tan B}$$

by dividing numerator and denominator by $\cos A \cos B$.
Similarly

$$\tan(A - B) = \frac{\sin(A - B)}{\cos(A - B)}$$

$$= \frac{\sin A \cos B - \cos A \sin B}{\cos A \cos B + \sin A \sin B}$$

$$= \frac{\tan A - \tan B}{1 + \tan A \tan B}.$$

Also by putting $B = A$ in the formulae for $\sin(A + B)$, $\cos(A + B)$ and $\tan(A + B)$ we get

$$\sin 2A = 2 \sin A \cos A$$

$$\begin{aligned}
\cos 2A &= \cos^2 A - \sin^2 A \\
&= 1 - 2 \sin^2 A \quad \text{since } \cos^2 A = 1 - \sin^2 A \\
&= 2 \cos^2 A - 1 \quad \text{since } \sin^2 A = 1 - \cos^2 A
\end{aligned}$$

$$\tan 2A = \frac{2 \tan A}{1 - \tan^2 A}$$

From the formulae for $\cos 2A$ we get

$$\sin^2 A = \tfrac{1}{2}(1 - \cos 2A)$$
and
$$\cos^2 A = \tfrac{1}{2}(1 + \cos 2A).$$

These results are of value in evaluating integrals involving $\sin^2 x$ and $\cos^2 x$ (see chapter 10).

Furthermore by putting θ for $2A$ in the $\tan 2A$ formula we get

$$\tan \theta = \frac{2 \tan \tfrac{1}{2}\theta}{1 - \tan^2 \tfrac{1}{2}\theta}$$

i.e. $$\tan \theta = \frac{2t}{1 - t^2}, \quad \text{where } t = \tan \tfrac{1}{2}\theta.$$

Putting $\theta = 2A$ in $\cos 2A = \cos^2 A - \sin^2 A$ gives

$$\begin{aligned}
\cos \theta &= \cos^2 \tfrac{1}{2}\theta - \sin^2 \tfrac{1}{2}\theta \\
&= \frac{\cos^2 \tfrac{1}{2}\theta - \sin^2 \tfrac{1}{2}\theta}{\cos^2 \tfrac{1}{2}\theta + \sin^2 \tfrac{1}{2}\theta} \\
&= \frac{1 - t^2}{1 + t^2} \quad \text{by dividing numerator and denominator by } \cos^2 \tfrac{1}{2}\theta.
\end{aligned}$$

Putting $\theta = 2A$ in $\sin 2A = 2 \sin A \cos A$ gives

$$\begin{aligned}
\sin \theta &= 2 \sin \tfrac{1}{2}\theta \cos \tfrac{1}{2}\theta \\
&= \frac{2 \sin \tfrac{1}{2}\theta \cos \tfrac{1}{2}\theta}{\cos^2 \tfrac{1}{2}\theta + \sin^2 \tfrac{1}{2}\theta} \\
&= \frac{2t}{1 + t^2}, \quad \text{by dividing numerator and denominator by } \cos^2 \tfrac{1}{2}\theta.
\end{aligned}$$

Figure 5.7 may be found useful as an aid to remembering these formulae for $\sin\theta$, $\cos\theta$ and $\tan\theta$ in terms of $t(=\tan\frac{1}{2}\theta)$.

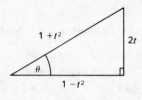

Fig. 5.7

The formulae for $\sin\theta$ and $\cos\theta$ in terms of t are useful for solving equations in $\sin\theta$ and $\cos\theta$ (see example 3 on page 114) and also for evaluating some integrals involving $\sin\theta$ and $\cos\theta$ (see chapter 10).

For easy reference all these formulae are now summarised.

Addition formulae

$$\sin(A+B) = \sin A \cos B + \cos A \sin B$$
$$\sin(A-B) = \sin A \cos B - \cos A \sin B$$

$$\cos(A+B) = \cos A \cos B - \sin A \sin B$$
$$\cos(A-B) = \cos A \cos B + \sin A \sin B$$

$$\tan(A+B) = \frac{\tan A + \tan B}{1 - \tan A \tan B}$$
$$\tan(A-B) = \frac{\tan A - \tan B}{1 + \tan A \tan B}$$

Double angle formulae

$$\sin 2A = 2\sin A \cos A$$
$$\cos 2A = \cos^2 A - \sin^2 A$$
$$= 2\cos^2 A - 1 \quad \text{or} \quad \cos^2 A = \tfrac{1}{2}(1 + \cos 2A)$$
$$= 1 - 2\sin^2 A \quad \text{or} \quad \sin^2 A = \tfrac{1}{2}(1 - \cos 2A)$$
$$\tan 2A = \frac{2\tan A}{1 - \tan^2 A}$$

Half angle formulae

$$\left. \begin{array}{l} \sin x = \dfrac{2t}{1+t^2} \\[2mm] \cos x = \dfrac{1-t^2}{1+t^2} \\[2mm] \tan x = \dfrac{2t}{1-t^2} \end{array} \right\} \qquad \text{where } t = \tan\tfrac{1}{2}x.$$

The use of these formulae is illustrated in the following examples.

Example 1
Express $\sin 3x$ in terms of $\sin x$.

Put $A = 2x$, $B = x$ in $\sin(A+B) = \sin A \cos B + \cos A \sin B$ to give

$$\begin{aligned}
\sin 3x &= \sin 2x \cos x + \cos 2x \sin x \\
&= (2 \sin x \cos x) \cos x + (1 - 2 \sin^2 x) \sin x \\
&= 2 \sin x \cos^2 x + \sin x - 2 \sin^3 x \\
&= 2 \sin x (1 - \sin^2 x) + \sin x - 2 \sin^3 x \\
&= 3 \sin x - 4 \sin^3 x
\end{aligned}$$

Example 2
Solve the equation $3 \cos 2x - \cos x + 2 = 0$ for $0° < x < 360°$.

$$\begin{aligned}
3 \cos 2x - \cos x + 2 &= 0 \\
\therefore \quad 3(2\cos^2 x - 1) - \cos x + 2 &= 0 \\
\therefore \quad 6 \cos^2 x - \cos x - 1 &= 0 \\
\therefore \quad (3 \cos x + 1)(2 \cos x - 1) &= 0
\end{aligned}$$

giving $\cos x = -\frac{1}{3}$ or $\frac{1}{2}$
and $x = 180° - 70\cdot5°$ or $180° + 70\cdot5°$ or $60°$ or $240°$
i.e. $x = 60°, 109\cdot5°, 240°, 250\cdot5°$.

Example 3
Find the general solution of the equation

$$2 \sin x - \cos x = 2.$$

$$\begin{aligned}
2 \sin x - \cos x &= 2 \\
\therefore \quad 2 \left(\frac{2t}{1+t^2} \right) - \frac{1-t^2}{1+t^2} &= 2 \\
\therefore \quad 4t - 1 + t^2 &= 2 + 2t^2 \\
\text{i.e.} \quad t^2 - 4t + 3 &= 0 \\
\text{giving} \quad (t-1)(t-3) &= 0 \\
\text{and} \quad \tan\tfrac{1}{2}x &= 1 \text{ or } 3 \\
\therefore \quad \tfrac{1}{2}x = 180n° + 45° \quad &\text{or} \quad 180n° + 71\cdot6° \\
\text{i.e.} \quad x = 360n° + 90° \quad &\text{or} \quad 360n° + 143\cdot1°.
\end{aligned}$$

This type of equation is often solved more easily by the methods of the next section (5.4).

Example 4
Show that $\dfrac{3 \sin x + \sin 2x}{1 + 3 \cos x + \cos 2x} = \tan x$ if $\cos x \neq 0$.

$$\frac{3\sin x + \sin 2x}{1 + 3\cos x + \cos 2x} = \frac{3\sin x + 2\sin x \cos x}{1 + 3\cos x + 2\cos^2 x - 1}$$

$$= \frac{\sin x\,(3 + 2\cos x)}{\cos x\,(3 + 2\cos x)}$$

$$= \frac{\sin x}{\cos x}$$

$$= \tan x.$$

Note that this result will not hold if $\cos x = 0$.

Exercises 5.3

1 If $\tan A = 3/4$ and $\tan B = 5/12$ and A and B are acute angles, find, without the use of tables or calculator,
   $\sin(A+B)$, $\cos(A+B)$, $\tan(A+B)$,
   $\sin(A-B)$, $\cos(A-B)$, $\tan(A-B)$.

2 If $\sin x = 8/17$, $\cos x = -15/17$, $\sin y = -24/25$, $\cos y = 7/25$, find, without the use of tables or calculator,
   $\sin(x+y)$, $\cos(x+y)$, $\tan(x+y)$,
   $\sin(x-y)$, $\cos(x-y)$, $\tan(x-y)$.

3 If α is an acute angle with $\sin\alpha = 4/5$ and β is an obtuse angle with $\sin\beta = 15/17$, find, without the use of tables or calculator,
   $\sin(\alpha+\beta)$, $\cos(\alpha+\beta)$, $\tan(\alpha+\beta)$,
   $\sin(\alpha-\beta)$, $\cos(\alpha-\beta)$, $\tan(\alpha-\beta)$.

4 If A and B are acute angles, $\sin A = 3/5$ and $\cos B = 8/17$, find, without the use of tables or calculator,
   $\sin 2A$, $\cos 2A$, $\tan 2A$,
   $\sin 2B$, $\cos 2B$, $\tan 2B$.

5 If A and B are acute angles, $\sin A = 0{\cdot}8$ and $\tan B = 2{\cdot}4$, find, without the use of tables or calculator,
   $\sin 2A$, $\cos 2A$, $\tan 2A$,
   $\sin 2B$, $\cos 2B$, $\tan 2B$.

6 If $0 < x < \frac{1}{2}\pi$, $\pi < y < 2\pi$, $\sin x = 5/13$ and $\tan y = 4/3$, find, without the use of tables or calculator,
   $\sin 2x$, $\cos 2x$, $\tan 2x$,
   $\sin 2y$, $\cos 2y$, $\tan 2y$.

7 Simplify
   $\sin(A+90°)$, $\sin(180°+A)$, $\sin(A-90°)$, $\sin(270°-A)$,
   $\cos(A+90°)$, $\cos(180°+A)$, $\cos(A-90°)$, $\cos(270°-A)$,
   $\tan(A+90°)$, $\tan(180°+A)$, $\tan(A-90°)$, $\tan(270°-A)$.

8 Without the use of tables or calculator, state the values of the sine, cosine and tangent of (a) $30°$ (b) $45°$ (c) $60°$.

   Using the addition formulae, find the sine, cosine, tangent, cotangent, secant and cosecant of (d) $75°$ (e) $105°$ (f) $15°$.

9 If $\tan\frac{1}{2}x = \frac{1}{2}$, find the possible values of $\sin x$, $\cos x$ and $\tan x$.

10 If $\tan y = \frac{2}{3}$, find the possible values of $\sin 2y$, $\cos 2y$ and $\tan 2y$.

11 If $\tan A = \frac{3}{4}$, find the possible values of $\tan \frac{1}{2}A$.

12 If $\tan A = \frac{4}{3}$, find the possible values of $\tan \frac{1}{2}A$.

13 Without using tables or calculator, find $\tan 22\frac{1}{2}°$, $\sin^2 22\frac{1}{2}°$ and $\cos^2 22\frac{1}{2}°$, leaving your answers in forms involving surds.

14 Express $\cos 3x$ in terms of $\cos x$.

15 Express $\tan 3x$ in terms of $\tan x$.

16 Express $\sin 4x$ in terms of $\sin x$ and $\cos x$.

17 Express $\cos 4x$ in terms of $\cos x$.

18 Solve, for $0° \leqslant x < 360°$, the equations
  (a) $2 \sin 2x + \cos x = 0$
  (b) $\sin 2x - \cos^2 x = 0$
  (c) $\sin 2x + \sin x = \tan x$
  (d) $\cos 2x - 3 \cos x - 1 = 0$
  (e) $3 \cos 2x + 4 \cos x + 1 = 0$.

19 Find the general solutions of the equations
  (a) $\sin 2x - 2 \cos x = 0$ in degrees
  (b) $\sin 2x + \sin^2 x = 0$ in radians
  (c) $2 \cos 2x + 4 \sin x - 3 = 0$ in degrees
  (d) $4 \cos 2x = 2 \sin x + 3$ in radians
  (e) $3 \cos 2x - 14 \cos x + 7 = 0$ in radians.

20 Solve, for $0° \leqslant x < 360°$,
  (a) $3 \sin x - 2 \cos x = 3$
  (b) $5 \sin x + 3 \cos x + 5 = 0$.

21 Find, in radians, the general solution of
  (a) $7 \sin x + \cos x + 5 = 0$
  (b) $\sin x + 8 \cos x + 4 = 0$.

22 Prove the identities
  (a) $\dfrac{\sin 2A}{1 + \cos 2A} \equiv \tan A$, $(A \neq \frac{1}{2}(2n+1)\pi)$,

  (b) $\dfrac{\sin 2A}{1 - \cos 2A} \equiv \cot A$, $(A \neq n\pi)$,

  (c) $\dfrac{1 + \cos x}{1 - \cos x} \equiv \cot^2 \frac{1}{2}x$, $(x \neq 2n\pi)$.

5.4 $a \cos x + b \sin x$

$a \cos x + b \sin x$ can be expressed in the forms $R \cos (x \pm \alpha)$ and $R \sin (x \pm \alpha)$. Whatever the signs of a and b it is possible to choose a form in which R is positive and α is acute.

For instance, consider $3 \cos x + 4 \sin x$. Compare this with $R \cos x \cos \alpha + R \sin x \sin \alpha \equiv R \cos (x - \alpha)$.

$$3 \cos x + 4 \sin x \equiv R \cos x \cos \alpha + R \sin x \sin \alpha \equiv R \cos (x - \alpha).$$

If $3 = R \cos \alpha$ and $4 = R \sin \alpha$, i.e. $R^2 = 3^2 + 4^2$ giving $R = 5$ and $\cos \alpha = 3/5$ and $\sin \alpha = 4/5$ giving $\alpha = 53 \cdot 1°$.

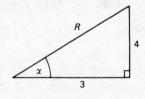

Fig. 5.8

$\therefore$ $\qquad$ $3 \cos x + 4 \sin x \equiv 5 \cos (x - 53 \cdot 1°)$.

Next consider $3 \cos x - 4 \sin x$. Compare this with
$R \cos x \cos \alpha - R \sin x \sin \alpha \equiv R \cos (x + \alpha)$.

$\qquad$ $3 \cos x - 4 \sin x \equiv R \cos x \cos \alpha - R \sin x \sin \alpha \equiv R \cos (x + \alpha)$.

If $3 = R \cos \alpha$ and $4 = R \sin \alpha$, i.e. $R^2 = 3^2 + 4^2$ giving $R = 5$ and $\cos \alpha = 3/5$, $\sin \alpha = 4/5$ giving $\alpha = 53 \cdot 1°$.

$\therefore$ $\qquad$ $3 \cos x - 4 \sin x \equiv 5 \cos (x + 53 \cdot 1°)$.

Next consider $3 \sin x - 4 \cos x$. Compare this with
$R \sin x \cos \alpha - R \cos x \sin \alpha \equiv R \sin (x - \alpha)$.

$\qquad$ $3 \sin x - 4 \cos x \equiv R \sin x \cos \alpha - R \cos x \sin \alpha \equiv R \sin (x - \alpha)$.

If $3 = R \cos \alpha$ and $4 = R \sin \alpha$, i.e. $R^2 = 3^2 + 4^2$ giving $R = 5$ and $\cos \alpha = 3/5$, $\sin \alpha = 4/5$ giving $\alpha = 53 \cdot 1°$.

$\therefore$ $\qquad$ $3 \sin x - 4 \cos x \equiv 5 \sin (x - 53 \cdot 1°)$.

Example 1
Express (a) $7 \cos x + 24 \sin x$ in the form $R \cos (x - \alpha)$, (b) $7 \sin x - 24 \cos x$ in the form $R \sin (x - \alpha)$, where $R > 0$ and $0 < \alpha < 90°$.

(a) $7 \cos x + 24 \sin x \equiv R \cos (x - \alpha) \equiv R \cos x \cos \alpha + R \sin x \sin \alpha$.
If $R \cos \alpha = 7$ and $R \sin \alpha = 24$, i.e. $R^2 = 7^2 + 24^2$, giving $R = 25$ and $\cos \alpha = 7/25$, $\sin \alpha = 24/25$ giving $\alpha = 73 \cdot 7°$.

$\therefore$ $\qquad$ $7 \cos x + 24 \sin x \equiv 25 \cos (x - 73 \cdot 7°)$

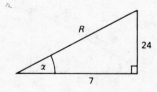

Fig. 5.9

(b) $7 \sin x - 24 \cos x \equiv R \sin (x - \alpha) \equiv R \sin x \cos \alpha - R \cos x \sin \alpha$.
If $R \cos \alpha = 7$ and $R \sin \alpha = 24$, i.e. $R^2 = 7^2 + 24^2$, giving $R = 25$ and $\cos \alpha = 7/25, \sin \alpha = 24/25$ giving $\alpha = 73.7°$.

$$\therefore \qquad 7 \sin x - 24 \cos x \equiv 25 \sin (x - 73.7°).$$

The sine or cosine of an angle has a maximum value of $+1$ and a minimum value of -1.

$$\therefore \qquad -1 \leqslant \cos(x \pm \alpha) \leqslant 1 \quad \text{and} \quad -1 \leqslant \sin(x \pm \alpha) \leqslant 1$$
$$\therefore \qquad -R \leqslant R \cos(x \pm \alpha) \leqslant R \quad \text{and} \quad -R \leqslant R \sin(x \pm \alpha) \leqslant R.$$

The maximum value of $a \cos x + b \sin x$ is R and the minimum value of $a \cos x + b \sin x$ is $-R$, where $R = \sqrt{(a^2 + b^2)}$.

Expressing $a \cos x + b \sin x$ in the forms $R \cos (x \pm \alpha)$ and $R \sin (x \pm \alpha)$ thus enables us to find the maximum and minimum values of $a \cos x + b \sin x$. It also provides a convenient way of solving equations of the type

$$a \cos x + b \sin x = c \quad \text{where } a, b \text{ and } c \text{ are constants.}$$

Since we can always express $a \cos x + b \sin x$ in the forms

$$\sqrt{(a^2 + b^2)} \cos (x \pm \alpha) \quad \text{and} \quad \sqrt{(a^2 + b^2)} \sin (x \pm \alpha)$$

it follows that the graph of $y = a \cos x + b \sin x$ will always be a periodic wave form with

$$-\sqrt{(a^2 + b^2)} \leqslant y \leqslant + \sqrt{(a^2 + b^2)},$$

i.e. with amplitude $\sqrt{(a^2 + b^2)}$ and with period 2π (see section 4.7).

Example 2
Find the maximum and minimum values of
(a) $3 \cos x - 4 \sin x$ (b) $5 \cos x + 12 \sin x$.

(a) Earlier in this section we have shown that

$$3 \cos x - 4 \sin x \equiv 5 \cos (x + 53.1°).$$

The maximum value of $3 \cos x - 4 \sin x$ is $+5$ and its minimum value is -5.
(b) $5 \cos x + 12 \sin x \equiv R \cos x \cos \alpha + R \sin x \sin \alpha \equiv R \cos (x - \alpha)$.
If $5 = R \cos \alpha$ and $12 = R \sin \alpha$, i.e. $R^2 = 5^2 + 12^2$, giving $R = 13$ and $\cos \alpha = 5/13, \sin \alpha = 12/13$ giving $\alpha = 67.4°$.

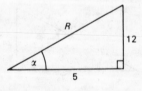

Fig. 5.10

$$\therefore \quad 5\cos x + 12\sin x \equiv 13\cos(x - 67 \cdot 4°)$$

and so the maximum value of $5\cos x + 12\sin x$ is $+13$ and the minimum value of $5\cos x + 12\sin x$ is -13.

Example 3
Find the general solution of the equation

$$2\sin x - \cos x = 2.$$

(This example was dealt with by a different method in section 5.3.)

$$2\sin x - \cos x \equiv R\sin x \cos \alpha - R\cos x \sin \alpha \equiv R\sin(x - \alpha).$$

If $2 = R\cos\alpha$ and $1 = R\sin\alpha$, i.e. $R^2 = 2^2 + 1^2$ giving $R = \sqrt{5}$ and $\cos\alpha = 2/\sqrt{5}$, $\sin\alpha = 1/\sqrt{5}$ giving $\alpha = 26 \cdot 6°$.

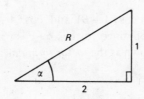

Fig. 5.11

$$\therefore \quad 2\sin x - \cos x \equiv \sqrt{5}\sin(x - 26 \cdot 6°).$$

Thus the equation becomes

$$2\sin x - \cos x = 2$$
$$\sqrt{5}\sin(x - 26 \cdot 6°) = 2$$

$$\therefore \quad \sin(x - 26 \cdot 6°) = \frac{2}{\sqrt{5}} = 0 \cdot 8945$$

$$\therefore \quad x - 26 \cdot 6° = 180n° + (-1)^n(63 \cdot 4°)$$
$$\therefore \quad x = 180n° + (-1)^n(63 \cdot 4°) + 26 \cdot 6°$$

which is the required general solution of the given equation.

Exercises 5.4
1 Find the maximum and minimum values of
  (a) $6\sin x + 8\cos x$ (d) $3\sin x - \cos x$
  (b) $12\sin x - 5\cos x$ (e) $\cos x - \sin x$.
  (c) $15\cos x - 8\sin x$
2 Find the general solutions of the following equations
  (a) $6\sin x + 8\cos x = 5$ (d) $2\sin x - 3\cos x = 2$
  (b) $5\cos x - 12\sin x = 4$ (e) $\cos 2x - 2\sin 2x = 1$.
  (c) $\cos x + 2\sin x = 1$

5.5 Factor formulae

In section 5.3 we had

$$\sin(A+B) = \sin A \cos B + \cos A \sin B \qquad \ldots(1)$$
and
$$\sin(A-B) = \sin A \cos B - \cos A \sin B \qquad \ldots(2)$$

By adding (1) and (2) we get

$$\sin(A+B) + \sin(A-B) = 2\sin A \cos B.$$

Putting $A+B = x$ and $A-B = y$ we get $A = \frac{1}{2}(x+y)$, $B = \frac{1}{2}(x-y)$

and
$$\sin x + \sin y = 2\sin\tfrac{1}{2}(x+y)\cos\tfrac{1}{2}(x-y).$$

By subtracting (2) from (1) we get

$$\sin(A+B) - \sin(A-B) = 2\cos A \sin B,$$
and
$$\sin x - \sin y = 2\cos\tfrac{1}{2}(x+y)\sin\tfrac{1}{2}(x-y).$$

From the formulae for $\cos(A+B)$ and $\cos(A-B)$ given in section 5.3 formulae can be similarly obtained for $\cos x + \cos y$ and $\cos x - \cos y$.

These four results are known as the *factor formulae* and are now listed for easy reference.

$$\sin x + \sin y = 2\sin\tfrac{1}{2}(x+y)\cos\tfrac{1}{2}(x-y)$$
$$\sin x - \sin y = 2\cos\tfrac{1}{2}(x+y)\sin\tfrac{1}{2}(x-y)$$
$$\cos x + \cos y = 2\cos\tfrac{1}{2}(x+y)\cos\tfrac{1}{2}(x-y)$$
$$\cos x - \cos y = -2\sin\tfrac{1}{2}(x+y)\sin\tfrac{1}{2}(x-y)$$

In using these formulae we try to have $A > B$ and hence $x > y$. Thus we would treat $\sin\theta + \sin 3\theta$ as $\sin 3\theta + \sin\theta$ with $x = 3\theta$ and $y = \theta$ so that $x - y$ is positive giving

$$\sin 3\theta + \sin\theta = 2\sin 2\theta \cos\theta.$$

Example 1

Solve, for $0° \leqslant x \leqslant 180°$, the equation

$$\sin x + \sin 3x + \sin 5x = 0.$$

Using the factor formulae

$$\sin 5x + \sin x = 2\sin 3x \cos 2x.$$

$\therefore$ $\sin x + \sin 3x + \sin 5x = 0$

becomes $2\sin 3x \cos 2x + \sin 3x = 0$

$\therefore$ $\sin 3x(2\cos 2x + 1) = 0$

$\therefore$ $\sin 3x = 0$ or $\cos 2x = -\tfrac{1}{2}$

$\therefore$ $3x = 0°, 180°, 360°, 540°$ (Note that for values of x between $0°$ and $180°$ we require values of $3x$ between $0°$ and $540°$.)

or $2x = 120°, 240°$ (i.e. between $0°$ and $360°$)

$$\therefore \quad x = 0°, 60°, 120°, 180°$$
$$\text{or} \quad x = 60°, 120°$$

i.e. the required solutions are $x = 0°, 60°, 120°, 180°$.

Example 2
Show that $\sin^2 A - \sin^2 B \equiv \sin(A + B)\sin(A - B)$.

$$\begin{aligned}
\sin^2 A - \sin^2 B &\equiv (\sin A + \sin B)(\sin A - \sin B) \\
&\equiv (2\sin\tfrac{1}{2}(A+B)\cos\tfrac{1}{2}(A-B))(2\cos\tfrac{1}{2}(A+B)\sin\tfrac{1}{2}(A-B)) \\
&\equiv (2\sin\tfrac{1}{2}(A+B)\cos\tfrac{1}{2}(A+B))(2\sin\tfrac{1}{2}(A-B)\cos\tfrac{1}{2}(A-B)) \\
&\equiv \sin(A+B)\sin(A-B).
\end{aligned}$$

Exercises 5.5
1. Express in factor form
   (a) $\sin 3x + \sin 5x$
   (b) $\sin 4x + \sin 8x$
   (c) $\sin (x + 30°) + \sin (x + 60°)$
   (d) $\sin 50° + \sin 40°$
   (e) $\sin (\alpha + \beta) + \sin (\alpha - \beta)$.

2. Express in factor form
   (a) $\sin 3x - \sin x$
   (b) $\sin 4x - \sin 2x$
   (c) $\sin 2x - \sin x$
   (d) $\sin 60° - \sin 30°$
   (e) $\sin (60° - x) - \sin (30° - x)$
   (f) $\sin (\alpha + \beta) - \sin (\alpha - \beta)$.

3. Express in factor form
   (a) $\cos x + \cos 5x$
   (b) $\cos 2x + \cos x$
   (c) $\cos (x + 45°) + \cos (x - 45°)$
   (d) $\cos 30° + \cos 60°$.

4. Express in factor form
   (a) $\cos x - \cos 5x$
   (b) $\cos 2x - \cos 6x$
   (c) $\cos (x + 60°) - \cos (x - 60°)$
   (d) $\cos 3x - \cos 2x$.

5. Solve, for $0° \leqslant x \leqslant 360°$, the equations
   (a) $\sin 2x - \sin 4x = \cos 3x$
   (b) $\cos x + \cos 3x = \cos 2x$
   (c) $\sin 3x + \sin x = \sin 2x$
   (d) $\cos x + \cos 5x = \cos 2x$.

6. Find the general solutions of the equations
   (a) $\cos x + \cos 3x + \cos 5x = 0$ in degrees
   (b) $\cos 2x - \cos 4x = \sin x$ in radians
   (c) $\cos 3x + \cos 5x = \sin 6x - \sin 2x$ in degrees.

7. Show that
   (a) $\dfrac{\cos 4x - \cos 2x}{\sin 4x - \sin 2x} \equiv -\tan 3x$ if $\sin 4x \neq \sin 2x$
   (b) $\dfrac{\sin 5x - \sin 3x}{\cos 5x + \cos 3x} = \tan x$ if $\cos 5x + \cos 3x \neq 0$
   (c) $\cos 2A + 2\cos 4A + \cos 6A = 4\cos^2 A \cos 4A$.

5.6 Area of a triangle

In the triangle ABC (Fig. 5.12), AD is an altitude such that $AD = h$.
Area of $\triangle ABC = \frac{1}{2}ah$ and from the triangle ADC, $h = b \sin C$.
$\therefore$ Area of $\triangle ABC = \frac{1}{2}ab \sin C$.
 Similarly the area can also be shown to be $\frac{1}{2}bc \sin A$ and $\frac{1}{2}ca \sin B$,

i.e.   area of $\triangle ABC = \frac{1}{2}ab \sin C = \frac{1}{2}bc \sin A = \frac{1}{2}ca \sin B$.

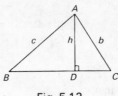

Fig. 5.12

The area of the triangle ABC can also be expressed in terms of the sides a, b, c.

Let $\qquad\qquad\qquad \frac{1}{2}(a+b+c) = s.$

Then $\qquad\qquad\qquad \frac{1}{2}(b+c-a) = s-a$
$\qquad\qquad\qquad\qquad \frac{1}{2}(c+a-b) = s-b$
$\qquad\qquad\qquad\qquad \frac{1}{2}(a+b-c) = s-c.$

Since $\qquad\qquad\qquad 2bc \cos A = b^2+c^2-a^2,$
$$2bc(1+\cos A) = 2bc+b^2+c^2-a^2$$
$$= (a+b+c)(b+c-a)$$
$$= 4s(s-a).$$

Also $\qquad\qquad 2bc(1-\cos A) = 2bc-b^2-c^2+a^2$
$$= (c+a-b)(a+b-c)$$
$$= 4(s-b)(s-c).$$

This gives $\qquad 4b^2c^2(1-\cos^2 A) = 16s(s-a)(s-b)(s-c)$
$\Rightarrow \qquad\qquad \frac{1}{4}b^2c^2 \sin^2 A = s(s-a)(s-b)(s-c).$

Then $\qquad\qquad$ area of $\triangle ABC = \frac{1}{2}bc \sin A$
$$= \sqrt{[s(s-a)(s-b)(s-c)]}$$

where $s = \frac{1}{2}(a+b+c)$.
 This is known as Heron's formula.

Example

Find, to 3 significant figures, the area of the triangle ABC given that

(a) $a = 27$, $c = 35$, $\angle B = 40°$
(b) $a = 27$, $b = 35$, $c = 47$.

(a) Area of $\triangle ABC = \frac{1}{2} ca \sin B$
$= \frac{1}{2}(35)(27) \sin 40°$
$= 304$

(b) $\quad\quad\quad 2s = 27 + 35 + 47 = 109$
∴ $\quad\quad\quad s = 54 \cdot 5$, $\; s - a = 27 \cdot 5$, $\; s - b = 19 \cdot 5$, $\; s - c = 7 \cdot 5$.
∴ $\quad$ Area of $\triangle ABC = \sqrt{s(s-a)(s-b)(s-c)}$
$= \sqrt{[(54 \cdot 5)(27 \cdot 5)(19 \cdot 5)(7 \cdot 5)]}$
$= 468$

Exercises 5.6

1 Find, to 3 significant figures, the area of triangle ABC given that
(a) $a = 7$, $b = 8$, $\angle C = 50°$ (c) $\angle A = 62 \cdot 3°$, $b = 6 \cdot 34$, $c = 8 \cdot 21$
(b) $a = 27$, $\angle B = 43 \cdot 5°$, $c = 16$ (d) $a = 3 \cdot 148$, $b = 2 \cdot 63$, $\angle C = 104 \cdot 3°$.
2 Find, to 3 significant figures, the area of the triangle ABC given that
(a) $a = 8$, $\quad\quad b = 6$, $\quad\quad c = 11$
(b) $a = 73$, $\quad\quad b = 59$, $\quad\quad c = 44$
(c) $a = 6 \cdot 7$, $\quad\quad b = 9 \cdot 3$, $\quad\quad c = 12 \cdot 8$
(d) $a = 145$, $\quad\quad b = 97$, $\quad\quad c = 168$
(e) $a = 16 \cdot 34$, $\quad b = 21 \cdot 07$, $\quad c = 14 \cdot 19$.

5.7 Problems in two and three dimensions

The following examples illustrate the use of trigonometrical formulae in the solution of problems in two and three dimensions.

Example 1

From a point A the angle of elevation to P, the top of a flag pole on the roof of a building, is 18°. From a point B the angle of elevation is 23° and $AB = 150$ m. If A and B are on level ground and A, B and P are in a vertical plane, find the height of P above the ground.

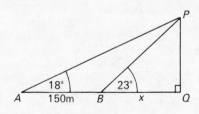

Fig. 5.13

Using the sine formula for the triangle APB

$$PB = \frac{150 \sin 18°}{\sin 5°}$$

In $\triangle PBQ$, $\sin 23° = \dfrac{PQ}{PB}$, where Q is at ground level vertically below P.

$$\therefore \qquad PQ = \frac{150 \sin 18° \sin 23°}{\sin 5°} = 207{\cdot}8 \text{ m},$$

i.e. the height of P above the ground is $207{\cdot}8$ m.

Alternative method:

$$h \cot 18° = 150 + x \quad \text{from } \triangle APQ$$
$$h \cot 23° = x \qquad\quad \text{from } \triangle BPQ$$

Subtracting $\qquad\qquad h(\cot 18° - \cot 23°) = 150$

$$\therefore \qquad PQ = h = \frac{150}{\cot 18° - \cot 23°} = 207{\cdot}8 \text{ m}.$$

Example 2

A vertical radio mast ST is of height 70 m. From points P and Q, both on the same level as S, the foot of the mast, the angles of elevation of T, the top of the mast, are $20°$ and $35°$ respectively. The bearings of S from P and Q are $062°$ and $304°$ respectively. Calculate the length PQ and the bearing from P to Q.

The situation is illustrated in Fig. 5.14.

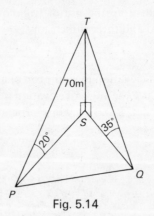

Fig. 5.14

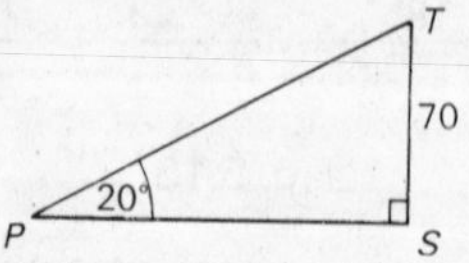

Fig. 5.15

In the triangle *PST*

$$\tan 20° = \frac{70}{PS}$$

∴ $$PS = \frac{70}{\tan 20°} \quad \text{or } 70 \cot 20°$$

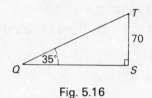

Fig. 5.16

In the triangle *QST*

$$\tan 35° = \frac{70}{QS}$$

∴ $$QS = 70 \cot 35°$$

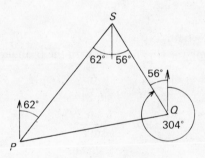

Fig. 5.17

From the triangle *PSQ* (Fig. 5.17) it can be seen that $\angle PSQ = 118°$. Using the cosine formula with the triangle *PSQ* we get

$$PQ^2 = 70^2(\cot^2 20° + \cot^2 35° - 2 \cot 20° \cot 35° \cos 118°)$$

⇒ $$PQ = 255 \text{ m to the nearest metre.}$$

Using the sine formula for the triangle *PSQ* we get

$$\frac{PQ}{\sin 118°} = \frac{QS}{\sin SPQ}$$

$$\therefore \qquad \sin SPQ = \frac{QS \sin 62°}{255}$$

$$\therefore \qquad \angle SPQ = 20\cdot3°$$

$\therefore$ Bearing of Q from $P = 62° + 20\cdot3° = 82\cdot3°$

Example 3
Two coastguard stations A and B are 6 km apart and B is due east of A. From A the bearings of two ships P and Q are 020° and 051° respectively and from B the bearings are 299° and 342° respectively. Find the distance between the ships.

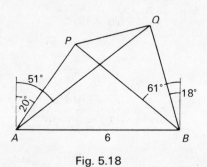

Fig. 5.18

The situation is shown in Fig. 5.18. $\angle PAB = 70°$ and $\angle PBA = 29°$, $\therefore \angle APB = 81°$. Using the sine formula with the triangle PAB we get

$$\frac{PA}{\sin 29°} = \frac{6}{\sin 81°}$$

$$\therefore \qquad PA = \frac{6 \sin 29°}{\sin 81°}$$

$$= 2\cdot95 \text{ km.}$$

$\angle QAB = 39°$ and $\angle QBA = 72°$ $\therefore \angle AQB = 69°$.
Using the sine formula with the triangle QAB we get

$$\frac{QA}{\sin 72°} = \frac{6}{\sin 69°}$$

$$QA = \frac{6 \sin 72°}{\sin 69°}$$

$$= 6\cdot11 \text{ km.}$$

$\angle PAQ = 51° - 20° = 31°$.
Using the cosine formula with the triangle PAQ we get

$$PQ^2 = PA^2 + QA^2 - 2(PA)(QA) \cos 31°$$

$$\therefore \qquad PQ = 3\cdot90 \text{ km to 2 decimal places.}$$

Exercises 5.7

1 A man walking due north along a straight road sights a chimney on a bearing of 033°. After walking a distance of 500 m, the bearing of the chimney from the man's new position is 69°. Find the distance of the chimney from the road.

2 In a tetrahedron $ABCD$ the edge AB is perpendicular to the face BCD, $AB = 30$ cm, $\angle CBD = 72°$, $\angle ACB = 42°$ and $\angle ADB = 53°$. Find the length of the edge CD.

3 In a quadrilateral $ABCD$ the length of the side AB is 34 cm and the angles DAB, CAB, DBA and CBA are 71·5°, 47·3°, 50·2° and 100·7° respectively. Find the length of the side CD.

5.8 Length of an arc and area of a sector of a circle

In Fig. 5.19, AB is an arc of length s of a circle, centre O and radius r, and $\angle AOB = \theta$ radians.

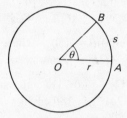

Fig. 5.19

From the definition of the radian (see section 4.6) we have

$$s = r\theta,$$

i.e. Length of the arc $AB = s = r\theta$.

Also

$$\frac{\text{area of sector } AOB}{\text{area of whole circle}} = \frac{\theta}{2\pi}$$

$\therefore$ $\text{area of sector } AOB = \left(\frac{\theta}{2\pi}\right)\pi r^2$

$$= \tfrac{1}{2}r^2\theta, \text{ where } \theta \text{ is in radians.}$$

Example 1

If AB is an arc of a circle, centre O and radius 8 cm, find the length of the arc AB and the area of the sector AOB given that

(a) $\angle AOB = 2\cdot5$ radians   (b) $\angle AOB = 70°$.

(a) Arc $AB = r\theta = 8 \times 2\cdot5 = 20$ cm.

Area of sector $AOB = \tfrac{1}{2}r^2\theta$

$$= \tfrac{1}{2}(64)(2\cdot5) = 80 \text{ cm}^2.$$

(b) $\angle AOB = 70° = \left(\dfrac{70}{180}\right)\pi \text{ radians} = \dfrac{7\pi}{18} \text{ radians}$

$\therefore \qquad\qquad \text{arc } AB = r\theta = 8\left(\dfrac{7\pi}{18}\right) = \dfrac{28\pi}{9} = 9\cdot774 \text{ cm}$

or $\quad \angle AOB = 70° = 1\cdot2217 \text{ radians}.$

$\therefore \qquad\qquad\qquad \text{arc } AB = 8 \times 1\cdot2217 = 9\cdot774 \text{ cm}.$

$$\text{Area of sector } AOB = \tfrac{1}{2}r^2\theta$$
$$= \tfrac{1}{2}(64)\dfrac{7\pi}{18} \text{ cm}^2$$
$$= 39\cdot1 \text{ cm}^2$$

Example 2
AB is an arc of a circle with centre O. The area of the sector AOB is 150 cm².
(a) Find the angle AOB in degrees if $AO = 10$ cm.
(b) Find the radius of the circle if $\angle AOB = 42\cdot5°$

(a) Area of sector $AOB = \tfrac{1}{2}(100)\theta = 150, \quad$ where $\theta = \angle AOB.$
$\therefore \qquad\qquad \theta = \angle AOB = 3 \text{ radians} = 171\cdot9°$

(b) Let $AO = r,\ \angle AOB = 42\cdot5° = 0\cdot7417$ radians.
Area of sector $AOB = \tfrac{1}{2}r^2(0\cdot7417) = 150 \text{ cm}^2.$
$\therefore \qquad\qquad r^2 = \dfrac{300}{0\cdot7417} \text{ cm}^2 = 404\cdot5 \text{ cm}^2$
$\therefore \qquad\qquad r = 20\cdot11 \text{ cm}.$

Example 3
An arc AB of a circle, centre O and radius 24 cm, subtends an angle of 110·7° at O.
Find the area of the segment formed by the chord AB and the minor arc AB.

The required area is shaded in Fig. 5.20.

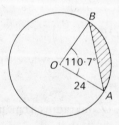

Fig. 5.20

Area of the segment = area of the sector AOB – area of $\triangle AOB$
$$= \tfrac{1}{2}r^2\theta - \tfrac{1}{2}r^2 \sin\theta$$
$$= \tfrac{1}{2}(24)^2(1\cdot9327 - \sin 110\cdot7°) \text{ cm}^2$$

Area of the segment $= 288(1{\cdot}9327 - \sin 69{\cdot}6°)$ cm²
$= 288(1{\cdot}9327 - 0{\cdot}9352)$ cm²
$= 287{\cdot}3$ cm².

Exercises 5.8

1 Find the length of the arc which subtends an angle θ at the centre of a circle of radius r when
(a) $r = 12$ cm, $\theta = 3$ radians   (c) $r = 8{\cdot}6$ cm, $\theta = 60°$
(b) $r = 25$ cm, $\theta = 4{\cdot}2$ radians   (d) $r = 136$ cm, $\theta = 100°$.
2 Find the areas of the four sectors defined by the data of question 1.
3 If AB is an arc of a circle centre O, find the angle AOB in both radians and degrees when
(a) $OA = 8$ cm and the area of the sector $AOB = 30$ cm²
(b) $OA = 12$ cm and the area of the sector $AOB = 100$ cm²
(c) $OA = 2$ m and the area of the sector $AOB = 1{\cdot}45$ m².
4 If AB is an arc of a circle centre O, find the radius OA when
(a) $\angle AOB = 2{\cdot}5$ radians and the area of the sector $AOB = 20$ cm²
(b) $\angle AOB = 110°$ and the area of the sector $AOB = 65$ cm²
(c) $\angle AOB = 75{\cdot}2°$ and the area of the sector $AOB = 4{\cdot}32$ m².
5 If an arc of length s subtends an angle θ at the centre of a circle of radius r,
(a) find r when $s = 40$ cm and $\theta = 2{\cdot}1$ radians
(b) find r when $s = 125$ cm and $\theta = 160°$
(c) find θ in radians when $s = 95$ cm and $r = 30$ cm
(d) find θ in degrees when $s = 3{\cdot}79$ m and $r = 1{\cdot}6$ m.
6 AB is an arc of a circle, centre O and radius r. Find the areas of the two segments into which the chord AB divides the circle when
(a) $\angle AOB = 65°$, $r = 15$ cm
(b) $\angle AOB = 136°$, $r = 1{\cdot}4$ m
(c) $\angle AOB = 89°$, $r = 28{\cdot}6$ cm.

5.9 The circular functions

In chapter 1 we considered functions which map the elements of one set to the elements of another set. For instance the function $d : \mathbb{R} \to \mathbb{R}$ defined by $d(x) = 2x$ or $x \mapsto 2x$ was considered.

Now consider the function which maps each element x of $\mathbb{R}$, the set of real numbers, to the sine of the angle x radians. This function is called the sine function and if we denote it by s, then

$$s : \mathbb{R} \to \mathbb{R} \text{ defined by } s(x) = \sin x \quad \text{or} \quad x \mapsto \sin x,$$

where $\sin x$ is the sine of the angle x radians.
Thus we have $\sin 1 = 0{\cdot}8415$
and $\sin 2{\cdot}3 = 0{\cdot}7457$.
Since, for all real values of x, $-1 \leqslant \sin x \leqslant 1$, the range of the sine function will be the set $\{y : y \in \mathbb{R} \text{ and } -1 \leqslant y \leqslant 1\}$.

The sine function can be illustrated by the graph in Fig. 5.21 in which each element x of the domain $\mathbb{R}$, the set of real numbers, and its image $s(x)$ are represented by the point with coordinates (x, y), where $y = s(x)$.

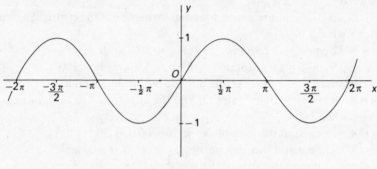

Fig. 5.21

The sine function is an odd function since $s(-x) = -s(x)$.

Does an inverse function s^{-1} exist?

When $x = \pi/6$, $\sin x = \sin \pi/6 = \frac{1}{2}$, i.e. $s(\pi/6) = \frac{1}{2}$. Does $s^{-1}(\frac{1}{2}) = \pi/6$? From Fig. 5.21 it can be seen that there is not just one value of x which is mapped by the sine function to the value $\frac{1}{2}$. Indeed $\sin x = \frac{1}{2}$ for an infinite set of values of x and so an inverse function s^{-1} does not exist.

Although no inverse function s^{-1} exists it is still possible to relate each element y of the range to the elements x of the domain such that $\sin x = y$. This inverse relation will map each element y of the range to an infinite set of numbers in the domain $\mathbb{R}$. For any given y, where $-1 \leqslant y \leqslant 1$, there is an infinite number of values of x which satisfy $\sin x = y$.

However if $-\frac{1}{2}\pi \leqslant x \leqslant \frac{1}{2}\pi$, there will be only one value of x for each value of y.

Consider a function f where

$$f\colon D \to \mathbb{R} \text{ defined by } f(x) = \sin x,$$

and $D = \{x\colon x \in \mathbb{R}, -\frac{1}{2}\pi \leqslant x \leqslant \frac{1}{2}\pi\}$. This set D is the domain and the range is the set $\{y\colon y \in \mathbb{R}, -1 \leqslant y \leqslant 1\}$. Since there is now only one element x of the domain for each element y of the range, an inverse function f^{-1} exists. The image of y under this inverse function is denoted by arcsin y or by $\sin^{-1} y$,

i.e. $\qquad\qquad f^{-1}(y) = \arcsin y \quad \text{or} \quad \sin^{-1} y, \quad -1 \leqslant y \leqslant 1.$

This inverse function maps each member of the set $\{y\colon y \in \mathbb{R}, -1 \leqslant y \leqslant 1\}$ to one member of the set $\{x\colon x \in \mathbb{R}, -\frac{1}{2}\pi \leqslant x \leqslant \frac{1}{2}\pi\}$.

This inverse function is represented in Fig. 5.22 by the graph $y = \arcsin x$. It has been shown that $\sin(\pi/6) = \frac{1}{2}$ and so we have

$$\arcsin(\tfrac{1}{2}) = \pi/6 \quad \text{or} \quad \sin^{-1}(\tfrac{1}{2}) = \pi/6.$$

Similarly $\qquad\qquad \arcsin(0\cdot6) = 0\cdot6435 \quad \text{or} \quad \sin^{-1}(0\cdot6) = 0\cdot6435.$

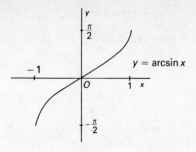

Fig. 5.22

The cosine function maps each element x of $\mathbb{R}$ to the cosine of the angle x radians. As with the sine function it can be shown that no inverse function exists. However a function g can be defined where

$$g:D_1 \to \mathbb{R} \text{ defined by } g(x) = \cos x,$$

and $D_1 = \{x : x \in \mathbb{R},\ 0 \leqslant x \leqslant \pi\}$. The range is again the set $\{y : y \in \mathbb{R}, -1 \leqslant y \leqslant 1\}$. An inverse function g^{-1} exists and the image of y under this inverse function is denoted by arccos y or by $\cos^{-1} y$,

i.e. $\qquad g^{-1}(y) = \arccos y \quad \text{or} \quad \cos^{-1} y, \quad -1 \leqslant y \leqslant 1.$

This inverse function maps each element of the set $\{y : y \in \mathbb{R}, -1 \leqslant y \leqslant 1\}$ to one element of the set $\{x : x \in \mathbb{R}, 0 \leqslant x \leqslant \pi\}$.

This inverse function is represented in Fig. 5.23 by the graph of $y = \arccos x$.

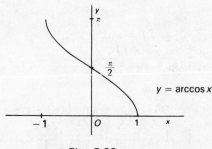

Fig. 5.23

The tangent function maps each element x of $\mathbb{R}$ to the tangent of the angle x radians. As with the sine and cosine functions no inverse function exists for the tangent function. However a function h can be defined where

$$h:D_2 \to \mathbb{R} \text{ defined by } h(x) = \tan x,$$

where $D_2 = \{x : x \in \mathbb{R}, -\frac{1}{2}\pi < x < \frac{1}{2}\pi\}$. For this function the range is $\mathbb{R}$, the set of real numbers. An inverse function h^{-1} exists and the image of y under this inverse function is denoted by arctan y or by $\tan^{-1} y$, i.e. $$h^{-1}(y) = \arctan y \quad \text{or} \quad \tan^{-1} y, \quad y \in \mathbb{R}.$$

This inverse function maps each element of $\mathbb{R}$, the set of real numbers, to one element of the set $\{x : x \in \mathbb{R}, \ -\frac{1}{2}\pi < x < \frac{1}{2}\pi\}$.

Example 1
Evaluate (a) $\arcsin \frac{1}{2}$ (b) $\cos^{-1}(1/\sqrt{2})$ (c) $\arctan(-\sqrt{3})$.

(a) Let $x = \arcsin \frac{1}{2}$, then $\sin x = \frac{1}{2}$, giving $x = \pi/6 = 0.5236$.
(b) Let $x = \cos^{-1}(1/\sqrt{2})$, then $\cos x = 1/\sqrt{2}$, giving $x = \pi/4 = 0.7854$.
(c) Let $x = \arctan(-\sqrt{3})$, then $\tan x = -\sqrt{3}$, giving $x = -\pi/3 = -1.0472$.

Example 2
(a) Find $\tan \theta$ if $\theta = \tan^{-1}(3/4) + \tan^{-1}(4/5)$.
(b) Find $\sin \theta$ if $\theta = \arcsin(5/13) - \arccos(8/17)$.

(a) Let $\alpha = \tan^{-1}(3/4)$ and $\beta = \tan^{-1}(4/5)$.
Then $\tan \alpha = 3/4$ and $\tan \beta = 4/5$.

$$\theta = \alpha + \beta$$

$$\tan \theta = \tan(\alpha + \beta) = \frac{\tan \alpha + \tan \beta}{1 - \tan \alpha \tan \beta}$$

$$= \frac{\dfrac{3}{4} + \dfrac{4}{5}}{1 - \left(\dfrac{3}{4}\right)\left(\dfrac{4}{5}\right)}$$

$$= 31/8.$$

(b) Let $\alpha = \arcsin(5/13)$ and $\beta = \arccos(8/17)$.
Then $\sin \alpha = 5/13$ and $\cos \beta = 8/17$.
and $\cos \alpha = 12/13$ and $\sin \beta = 15/17$.

$$\theta = \alpha - \beta$$
$$\sin \theta = \sin(\alpha - \beta) = \sin \alpha \cos \beta - \cos \alpha \sin \beta$$
$$= \left(\frac{5}{13}\right)\left(\frac{8}{17}\right) - \left(\frac{12}{13}\right)\left(\frac{15}{17}\right)$$
$$= -140/221.$$

Exercises 5.9
1 Evaluate
    (a) $\arcsin(3/4)$ (d) $\frac{1}{2}\sin^{-1}(-0.64)$ (g) $2\cos^{-1}(-0.234)$
    (b) $\arcsin(0.34)$ (e) $\arccos(4/5)$ (h) $\arctan(2.4)$
    (c) $\sin^{-1}(-\frac{1}{4})$ (f) $\arccos(0.321)$ (i) $\frac{1}{3}\tan^{-1}(-0.789)$.
2 Find $\tan \theta$ if (a) $\theta = \arctan 2 + \arctan \frac{2}{3}$ (b) $\theta = \tan^{-1}(1.5) - \tan^{-1}(0.8)$.
3 Find $\sin \theta$ when
    (a) $\theta = \arcsin(3/5) + \arcsin(12/13)$ (b) $\theta = \cos^{-1}(8/17) - \cos^{-1}(4/5)$.

4 Find $\cos \theta$ when

(a) $\theta = \arcsin (4/5) + \arccos (15/17)$ (b) $\theta = \cos^{-1} (1/2) - \sin^{-1} (1/3)$.

Miscellaneous exercises 5

1 Prove that in *any* triangle ABC, with the usual notation,
$$a^2 = b^2 + c^2 - 2bc \cos A.$$

A convex quadrilateral $ABCD$ has $AB = 5$ cm, $BC = 8$ cm, $CD = 3$ cm, $DA = 3$ cm and $BD = 7$ cm. Find the angles DAB, BCD and show that the quadrilateral is cyclic. Show also that $AC = 39/7$ cm. [L]

2 Show that, in the triangle ABC, the length of the perpendicular from C to AB is given by

$$\frac{ab}{c} \sin C.$$

The triangle ABC, in which $\angle ACB = 150°$, lies in a horizontal plane and D is the point at a distance 2 units vertically above C. Given that $\angle DBC = 30°$ and $\angle DAC = 45°$, find the lengths of the sides of the triangle ABC and the length of the perpendicular from C to AB. [JMB]

3 In the triangle ABC, X, Y and Z are the mid-points of BC, CA and AB, respectively. Using the cosine rule or otherwise, prove that
$$b^2 + c^2 = 2AX^2 + 2BX^2.$$

Write down two other similar results, and hence show that

$$AX^2 + BY^2 + CZ^2 = \tfrac{3}{4}(a^2 + b^2 + c^2).$$ [JMB]

4 Show that for any triangle ABC

$$\tan \frac{A}{2} \tan \frac{B-C}{2} = \frac{b-c}{b+c}.$$

Three towns A, B and C are all at sea level. The bearings of towns B and C from A are 36° and 247° respectively. If B is 120 km from A and C is 234 km from A, calculate the distance and bearing of town B from C. [AEB]

5 Prove that, for a triangle PQR,

$$\cot P = \frac{q}{p} \operatorname{cosec} R - \cot R.$$

To an observer standing at the top of a cliff the bearings of two ships P and Q at sea are 5·6° and 127·2° respectively. If to the observer the angle of depression of ship P 14·4° whilst that of ship Q is 10·1°, calculate the bearing of ship Q from ship P. [AEB]

6 In the tetrahedron $VABC$ the angles ACB, VAC and VBC are 60°, α and β respectively, $AB = d$ and VC is perpendicular to the plane ABC. Show that

$$VC = \frac{d}{(\cot^2 \alpha + \cot^2 \beta - \cot \alpha \cot \beta)^{\frac{1}{2}}}.$$

If $\tan \alpha = \frac{3}{4}$ and $\tan \beta = \frac{1}{2}$, find

  (i) the area of the triangle ABC

  (ii) the volume of the tetrahedron $VABC$

  (iii) the tangent of the angle between the planes VAB and ABC.   [AEB]

7   An equilateral triangle ABC of side 6 cm lies in a horizontal plane. Points D, E and F are vertically above A, B and C respectively at heights above the plane of 1 cm, 2 cm and 2 cm respectively. A point P is taken on BC, distant x cm from B. If the angle $EPD = \theta$, show that

$$\cos \theta = \frac{x^2 - 3x + 2}{\sqrt{(x^2 + 4)} \sqrt{(x^2 - 6x + 37)}}.$$

   Find the points P_1 and P_2 on BC at which ED subtends an angle of 90°. Show that the ratio of the area of the triangle EP_1D to that of EP_2D is equal to the ratio of FP_2 to FP_1.   [AEB]

8   Three vertical wires are such that they cut any horizontal plane in the vertices of an equilateral triangle of side $2a\sqrt{6}$. A plane cuts the wires at points P, Q and R so that the horizontal level of Q is $5a$ below that of P, and that of R is a above the level of P. Calculate the angle P of the triangle PQR.   [AEB]

9   The plane π is at a distance $r/2$ from a point O, and P and Q are points on the intersection of π with the surface of a sphere of centre O and radius r. The angle POQ is θ. Find, in terms of r and θ, the length of the minor arc PQ of the circle in which π intersects the sphere.   [JMB]

10   In the triangle ABC, $BC = a$, $CA = b$, $AB = c$ and $c^2 = a^2 + b^2$. Show that the area of the region that is common to the two circles, one of centre A and radius b and the other of centre B and radius a, is

$$a^2 \sin^{-1} \left(\frac{b}{c} \right) + b^2 \sin^{-1} \left(\frac{a}{c} \right) - ab. \qquad \text{[JMB]}$$

11   (a) Without the use of tables or calculator, evaluate $\tan (A + B + C)$, given that $\tan A = \frac{1}{2}$, $\tan B = \frac{1}{3}$ and $\tan C = \frac{1}{4}$.

   (b) Find all values of θ between 0° and 360° such that $5(\cos \theta + \sin \theta) = 1$.
   [C]

12   (i) If A is the acute angle such that $\sin A = 3/5$ and B is the obtuse angle such that $\sin B = 5/13$, find the exact values of $\cos (A + B)$ and $\tan (A - B)$.

   (ii) Find the solutions of the equation

$$\tan \theta + 3 \cot \theta = 5 \sec \theta$$

for which $0 < \theta < 2\pi$.   [L]

13   (a) Prove that

$$\cot (A + B) - \tan (A - B) = \frac{2 \cos 2A}{\sin 2A + \sin 2B}.$$

   (b) Given that $\operatorname{cosec} x = \sin \alpha - \cos \alpha$, prove that

(i) $\tan^2 x = -\operatorname{cosec} 2\alpha,$

(ii) $\cos 2x = -\left(\dfrac{\sin\alpha+\cos\alpha}{\sin\alpha-\cos\alpha}\right)^2.$ [C]

14 Find the values of x, for angles between $0°$ and $360°$ inclusive, for which

$$2\sin 3x + \cos 2x = 1.$$ [C]

15 (a) Solve the equation $1 - 3\cos^2\theta = 5\sin\theta$ for $0° \leqslant \theta \leqslant 360°$.

(b) State the formula for $\tan(x+y)$ in terms of $\tan x$ and $\tan y$.

If $2x + y = \dfrac{\pi}{4}$, show that

$$\tan y = \frac{1 - 2\tan x - \tan^2 x}{1 + 2\tan x - \tan^2 x}.$$

Deduce that $\tan\dfrac{\pi}{8}$ is a root of the equation $t^2 + 2t - 1 = 0$ and that its value is $\sqrt{2}-1$. [AEB]

16 Find in radians the general solutions of the equations

(a) $\cos x = \sin 3x$

(b) $\cos x + \cos 7x = \cos 4x$

(c) $\sin x - \cos x = 1.$ [L]

17 Show that the function $25\cos^2\theta - 4\sin^2\theta - 20\cos\theta - 8\sin\theta$ may be expressed as the difference of two squares and hence or otherwise, find the general solution of the equation

$$25\cos^2\theta - 4\sin^2\theta - 20\cos\theta - 8\sin\theta = 0.$$ [AEB]

18 (i) Find the values of x between $0°$ and $180°$ inclusive for which

$$\sin x + 2\sin 2x + \sin 3x = 0.$$

(ii) Find the values of x between $0°$ and $360°$ for which

$$2\cos x + 3\sin x = 3.$$ [C]

19 Calculate the values of θ in the interval $0 \leqslant \theta \leqslant 180°$ which satisfy the equations

(i) $2\sin\theta + \cos\theta = 1$

(ii) $2\sin\theta + \cos 2\theta = 1.$

By using a graphical method, find approximate solutions to the equation $2\sin\theta + \cos 3\theta = 1$ for values of θ in the interval $0 \leqslant \theta \leqslant 90°$. [AEB]

20 (a) Prove that $(\cot\theta + \operatorname{cosec}\theta)^2 = \dfrac{1+\cos\theta}{1-\cos\theta}$ and hence, or otherwise, solve the equation $(\cot 2\theta + \operatorname{cosec} 2\theta)^2 = \sec 2\theta$ for values of θ between $0°$ and $180°$.

(b) Find the general solution of the equation

$$\sin 2x + \sin 3x + \sin 5x = 0.$$ [AEB]

21 (i) Solve the equation

$$\sin \theta + \sin 3\theta + \sin 5\theta + \sin 7\theta = 0$$

for $0° < \theta < 360°$.
(ii) Given that $3 \tan \theta - \sec \theta = 1$, find the possible values of

$$3 \sec \theta + \tan \theta. \qquad [\text{L}]$$

22 Find the general solution, in degrees, of the equation
$$2 \cos x + \sin x = 1. \qquad [\text{L}]$$

23 Express $\tan 2\theta$ in terms of t where $t = \tan \theta$. Derive expressions for $\cos 2\theta$ and $\sin 2\theta$ in terms of t. Hence, or otherwise, solve the equation

$$5 \cos 2\theta - 2 \sin 2\theta = 2 \quad \text{for } 0° < \theta < 360°. \qquad [\text{AEB}]$$

24 (i) Show that $\cos^6 x + \sin^6 x = 1 - \frac{3}{4} \sin^2 2x$.
(ii) Solve, for $0° \leqslant x \leqslant 180°$, the equation

$$\sin x + \sin 5x = \sin 3x.$$

(iii) Find the general solution of the equation

$$3 \cos x + 4 \sin x = 2,$$

giving the answer in degrees. $\qquad [\text{L}]$

25 (i) Express $\sin 2\theta - \cos 2\theta$ in the form $r \cos (2\theta + \alpha)$, and hence or otherwise find the general solution of the equation

$$\sin 2\theta = \cos 2\theta - 1.$$

(ii) Show that $\cos 3\theta = 4 \cos^3 \theta - 3 \cos \theta$. Solve the equation $x^3 - 12x = 8$, by means of the substitution $x = 4 \cos \theta$. $\qquad [\text{L}]$

26 (a) Solve, for values of x between $0°$ and $360°$ inclusive, the equations
(i) $\cos 2x = \cos x$,
(ii) $\sin x = 2 \sin (60° - x)$.
(b) Given that $p \cos 2x + q \sin 2x + r = 0$, where $p \neq 0$ and $p \neq r$, find an equation for $\tan x$.
Deduce that, if the roots of this equation are $\tan x_1$ and $\tan x_2$, then

$$\tan (x_1 + x_2) = \frac{q}{p}. \qquad [\text{C}]$$

27 Find the values of θ, lying in the interval $-\pi \leqslant \theta \leqslant \pi$, for which $\sin 2\theta = \cos \theta$.
Sketch on the same axes the graphs of $y = \sin 2\theta$ and $y = \cos \theta$ in the above interval and deduce the set of values of θ in this interval for which $\sin 2\theta \geqslant \cos \theta$. $\qquad [\text{JMB}]$

28 Express $7 \sin x - 24 \cos x$ in the form $R \sin (x - \alpha)$, where R is positive and α is an acute angle.
Hence or otherwise solve the equation

$$7 \sin x - 24 \cos x = 15, \quad \text{for } 0° < x < 360°. \qquad \text{[AEB]}$$

29 Express $(\sqrt{3}) \sin \theta - \cos \theta$ in the form $R \sin (\theta - \alpha)$ where R is positive. Find all values of θ in the interval $0° \leqslant \theta \leqslant 360°$ which satisfy the equation
$$4 \sin \theta \cos \theta = (\sqrt{3}) \sin \theta - \cos \theta. \qquad \text{[JMB]}$$

30 (i) Find, in radians, the general solution of the equation
$$2 \sin \theta = \sqrt{3} \tan \theta.$$

(ii) Express $4 \sin \theta - 3 \cos \theta$ in the form $R \sin (\theta - \alpha)$, where α is an acute angle.
(a) Solve the equation
$$4 \sin \theta - 3 \cos \theta = 3,$$
giving all solutions between $0°$ and $360°$.
(b) Find the greatest and least values of
$$\frac{1}{4 \sin \theta - 3 \cos \theta + 6}. \qquad \text{[L]}$$

31 (a) Prove that
$$\cos 3\theta - \sin 3\theta = (\cos \theta + \sin \theta)(1 - 4 \cos \theta \sin \theta).$$

(b) Prove that if $\sec A = \cos B + \sin B$,
(i) $\tan^2 A = \sin 2B$ (ii) $\cos 2A = \tan^2 (\frac{1}{4}\pi - B)$. $\qquad \text{[C]}$

32 (i) Solve the equations
(a) $5 \sin \theta = 2 \sec \theta$
(b) $2 \tan^2 \theta = 3 \sec \theta$,
giving, in each case, all solutions between $0°$ and $360°$.
(ii) Find all values of θ between $0°$ and $180°$ which satisfy the equation
$$\sin 2\theta + 3 \cos 2\theta = 2. \qquad \text{[C]}$$

33 (a) Prove that
$$\sin^4 \theta + \cos^4 \theta = \tfrac{1}{4}(3 + \cos 4\theta).$$

Hence, or otherwise, find the greatest and least possible values of $\sin^4 \theta + \cos^4 \theta$.
(b) Using the identity $\operatorname{cosec}^2 \theta - \cot^2 \theta = 1$, or otherwise, find the exact value of $\tan \theta$ if
$$\operatorname{cosec} \theta + \cot \theta = 4. \qquad \text{[C]}$$

34 (a) Find all angles θ between $0°$ and $360°$ such that
$$2 \cos 2\theta = 7 \sin \theta.$$

(b) Find all angles θ between $0°$ and $360°$ such that
$$2 \cos \theta - 4 \sin \theta = 3. \qquad \text{[C]}$$

35 Draw the graph of $f(\theta) = 3 \sin \theta + 4 \cos \theta$ for $0° \leqslant \theta \leqslant 360°$. Hence find approximate values of the roots of $3 \sin \theta + 4 \cos \theta = 1$ which lie in this interval.

Express $f(\theta)$ in the form $R \sin (\theta + \alpha)$. Hence calculate the values of the roots of the equation $f(\theta) = 1$ in the interval $0° \leqslant \theta \leqslant 360°$.

If the equation is rewritten as $3 \sin \theta = 1 - 4 \cos \theta$ and both sides are then squared, the subsequent solution gives rise to values additional to those obtained above. Explain the significance of these additional values.

[AEB]

36 (a) Prove the identity $\sec 2\theta + \tan 2\theta = \dfrac{1 + \tan \theta}{1 - \tan \theta}$ and hence, or otherwise, find the general solution of the equation $\sec 2\theta + \tan 2\theta = 3$.

(b) Show that $8 \cos x - 6 \sin x \leqslant 10$. Determine the general values of x when $8 \cos x - 6 \sin x = 10$. [AEB]

37 (a) Solve the equation $\sin 4x = \cos x$ for values of x between $0°$ and $180°$.

(b) Given that $2 \sin 2x + \cos 2x = k$, show that

$$(1 + k) \tan^2 x - 4 \tan x - 1 + k = 0.$$

Hence, or otherwise, show that if $\tan x_1$ and $\tan x_2$ are the roots of this quadratic equation in $\tan x$, then

$$\tan (x_1 + x_2) = 2.$$

[AEB]

38 (a) Solve the equation

$$\sin x - \sin 4x + \sin 7x = 0 \quad \text{for } 0° \leqslant x \leqslant 180°.$$

(b) Verify that the equation $\sin 3\theta = 2 \cos 2\theta$ is satisfied by $\theta = 30°$.
Find the general solution of this equation. [AEB]

39 Find all the values of θ in the interval $0 \leqslant \theta \leqslant 2\pi$ for which

$$\sin \theta + \sin 3\theta = \cos \theta + \cos 3\theta.$$

[JMB]

40 (a) Prove the identity

$$\sin \theta + 2 \sin 3\theta + \sin 5\theta = 2 \cos \theta (\sin 2\theta + \sin 4\theta).$$

(b) Express $\sin \theta$ and $\cos \theta$ in terms of $t = \tan (\theta/2)$. By expressing $\dfrac{3 - \sin \theta}{\cos \theta}$ in terms of t, or otherwise, show that this expression cannot have any value between $-2 \sqrt{2}$ and $+2 \sqrt{2}$. [AEB]

41 (i) Find the values of x between $0°$ and $360°$ which satisfy the equation $\sin 2x + 2 \cos 2x = 1$.

(ii) Find the general solution of the equation

$$\cos 3x + \cos x = \sin 2x.$$

(iii) If $\tan^{-1} \tfrac{1}{3} + \tan^{-1} \tfrac{1}{4} = \tan^{-1} (a/b)$, where a and b are positive integers, find the least values of a and b. [L]

42 Express $\cot(A+B)$ in terms of $\cot A$ and $\cot B$ and hence express $\cot 22\frac{1}{2}°$ in surd form. Deduce $\cot 67\frac{1}{2}°$ in surd form.

Show that $\cot^{-1}(1+x)+\cot^{-1}(1-x)=\cot^{-1}(-x^2/2)$.　　[AEB]

43 (a) Use the relationship $\tan^{-1}x+\tan^{-1}y=\tan^{-1}\dfrac{x+y}{1-xy}$ to show that, if

$\tan^{-1}x+\tan^{-1}y+\tan^{-1}z=\pi/2$, then $xy+yz+zx=1$.

(b) Solve the equations

　(i) $\cos 6\theta+\sin 6\theta+\cos 4\theta+\sin 4\theta=0$, for $0\leqslant\theta\leqslant\pi/2$,

　(ii) $\cos 2\theta+3\sin\theta+1=0$, for $0\leqslant\theta\leqslant 2\pi$.　　[AEB]

44 (a) Find, without using tables, the value of x when

$$\tan^{-1}\tfrac{1}{2}-\tan^{-1}\tfrac{1}{3}=\sin^{-1}x.$$

(b) The length of an edge of a regular tetrahedron $PQRS$ is a. Show that the perpendicular distance of a vertex from the opposite face is $a\sqrt{(2/3)}$.

A point X lies on the edge PQ and θ is the acute angle which SX makes with the plane PQR. Prove that as X moves on the edge PQ the greatest and least values of θ are 70·5° and 54·7° approximately.　　[AEB]

45 (a) An object is descending in a straight line making an angle of 30° with a horizontal plane p so as to land at a point which is the centre of a horizontal circle of diameter 1000 metres lying in p. When the object is still at a height of 1000 metres, calculate the angles which are subtended at the object by the two mutually perpendicular diameters of the circle, one in the vertical plane containing the path of descent and the other at right angles to this.

(b) Given that $\tan^{-1}A+\tan^{-1}B=\tan^{-1}\dfrac{A+B}{1-AB}$, show that

$$\pi/4=\tan^{-1}\tfrac{1}{2}+\tan^{-1}\tfrac{1}{3}=2\tan^{-1}\tfrac{1}{2}-\tan^{-1}\tfrac{1}{7}.　　[AEB]$$

6 Differentiation 1

6.1 The idea of a limit

Let n denote a positive integer. The larger the value of n, the smaller will be the value of $1/n$. If we keep on increasing n, then $1/n$ will tend to zero. We express this by saying that 'as n tends to infinity, $1/n$ tends to the limit zero'. Note that infinity, which is represented by the symbol ∞, is not a number. The statement 'n tends to infinity' is just a brief way of saying that n increases without there being any ceiling to its value. Note also that while $1/n$ gets closer and closer to zero, it never reaches zero.

In the sequence of fractions

$$\frac{1}{2}, \frac{2}{3}, \frac{3}{4}, \frac{4}{5}, \ldots, \frac{n}{n+1}, \ldots$$

each fraction is larger than the one which precedes it, but every fraction is less than 1. By increasing n, we can bring the value of $\dfrac{n}{n+1}$ as close to 1 as we please. We say that, as n tends to infinity, $\dfrac{n}{n+1}$ tends to the limit 1.

In the sequence

$$\frac{2}{3}, \frac{3}{5}, \frac{4}{7}, \frac{5}{9}, \ldots, \frac{n+1}{2n+1}, \ldots$$

each term is less than the one which precedes it. By increasing n we can make the value of $\dfrac{n+1}{2n+1}$ as close to $\dfrac{1}{2}$ as we please. In this case, as n tends to infinity, $\dfrac{n+1}{2n+1}$ tends to the limit $\dfrac{1}{2}$.

Consider now f(x), where f$(x) = \dfrac{x^3 - a^3}{x - a}$, where a is a constant. Provided that x does not equal a, we have

$$f(x) = x^2 + ax + a^2,$$

As x approaches a, the value of $x^2 + ax + a^2$ approaches $3a^2$, so that f(x) tends to the limit $3a^2$. There are two points to notice. First, we do not let x equal a, and so the limiting value $3a^2$ is never reached; and second, f(x) tends to $3a^2$ whether x increases or decreases towards a. The statement 'f(x) tends to the limit $3a^2$ as x tends to a' can be expressed briefly as

$$f(x) \to 3a^2 \text{ as } x \to a,$$

or as
$$\lim_{x \to a} f(x) = 3a^2.$$

This means that $f(x)$ will be as close to $3a^2$ as we require for all values of x sufficiently close to a.

The following theorems on limits will be needed, but it is not within the scope of this book to prove them.

If $f(x) \to l_1$ and $g(x) \to l_2$ as $x \to a$,

$$f(x) + g(x) \to l_1 + l_2,$$
$$cf(x) \to cl_1, \text{ where } c \text{ is a constant,}$$
$$f(x)g(x) \to l_1 l_2,$$

and
$$f(x)/g(x) \to l_1/l_2, \text{ if } l_2 \text{ is not zero.}$$

Example 1

Find the limit as x tends to 2 of $f(x)$ where

$$f(x) = \frac{3x^2 - 12}{2x - 4}.$$

Provided that x is not equal to 2, we can divide the numerator and the denominator by $(x - 2)$. Then

$$f(x) = 3(x + 2)/2,$$

and so $f(x)$ tends to 6 as x tends to 2.

Example 2

Find the limit as n tends to infinity of $(2n + 3)/n^2$. (Use is made here of two of the theorems on limits quoted above.)

As n tends to infinity, $1/n$ and $1/n^2$ both tend to zero, and hence $2/n$ and $3/n^2$ both tend to zero. Therefore $(2/n + 3/n^2)$ tends to zero, i.e.

$$\lim_{n \to \infty} \frac{2n + 3}{n^2} = 0.$$

Example 3

Find the limit of $(n + 2)/(2n + 1)$ as n tends to infinity. Find also the smallest positive integer n for which the expression differs from its limit by less than 10^{-3}.

By dividing numerator and denominator by n we have

$$\frac{n + 2}{2n + 1} = \frac{1 + 2/n}{2 + 1/n}.$$

As n tends to infinity, $2/n$ and $1/n$ each tend to zero, so that the expression tends to the limit $1/2$.

Now

$$\frac{n+2}{2n+1} - \frac{1}{2} = \frac{3}{4n+2}.$$

This is less than 10^{-3} when $(4n+2)$ is greater than 3000. Hence the smallest integer is 750.

Exercises 6.1

1 Find the smallest positive integer n such that
   (a) $1/n < 10^{-6}$ (b) $1/2^n < 10^{-6}$ (c) $(2/3)^n < 10^{-6}$.
2 Find the smallest positive integer n such that $(2n+5)/(n+2)$ differs from 2 by less than 10^{-2}.
3 Write down the first five members of the sequence in which the nth member is $[n+(-1)^n]/n$. Find the limit of the nth member as n tends to infinity.
4 Find the limit as n tends to infinity of
   (a) $\dfrac{2n-1}{3n+1}$ (b) $\dfrac{2n+1}{3n-1}$ (c) $\dfrac{4+2n}{2+4n}$ (d) $\dfrac{n+2n^2}{n^2+2n}$.
5 As $x \to 1$, $f(x) \to 4$ and $g(x) \to 2$. Find the limits as $x \to 1$ of
   (a) $2f(x) - 3g(x)$ (b) $f(x)\,g(x)$ (c) $f(x)/g(x)$.
6 Find the limit as x tends to zero of
   (a) $\dfrac{3x+x^2}{x}$ (b) $\dfrac{2x^2+3x}{x^2+x}$ (c) $\dfrac{x^2-x}{2x-x^2}$ (d) $\dfrac{x+2}{x} - \dfrac{x+4}{2x}$.
7 Show that
   (a) $\displaystyle\lim_{x \to -1} \frac{x^3+1}{x+1} = 3$ (b) $\displaystyle\lim_{x \to 3} \frac{3x-9}{2x^2-18} = 1/4$.
8 Show that as x tends to 1
   (a) $\dfrac{x^3-1}{x^2-1} \to \dfrac{3}{2}$ (b) $\dfrac{x^3-x^2}{x^2-1} \to \dfrac{1}{2}$.

6.2 The gradient of a curve

Let A and B be the points on the curve $y = f(x)$ at which $x = a$ and $x = b$ respectively. The gradient of the chord AB will be

$$\frac{f(b)-f(a)}{b-a}.$$

Now let the point B move along the curve towards the point A which is kept fixed. If the gradient of AB tends to a limit m as b approaches a, we define the gradient of the curve at A to be m. This will be the gradient of the tangent to the curve at A. If the gradient of AB tends to one value when B approaches A from the right, and to a different value when B approaches A from the left, there will not be a tangent to the curve at A.

To find the gradient of the curve $y = x^2$ at the point $A(a, a^2)$ we consider the chord joining A to the point $B(b, b^2)$ (Fig. 6.1). The gradient of the chord AB is $(b^2 - a^2)/(b - a)$, which equals $(b + a)$ provided that b does not equal a. As b tends to a, the gradient of AB tends to the limit $2a$, the gradient of the curve at A.

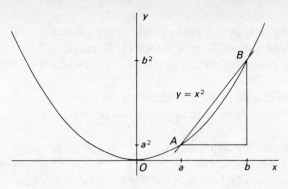

Fig. 6.1

Example 1
Show that the gradient of the curve $y = 3x^2 - 2x + 1$ at the point (1, 2) is 4.

The gradient of the chord joining the points on the curve given by $x = a$ and $x = b$ will be

$$\frac{(3b^2 - 2b + 1) - (3a^2 - 2a + 1)}{(b - a)}$$

$$= \frac{3(b^2 - a^2) - 2(b - a)}{(b - a)}.$$

Since $(b - a)$ is not zero, this simplifies to $3(b + a) - 2$. As b tends to a, this tends to the limit $(6a - 2)$. Hence when $a = 1$, the gradient of the curve is 4.

Example 2
Find the gradient of the curve $y = x^2 + 4x - 8$ at the point $P(2, 4)$.

Let Q be the point on the curve at which $x = 2 + h$. Then $y = 4 + 8h + h^2$ at Q, and the gradient of PQ is

$$\frac{(4 + 8h + h^2) - 4}{(2 + h) - 2},$$

which simplifies to $(8 + h)$ if h is not zero. When h tends to zero, Q moves up to P and the gradient of PQ tends to 8. This is the gradient of the curve at P.

Exercises 6.2
1   Find the gradients of the lines PQ, QR, RP where P, Q, R are the points (1, 1), (4, 3), $(-1, 5)$ respectively.
2   Find the gradient of the curve
   (a) $y = 3x^2$ at the point (1, 3)
   (b) $y = 2x^2 + 3$ at the point $(-1, 5)$
   (c) $y = 4x^2 - 6x - 1$ at the point (2, 3).

3 Find the gradient of the curve $y = x^2 - 4x + 5$ at the point given by $x = a$. Find the points on the curve at which the gradient is (a) 0 (b) 1 (c) -1. Check your answers from a sketch of the curve.

4 Find the points on the curve $y = 6x - 2x^2$ at which the gradient is (a) 2 (b) 0 (c) -2. Check your answers from a sketch of the curve.

5 Obtain the equation of the tangent to the curve $y = 3x^2 - 5x - 2$ at the point $(2, 0)$.

6 Find the gradient of the curve $y = x^3$ at the point $(1, 1)$.

7 Show that the gradient of the curve $y = 3x^2 - 2x + 2$ at the point $(0, 2)$ is -2. Find the value of c for which the line $y = 10x + c$ is a tangent to this curve.

8 Find the equation of the tangent to the curve $y = 3x^2 - 4x + 2$ which is parallel to the line $y = 2x$.

9 Sketch the curves (a) $y = |x|$ (b) $y = x + |x|$ (c) $y = x|x|$. Which of these curves has a tangent at the origin?

10 Show that the gradient of the curve $y = x^3 - 3x$ is negative when x lies between -1 and 1.

6.3 Derived functions

Let P and Q be the points on the continuous curve $y = f(x)$ given by $x = a$ and $x = a + h$ respectively. In Fig. 6.2, h is taken to be positive, but it can be either positive or negative. The gradient of the chord PQ is QR/PR, which equals

$$\frac{f(a+h) - f(a)}{h}.$$

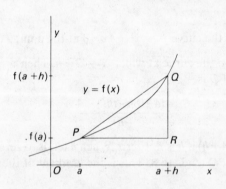

Fig. 6.2

Keep P fixed and let Q move along the curve towards P. As h tends to zero, $f(a+h) - f(a)$ also tends to zero. If the gradient of PQ tends to a limit, giving the gradient of the curve at P, this limit will be the value at $x = a$ of a new function. This is known as the derived function of the function f and is denoted by f′. The value of the derived function f′ at any point is called the derivative and is equal to the gradient of the curve $y = f(x)$ at that point. The derivative is defined by

$$f'(x) = \lim_{h \to 0} \frac{f(x+h) - f(x)}{h}.$$

There are three steps in the process of differentiating $f(x)$ i.e. of finding $f'(x)$.

Step 1   Form the difference $f(x+h) - f(x)$.
Step 2   Divide by h to obtain the gradient of the chord.
Step 3   Find the limit of this gradient as h tends to zero.

Example 1
Find the derivative $f'(x)$ when $f(x) = x^3$.

Step 1 $f(x+h) - f(x) = (x+h)^3 - x^3$
$$= 3x^2 h + 3xh^2 + h^3.$$
Step 2   Division by h gives $(3x^2 + 3xh + h^2)$.
Step 3   As $h \to 0$, $(3x^2 + 3xh + h^2) \to 3x^2$.

Therefore when $f(x) = x^3$, $f'(x) = 3x^2$.

Example 2
Find the derivative $f'(x)$ when $f(x) = 1/x$.

Step 1 $f(x+h) - f(x) = \dfrac{1}{x+h} - \dfrac{1}{x} = \dfrac{-h}{x(x+h)}.$
Step 2   Division by h gives $-1/(x^2 + xh)$.
Step 3   As $h \to 0$, $-1/(x^2 + xh) \to -1/x^2$.

Therefore when $f(x) = 1/x$, $f'(x) = -1/x^2$.

Example 3
If $f(x) = x^n$, where n is a positive integer, show that $f'(x) = nx^{n-1}$.

By the binomial expansion, $(x+h)^n - x^n$ equals

$$nx^{n-1}h + \binom{n}{2}x^{n-2}h^2 + \binom{n}{3}x^{n-3}h^3 + \ldots + h^n.$$

Division by h gives

$$nx^{n-1} + \binom{n}{2}x^{n-2}h + \binom{n}{3}x^{n-3}h^2 + \ldots + h^{n-1}.$$

As h tends to zero, each term except the first tends to zero, and so the limit is nx^{n-1}. Hence $f'(x) = nx^{n-1}$.

    Differentiating x^n gives nx^{n-1}. Differentiating nx^{n-1} will give $n(n-1)x^{n-2}$.

This is called the second derivative of x^n. For any function f, we define the second derivative $f''(x)$ by

$$f''(x) = \lim_{h \to 0} \frac{f'(x+h) - f'(x)}{h}.$$

Exercises 6.3

1 Showing the three steps in your working, differentiate
   (a) x^2 (b) $2x^2 + 3$ (c) x^4 (d) x^5.
2 Show that the derivative of $1/x^2$ is $-2/x^3$.
3 By using the identity $(b - a) \equiv (b^2 - a^2)/(b + a)$, show that the derivative of $\sqrt{x}$ is $1/(2\sqrt{x})$.
4 If $f(x) = |x|$, show that $f'(x) = 1$ when $x > 0$, and that $f'(x) = -1$ when $x < 0$. Draw the graphs of $f(x)$ and $f'(x)$.
5 Use the result of example 3 above to find $f'(x)$ and $f''(x)$ when $f(x)$ equals
   (a) x^6 (b) $3x^2 + 2$ (c) x^{10} (d) $2x^3 - 3x + 1$.
6 Show that the gradient of the curve $y = x^4 - x^2 + 1$ at the point $(1, 1)$ is 2.
7 Find the gradient of the curve $y = 2x^3 - x + 1$ at the points $(-1, 0)$, $(0, 1)$, $(1, 2)$. Sketch the curve.
8 If $f(x) = x^3 - x$, draw the curves $y = f(x)$ and $y = f'(x)$ in the same diagram.
9 Find the values of x at the points on the curve $y = 2x^3 - 3x^2 - 30x$ at which the gradient is 6.
10 Find $f'(t)$ and $f''(t)$ when $f(t)$ equals
   (a) t (b) t^2 (c) t^3 (d) $(t + 1)^2$.

6.4 Rates of change

If the relation $y = f(x)$ exists between the variables x and y, x is known as the independent variable and y as the dependent variable. The rate of change of y with respect to x is denoted by

$$\frac{dy}{dx} \quad \text{(dee } y \text{ by dee } x\text{)}.$$

This symbol is not a fraction; the d's cannot be cancelled.

Let δx (delta x) denote a small change in the value of x. Note that δx does not mean δ times x. If x changes from 3·2 to 3·3, then $\delta x = 0·1$; if x changes from 3·2 to 3·1, then $\delta x = -0·1$. Let δy denote the change in y corresponding to the change δx in x. Then

$$\delta y = f(x + \delta x) - f(x).$$

The fraction

$$\frac{\delta y}{\delta x} = \frac{f(x + \delta x) - f(x)}{\delta x}$$

gives the average rate of change of y with respect to x. The limit of this fraction as

δx tends to zero is denoted by $\dfrac{dy}{dx}$.

$$\frac{dy}{dx} = \lim_{\delta x \to 0} \frac{\delta y}{\delta x}.$$

The fraction $\dfrac{\delta y}{\delta x}$ corresponds to the gradient of a chord, while the limit $\dfrac{dy}{dx}$ corresponds to the gradient of a tangent. If we now compare the limit $\dfrac{dy}{dx}$, given by

$$\frac{dy}{dx} = \lim_{\delta x \to 0} \frac{f(x + \delta x) - f(x)}{\delta x},$$

with the derivative $f'(x)$, given by

$$f'(x) = \lim_{h \to 0} \frac{f(x + h) - f(x)}{h},$$

it is clear that they are equivalent to each other, so that if $y = f(x)$, $\dfrac{dy}{dx} = f'(x)$.

The rate of change of $\dfrac{dy}{dx}$ with respect to x is denoted by

$$\frac{d^2 y}{dx^2} \quad \text{(dee two } y \text{ by dee } x \text{ squared)},$$

and this is equal to the second derivative $f''(x)$.

It is often convenient to define a curve in terms of a parameter. Let $x = f(t)$ and $y = g(t)$, so that x and y are dependent variables and the parameter t is the independent variable. Let δx and δy be the changes in x and y when t becomes $(t + \delta t)$. Then provided that the denominator $\delta x/\delta t$ does not vanish

$$\frac{\delta y}{\delta x} = \frac{\delta y}{\delta t} \bigg/ \frac{\delta x}{\delta t}.$$

Hence, when δt tends to zero, we have

$$\frac{dy}{dx} = \frac{dy}{dt} \bigg/ \frac{dx}{dt}.$$

Consider now a point P moving in a straight line through a fixed point O. Take the time t as the independent variable and let the dependent variable be s, the directed length of OP. We take s to be positive in one direction and negative in the opposite direction. If s changes to $(s + \delta s)$ when t increases to $(t + \delta t)$, the average velocity over this time interval is $\delta s/\delta t$. The velocity v of the point P at time t is defined to be the limit of $\delta s/\delta t$ as δt tends to zero.

$$v = \lim_{\delta t \to 0} \frac{\delta s}{\delta t} = \frac{ds}{dt}.$$

The speed of the point P is the magnitude of its velocity. While v may be positive or negative, the speed is always positive (or zero).

If v changes to $(v + \delta v)$ when t increases to $(t + \delta t)$, the average acceleration over this time interval is $\delta v / \delta t$. The acceleration a at a given instant is the rate of change of v with respect to t at that instant.

$$a = \lim_{\delta t \to 0} \frac{\delta v}{\delta t} = \frac{dv}{dt}.$$

Since $v = \dfrac{ds}{dt}$, we have $\dfrac{dv}{dt} = \dfrac{d^2 s}{dt^2}$, so that the acceleration a equals $\dfrac{d^2 s}{dt^2}$.

Example 1

A particle starts from rest and travels a distance s metres in a straight line in t seconds. Given that $s = t^3 + 2t^2$, find the velocity v m s^{-1} and acceleration a m s^{-2} of the particle after 2 seconds.

Since $s = t^3 + 2t^2$, $v = \dfrac{ds}{dt} = 3t^2 + 4t$. When $t = 2$, the velocity is 20 m s^{-1}.

The acceleration a m s^{-2} is given by

$$a = \frac{d^2 s}{dt^2} = 6t + 4.$$

Hence when $t = 2$, the acceleration is 16 m s^{-2}.

Example 2

The coordinates of a point on a curve are given, in terms of a parameter t, by $x = t^2$, $y = t^3$. Find the gradient of the curve at the point at which $t = 2$.

Since $y = t^3$, $\dfrac{dy}{dt} = 3t^2$, and since $x = t^2$, $\dfrac{dx}{dt} = 2t$.

Gradient $= \dfrac{dy}{dx} = \dfrac{dy}{dt} \bigg/ \dfrac{dx}{dt} = 3t/2$.

When $t = 2$, the gradient will be 3.

Exercises 6.4

1 Find $\dfrac{dy}{dx}$ when y equals
   (a) $x^5 + 4x + 2$ (c) $3x^6 - 2x^3 + x$
   (b) $2x^4 + 4x^2 - 1$ (d) $x(x + 1)(x + 2)$.

2 If the radius r of a circle is increasing, show that the rate of change of its area with respect to its radius is $2\pi r$. Interpret this result geometrically.

3 A point moves in a straight line through a distance s metres in t seconds. Find its velocity and its acceleration when $t = 2$, given that
   (a) $s = t^3$ (b) $s = 3t^2 + t + 1$ (c) $s = 3t^4 - 4t^3$.

4 A ball is thrown vertically upwards, and after t seconds its height from the ground is $(30t - 5t^2)$ metres. Find its height when its speed is 10 m/s.

5 The coordinates of a point on a curve are $x = 2t$, $y = 2/t$. Find the points on the curve at which the gradient is $-1/4$. Sketch the curve.

6 The distance travelled in a straight line by a particle at time t is given by $(t^2 - t^3)$. Find the velocity of the particle at the instant when its acceleration is zero.

7 A solid circular cylinder has radius r and height h. If h is constant and r is increasing, find the rate of change of the volume with respect to the radius. Find also the rate of change of the total surface area with respect to the radius at the instant when $r = h$.

8 Find $\dfrac{dx}{dt}$ and $\dfrac{d^2x}{dt^2}$ when $t = 1$ given that

(a) $x = t^4$ (b) $x = (t+1)^3$ (c) $x = 1/t$.

6.5 Differentiation of sin x and cos x

First we show that, as θ tends to zero, $\sin\theta/\theta$ tends to the limit 1. In Fig. 6.3 the tangent at A to the circle with centre O and radius a meets OB produced at C. Let

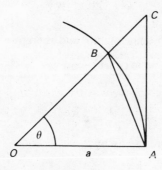

Fig. 6.3

the acute angle AOB equal θ radians, where $0 < \theta < \pi/2$. The following three areas are then in ascending order of magnitude:

(a) area of the triangle $AOB = \frac{1}{2}a^2 \sin\theta$
(b) area of the sector $AOB = \frac{1}{2}a^2\,\theta$
(c) area of the triangle $AOC = \frac{1}{2}a^2 \tan\theta$.
It follows that $\sin\theta < \theta < \tan\theta$.

Dividing by $\sin\theta$, which is positive, gives

$$1 < \frac{\theta}{\sin\theta} < \frac{1}{\cos\theta}.$$

Since $\cos\theta$ tends to 1 as θ tends to zero, $\theta/\sin\theta$ must also tend to 1. This result is true for positive values of θ, but it is also true for negative values, since the value of $\theta/\sin\theta$ does not change if θ is replaced by $-\theta$. It follows that

$$\lim_{\theta\to 0}\frac{\sin\theta}{\theta}=1,$$

where θ is in radians.

To obtain the derivative of $\sin x$, the limit as h tends to zero of

$$\frac{\sin(x+h)-\sin x}{h}.$$

must be found.

By putting $A=x+h$ and $B=x$ in the identity

$$\sin A-\sin B\equiv 2\cos\tfrac{1}{2}(A+B)\sin\tfrac{1}{2}(A-B),$$

and dividing by h, we get

$$\frac{\sin(x+h)-\sin x}{h}=\frac{2\cos(x+\tfrac{1}{2}h)\sin(\tfrac{1}{2}h)}{h}.$$

Now $\cos(x+\tfrac{1}{2}h)$ tends to $\cos x$ as h tends to zero, and if $h=2\theta$,

$$\frac{2\sin(\tfrac{1}{2}h)}{h}=\frac{\sin\theta}{\theta}.$$

This tends to 1 as θ and h tend to zero, and therefore

$$\lim_{h\to 0}\frac{\sin(x+h)-\sin x}{h}=\cos x.$$

This shows that when $y=\sin x$, $\dfrac{dy}{dx}=\cos x$.

This can be expressed more briefly by

$$\frac{d}{dx}(\sin x)=\cos x.$$

(Note that x is in radians.)

The derivative of $\cos x$ is given by the limit as h tends to zero of

$$\frac{\cos(x+h)-\cos x}{h}.$$

By putting $A=x+h$ and $B=x$ in the identity

$$\cos A-\cos B\equiv -2\sin\tfrac{1}{2}(A+B)\sin\tfrac{1}{2}(A-B),$$

and dividing by h, we get

$$\frac{\cos(x+h)-\cos x}{h}=-\frac{2\sin(x+\tfrac{1}{2}h)\sin(\tfrac{1}{2}h)}{h}.$$

As h tends to zero, $\sin(x + \tfrac{1}{2}h)$ tends to $\sin x$, while $\dfrac{[2\sin(\tfrac{1}{2}h)]}{h}$ tends to 1. It follows that the derivative of $\cos x$ is $-\sin x$. When $y = \cos x$, $\dfrac{dy}{dx} = -\sin x$, or

$$\frac{d}{dx}(\cos x) = -\sin x.$$

(Again note that x is in radians.)

The second derivatives of $\sin x$ and $\cos x$ can be found at once.

When $y = \sin x$, $\dfrac{dy}{dx} = \cos x$ and so $\dfrac{d^2 y}{dx^2} = -\sin x$.

When $y = \cos x$, $\dfrac{dy}{dx} = -\sin x$ and so $\dfrac{d^2 y}{dx^2} = -\cos x$.

Exercises 6.5

1 Sketch the curves $y = \sin x$, $y = x$, $y = \tan x$ on the same axes for values of x between $-\pi/4$ and $\pi/4$.

2 Find $\dfrac{dy}{dx}$ and $\dfrac{d^2 y}{dx^2}$ when
   (a) $y = \cos x - \sin x$
   (b) $y = 2\sin x + 3\cos x$.

3 Show that the gradient of the curve $y = x - \sin x$ is never negative. Sketch the curve for $-\pi \leqslant x \leqslant \pi$.

4 Show that

$$\sin(2x + 2h) - \sin(2x) \equiv 2\cos(2x + h)\sin h.$$

Use this identity to show that if $y = \sin(2x)$, $\dfrac{dy}{dx} = 2\cos(2x)$.

5 Show that

$$\cos(2x + 2h) - \cos(2x) \equiv -2\sin(2x + h)\sin h.$$

Hence show that if $y = \cos(2x)$, $\dfrac{dy}{dx} = -2\sin(2x)$.

6 At time t the coordinates of the point P are $x = 2\cos t$, $y = 2\sin t$. Show that the path of P is a circle and that $\dfrac{dy}{dx} = -\cot t$.

6.6 The product rule

The derivative of the product $u(x)v(x)$ is given by the rule

$$\frac{d}{dx}(u \times v) = u\frac{dv}{dx} + v\frac{du}{dx}.$$

Let δu and δv be the changes in $u(x)$ and $v(x)$ due to the change δx in the value of x.

Then since

$$(u + \delta u)(v + \delta v) - uv = u\,(\delta v) + v\,(\delta u) + (\delta u)\,(\delta v),$$

$$\frac{(u + \delta u)(v + \delta v) - uv}{\delta x} = u\frac{\delta v}{\delta x} + v\frac{\delta u}{\delta x} + \delta u \times \frac{\delta v}{\delta x}.$$

The limit of the left-hand side as δx tends to zero is the derivative of the product $u \times v$. On the right-hand side, as $\delta x \to 0$,

$$\frac{\delta v}{\delta x} \to \frac{dv}{dx}, \quad \frac{\delta u}{\delta x} \to \frac{du}{dx} \quad \text{and} \quad \delta u \times \frac{\delta v}{\delta x} \to 0.$$

This shows that

$$\frac{d}{dx}(u \times v) = u\frac{dv}{dx} + v\frac{du}{dx}.$$

This rule is known as the *product rule*.

Example 1

Find the derivative of $x \cos x$.

Put $u = x$ and $v = \cos x$. Then

$$\frac{du}{dx} = 1 \quad \text{and} \quad \frac{dv}{dx} = -\sin x.$$

Hence

$$\frac{d}{dx}(x \cos x) = (x)(-\sin x) + (\cos x)(1)$$

$$= \cos x - x \sin x.$$

Example 2

Use the product rule to differentiate $\sqrt{x}$.

Take $u = v = \sqrt{x}$, so that $u \times v = x$ and $\dfrac{d}{dx}(u \times v) = 1$.

By the product rule,

$$1 = \sqrt{x}\frac{dv}{dx} + \sqrt{x}\frac{du}{dx} = 2\sqrt{x}\frac{d}{dx}(\sqrt{x}).$$

It follows that $\dfrac{d}{dx}(\sqrt{x}) = \dfrac{1}{2\sqrt{x}}.$

Example 3

Find the derivative of the product $u(x)v(x)w(x)$.

By treating $u \times v \times w$ as the product of u and $v \times w$, we get

$$\frac{d}{dx}(u \times v \times w) = u\frac{d}{dx}(v \times w) + (v \times w)\frac{du}{dx}$$

$$= (u \times v)\frac{dw}{dx} + (u \times w)\frac{dv}{dx} + (v \times w)\frac{du}{dx}$$

$$= (v \times w)\frac{du}{dx} + (w \times u)\frac{dv}{dx} + (u \times v)\frac{dw}{dx}.$$

Exercises 6.6
1 Differentiate with respect to x
   (a) $x \sin x$ (b) $(\sin x)(\cos x)$ (c) $x^2 \cos x$.
2 Differentiate with respect to t
   (a) $\cos^2 t$ (b) $t^2 \sin t$ (c) $t (\sin t) (\cos t)$.
3 Use the product rule to find the derivative of $(x + 2x^2)(2 - x^3)$. Check your answer by multiplying out before differentiating.
4 Differentiate with respect to x
   (a) $x \sqrt{(x + 1)}$ (b) $x^2 (\sin x + \cos x)$.
5 By putting $u(x) = v(x) = w(x) = x^{\frac{1}{3}}$ in the product rule, show that the derivative of $x^{\frac{1}{3}}$ is $1/(3x^{\frac{2}{3}})$.

6.7 The quotient rule
The derivative of the quotient $u(x)/v(x)$ is given by the rule

$$\frac{d}{dx}\left(\frac{u}{v}\right) = \frac{v\dfrac{du}{dx} - u\dfrac{dv}{dx}}{v^2}.$$

When x becomes $(x + \delta x)$, let u and v become $(u + \delta u)$ and $(v + \delta v)$ respectively. Now

$$\frac{u + \delta u}{v + \delta v} - \frac{u}{v} = \frac{v(\delta u) - u(\delta v)}{v(v + \delta v)},$$

and so

$$\frac{\dfrac{u + \delta u}{v + \delta v} - \dfrac{u}{v}}{\delta x} = \frac{v\dfrac{\delta u}{\delta x} - u\dfrac{\delta v}{\delta x}}{v(v + \delta v)}.$$

When δx tends to zero, the limit of the left-hand side is the derivative of u/v, while on the right-hand side

$$\frac{\delta u}{\delta x} \to \frac{du}{dx}, \quad \frac{\delta v}{\delta x} \to \frac{dv}{dx} \quad \text{and} \quad v(v + \delta v) \to v^2.$$

It follows that

$$\frac{d}{dx}\left(\frac{u}{v}\right) = \frac{v\frac{du}{dx} - u\frac{dv}{dx}}{v^2}.$$

This rule is known as the *quotient rule*.

Example 1

Differentiate $\frac{x^2}{x+1}$.

Put $u = x^2$ and $v = x + 1$. Then $\frac{du}{dx} = 2x$ and $\frac{dv}{dx} = 1$.

By the quotient rule,

$$\frac{d}{dx}\left(\frac{x^2}{x+1}\right) = \frac{(x+1)(2x) - x^2}{(x+1)^2} = \frac{x^2 + 2x}{(x+1)^2}.$$

Example 2

Differentiate $\tan x$.

Put $u = \sin x$ and $v = \cos x$, so that $\frac{du}{dx} = \cos x$ and $\frac{dv}{dx} = -\sin x$. Then

$$\frac{d}{dx}(\tan x) = \frac{d}{dx}\left(\frac{\sin x}{\cos x}\right) = \frac{(\cos x)(\cos x) - (\sin x)(-\sin x)}{(\cos x)^2}$$

$$= \frac{\cos^2 x + \sin^2 x}{\cos^2 x}$$

Hence $\frac{d}{dx}(\tan x) = \frac{1}{\cos^2 x} = \sec^2 x.$

Example 3

If k is a negative integer, show that $\frac{d}{dx}(x^k) = kx^{k-1}$.

It has already been shown that when n is a positive integer

$$\frac{d}{dx}(x^n) = nx^{n-1}.$$

Let $u = 1$ and $v = x^n$, where n is a positive integer.
Then $\frac{du}{dx} = 0$ and $\frac{dv}{dx} = nx^{n-1}$, so that

$$\frac{d}{dx}\left(\frac{u}{v}\right) = \frac{d}{dx}\left(\frac{1}{x^n}\right) = \frac{0 - nx^{n-1}}{x^{2n}} = -nx^{-n-1}.$$

Now put $n = -k$ so that k is a negative integer.

Since $\dfrac{d}{dx}\left(\dfrac{1}{x^n}\right) = -nx^{-n-1}$, $\dfrac{d}{dx}(x^k) = kx^{k-1}$.

Exercises 6.7

1 Differentiate with respect to x

   (a) $\dfrac{x}{x+1}$ (b) $\dfrac{2x+3}{x-1}$ (c) $\dfrac{x-1}{x+1}$ (d) $\dfrac{x^2}{x^2+1}$.

2 By taking $u = 1$ and $v = \cos x$, show that the derivative of $\sec x$ is $\sec x \tan x$.

3 Differentiate with respect to t

   (a) $\dfrac{\sin t}{t}$ (b) $\dfrac{\cos t}{t}$ (c) $\dfrac{t}{\sin t}$ (d) $\dfrac{\sin t}{\sin t + \cos t}$.

4 Show that
   (a) if $f(x) = \cot x$, then $f'(x) = -\operatorname{cosec}^2 x$
   (b) if $f(x) = \operatorname{cosec} x$, then $f'(x) = -\operatorname{cosec} x \cot x$.

5 Obtain the quotient rule by applying the product rule to the product of u and $(1/v)$.

6 Differentiate with respect to x
   (a) $1/x^2$ (b) $1/x^4$ (c) $(x^3+1)/x^3$ (d) $1/(x+1)^3$.

6.8 The chain rule

The chain rule is for differentiating a composite function, i.e. a function of a function. It can be stated either in the form

$$\frac{dy}{dx} = \frac{dy}{du} \times \frac{du}{dx},$$

or in the form

$$(fg)'(x) = f'[g(x)]x(x).$$

Let $y = f(u)$ where $u = g(x)$, so that $y = fg(x)$. When x becomes $(x + \delta x)$, let u change to $(u + \delta u)$ and y change to $(y + \delta y)$. The rate of change of y with respect to x will be given by the limit of $\delta y/\delta x$ as δx tends to zero. Provided that δu does not become zero for any value of δx as δx approaches zero, the equation

$$\frac{\delta y}{\delta x} = \frac{\delta y}{\delta u} \times \frac{\delta u}{\delta x}$$

will remain true as δx tends to zero. If $\delta y/\delta u$ tends to a limit as δu tends to zero, this limit will be dy/du, and if $\delta u/\delta x$ tends to a limit as δx tends to zero, this limit will be du/dx. Hence by a theorem on limits stated in section 6.1,

$$\frac{dy}{dx} = \frac{dy}{du} \times \frac{du}{dx}.$$

In the case when $y = f(u)$, $u = g(v)$ and $v = h(x)$, so that $y = fgh(x)$,

$$\frac{dy}{dx} = \frac{dy}{du} \times \frac{du}{dv} \times \frac{dv}{dx}.$$

Notice that the links in the chain, $\frac{dy}{du}, \frac{du}{dv}, \frac{dv}{dx}$, are not fractions. The 'du' in $\frac{dy}{du}$ cannot be cancelled with the 'du' in $\frac{du}{dv}$.

Example 1

Find the derivative of $(3x+4)^5$.

Let $y = u^5$ and $u = 3x + 4$.

Then $\frac{dy}{du} = 5u^4$ and $\frac{du}{dx} = 3$.

Hence $\frac{dy}{dx} = (5u^4) \times (3) = 15(3x+4)^4$.

Example 2

If $x = t^2$ and $y = t^3$, show that $\frac{d^2 y}{dx^2} = 3/(4t)$.

$$\frac{dy}{dx} = \frac{dy}{dt} \bigg/ \frac{dx}{dt} = (3t^2)/(2t) = 3t/2.$$

$$\frac{d^2 y}{dx^2} = \frac{d}{dx}\left(\frac{dy}{dx}\right) = \frac{d}{dt}\left(\frac{dy}{dx}\right) \times \frac{dt}{dx} = \left(\frac{3}{2}\right)\left(\frac{1}{2t}\right) = 3/(4t).$$

Example 3

If $y = x^k$, where k is a rational number, show that $\frac{dy}{dx} = kx^{k-1}$.

Let $k = p/q$ where p and q are integers, and let $y = x^{p/q}$. Then if $z = x^{1/q}$, we have $x = z^q$ and $y = z^p$. By the chain rule,

$$\frac{dy}{dx} \times \frac{dx}{dz} = \frac{dy}{dz},$$

i.e.

$$\frac{dy}{dx}(qz^{q-1}) = pz^{p-1}.$$

This gives

$$\frac{dy}{dx} = (p/q)z^{p-q} = kx^{k-1}.$$

Example 4

If $y = \frac{1}{f(x)}$, show that $\frac{dy}{dx} = -\frac{f'(x)}{f^2(x)}$.

Put $y = 1/u$ and $u = f(x)$.

Then $\dfrac{dy}{du} = -\dfrac{1}{u^2}$ and $\dfrac{du}{dx} = f'(x)$.

Hence $\dfrac{dy}{dx} = -\dfrac{1}{u^2} \times f'(x) = -\dfrac{f'(x)}{f^2(x)}$.

Exercises 6.8

1. Find $f'(x)$ given that $f(x)$ equals
   (a) $(2x+1)^3$ (c) $\sin(3x)$ (e) $\sin(\frac{1}{2}x+1)$
   (b) $(3x-4)^2$ (d) $\cos(x^2)$ (f) $(3-2x)^3$.

2. Differentiate with respect to x
   (a) $x^{1/2}$ (b) $x^{3/2}$ (c) $x^{-1/2}$ (d) $x^{-3/2}$.

3. Differentiate with respect to t
   (a) $(2t+1)^{1/2}$ (b) $(3t-1)^{2/3}$ (c) $(t^2+2t)^{-2}$

4. Find the derivatives of
   (a) $\sin^3 x$ (b) $\tan^2 x$ (c) $\cos^2(2x)$ (d) $\sec(2x)$.

5. Differentiate with respect to x
   (a) $1/(2x+3)$ (b) $x^2 \sin(3x)$ (c) $2\tan(\frac{1}{2}x)$ (d) $(x^2+1)^3$.

6. The radius of a circle is increasing at the rate of 0.05 m s^1. Find the rate at which the area is increasing at the instant when the radius in 10 m.

7. The edges of a hollow cube are made of metal rods such that one metre length expands by 12×10^{-6} m for each degree Celsius (°C) rise in temperature. Find the rate of increase in m^3 s^{-1} of the volume of the cube when the temperature is rising by 2°C per minute and the length of an edge is 5 m.

8. The line AB joining the point A on the x-axis to the point B on the y-axis is of constant length. If A is moving away from the origin O with speed u, find the speed of B at the instant when $OA = 2OB$.

9. The radius of a circle is increasing at the constant rate of 0.5 m s^{-1}. Find the rate of increase of the area of the circle at the instant when its circumference is 12 m long.

10. Each edge of a cube of ice is decreasing at the constant rate of one centimetre per second. Find the rate of change of its volume at the instant when the volume is 8 m^3. Draw a graph showing how the volume changes with time from this instant onwards.

11. Given that $x = \cos t$ and $y = \sin t$, show that
    (a) $\dfrac{dy}{dx} = -\cot t$ (b) $\dfrac{d^2y}{dx^2} = -1/\sin^3 t$.

12. Show that at any point on the parabola given by the parametric equations $x = at^2$, $y = 2at$,
    (a) $\dfrac{d^2y}{dx^2} = -1/(2at^3)$ (b) $\dfrac{d^2x}{dy^2} = 1/(2a)$.

13 Given that $x = t - \sin t \cos t$, $y = 4 \cos t$, show that

(a) $\dfrac{dy}{dx} = -\dfrac{2}{\sin t}$ (b) $\dfrac{d^2 y}{dx^2} = \dfrac{\cos t}{\sin^4 t}$.

14 Show that on the curve $x = \sec t$, $y = \tan t$

(a) $\dfrac{dy}{dx} = \operatorname{cosec} t$ (b) $\dfrac{d^2 y}{dx^2} = -\cot^3 t$.

15 Given that $x = 2t + 1$, $y = 2/t$, express $\dfrac{d^2 y}{dx^2}$ in terms of t.

6.9 Differentiation of inverse functions

If the function g is the inverse of the function f, $g[f(x)] = x$. If $y = f(x)$, we have $g(y) = x$, so that

$$g'(y)\frac{dy}{dx} = 1.$$

since $\dfrac{dy}{dx} = f'(x)$, this gives $g'(y)f'(x) = 1$. Thus when the two derivatives both exist they are reciprocals of each other, i.e.

$$f'(x) = 1/g'(y) = 1/g'f(x).$$

Example 1

Show that $\dfrac{d}{dx}(\arctan x) = \dfrac{1}{1 + x^2}$.

Let $y = f(x) = \arctan x$,
and $x = g(y) = \tan y$, where $-\pi/2 < y < \pi/2$.
Then $g'(y) = \sec^2 y = 1 + \tan^2 y = 1 + x^2$.
Since $f'(x)$ is the reciprocal of $g'(y)$, we have

$$\frac{d}{dx}(\arctan x) = f'(x) = \frac{1}{1 + x^2}.$$

Example 2

Show that $\dfrac{d}{dx}(\arcsin x) = \dfrac{1}{\sqrt{(1 - x^2)}}$.

Let $y = f(x) = \arcsin x$,
and $x = g(y) = \sin y$, where $-\pi/2 \leqslant y \leqslant \pi/2$.
Then $g'(y) = \cos y = \sqrt{(1 - x^2)}$.
We take the positive square root because $\cos y$ is positive or zero in the interval $-\pi/2 \leqslant y \leqslant \pi/2$. Then

$$\frac{d}{dx}(\arcsin x) = f'(x) = \frac{1}{\sqrt{(1 - x^2)}}$$

where $-1 < x < 1$.

Exercises 6.9

1 Show that $\dfrac{d}{dx}(\arccos x) = -\dfrac{1}{\sqrt{(1-x^2)}}$, where $-1 < x < 1$.

2 Evaluate
   (a) $\arcsin(-1/2)$ (c) $\arctan(-1)$
   (b) $\arctan 1$ (d) $\arcsin(\sqrt{3}/2)$.

3 Differentiate with respect to x
   (a) $\arcsin(2x)$ (b) $\arctan(3x)$ (c) $(\arcsin x)^2$.

4 Find the constants A and B, given that for $-2 < x < 2$

$$A\frac{d}{dx}[x\sqrt{(4-x^2)}] + B\frac{d}{dx}[\arcsin(x/2)] \equiv \sqrt{(4-x^2)}.$$

5 Differentiate with respect to x
   (a) $x \arcsin x$ (c) $x^2 \arcsin(2x)$
   (b) $\arcsin(x^2)$ (d) $\tan(\arctan x)$.

Miscellaneous exercises 6

1 Find the limit as x tends to 1 of
   (a) $\dfrac{x^4-1}{x^3-1}$ (b) $\dfrac{1}{x-1} - \dfrac{2}{x^2-1}$.

2 Find the limit as n tends to infinity of
   (a) $\dfrac{n(n-1)}{2^n}$ (b) $\dfrac{3^n+2^n}{3^n-2^n}$.

3 Find the limit as θ tends to zero of
   (a) $\dfrac{\sin(\theta/2)}{2\theta}$ (b) $\dfrac{1-\cos(2\theta)}{\theta^2}$.

4 Find the set of values of x for which the gradient of the cubic curve $y = x^3 - 2x^2 + x + 4$ is positive.

5 A curve is given by the parametric equations $x = 2\cos^3 t$, $y = 2\sin^3 t$ where $0 \leqslant t < 2\pi$. Find the coordinates of the points on the curve at which the gradient is 1.

6 Show that the gradient of the curve $y = x^4 - 2x^2$ is positive when x is greater than 1. Find whether the gradient is positive for any other values of x.

7 At a certain instant the volume of a cube is increasing at the rate of $3 \text{ cm}^3 \text{ s}^{-1}$ while the area of each face is increasing at the rate of $0 \cdot 2 \text{ cm}^2 \text{ s}^{-1}$. Find the rate of increase of the length of an edge of the cube at this instant.

8 A conical vessel with a semi-vertical angle of $30°$ is fixed with its axis vertical and its vertex downwards. Water is poured into the vessel at the constant rate of one litre per minute. Find the rate at which the depth of the water is increasing at the instant when the depth is 5 cm.

9 A ladder rests against a vertical pole. The foot of the ladder is sliding away from the pole along horizontal ground. Find the inclination of the ladder to the horizontal at the instant when the top of the ladder is moving three times as fast as its foot. [L]

10 A horizontal circular disc of radius a is fixed at a height a above a horizontal floor. A small lamp vertically above the centre of the disc is moving downward with speed v. Find the rate of increase with respect to time of the area of the shadow of the disc at the instant when the height of the lamp above the floor is $2a$. [L]

11 A point moves on the positive x-axis in such a way that its distance from the origin at time t seconds is $4t/(1+t^2)$ metres. Find its distance from O when its speed is zero, and find its acceleration at this instant.

12 Find the values of x between 0 and 2π for which the gradient of the curve $y = \sin x + \cos x$ is zero. Sketch the curve.

13 Differentiate with respect to x (a) $\sin(x^2)$ (b) $4\cos^2 x$
(c) $\tan^2(2x)$.

14 Differentiate with respect to x (a) $x^2 \tan x$ (b) $x \sin x + \cos x$
(c) $\sin^2 x \cos^2 x$ (d) $x \cos x - \sin x$.

15 Differentiate with respect to x (a) $(x-1)/\sqrt{x}$ (b) $(x^2+3x)^{\frac{1}{2}}$
(c) $(x^3+3x)^{2/3}$ (d) $(x+1)^{-2/3}$.

16 Differentiate with respect to x (a) $(x-1/x)^8$ (b) $(3-4x^2)^{-5}$
(c) $(x^3+3)^{-1}$ (d) $(x^2+x+1)^4$.

17 Differentiate with respect to x (a) $\dfrac{x^2-1}{x^2-2}$ (b) $\dfrac{1-3x^2}{2+4x^2}$ (c) $\dfrac{2x+3}{(x+3)^2}$.

18 Given that $y = \arctan[(1-x)/(1+x)]$, find $\dfrac{dy}{dx}$ and $\dfrac{d^2y}{dx^2}$.

19 If $y = \dfrac{1-\cos x}{1+\cos x}$ show that $\dfrac{dy}{dx} = t+t^3$, where $t = \tan(x/2)$.

20 Given that $y = (\cos x)/x$, prove that

$$x\frac{d^2y}{dx^2} + 2\frac{dy}{dx} + xy = 0.$$

21 If $x = \dfrac{1+t}{1-2t}$ and $y = \dfrac{1+2t}{1-t}$, find the value of $\dfrac{dy}{dx}$ when $t = 0$.

22 Find $f'(x)$ given that $f(x) = 3x + \sin x - 8\sin \frac{1}{2}x$ and deduce that $f(x)$ is positive for $x > 0$.

23 A horizontal trough 2 m long is closed at both ends and is 1 m deep. Its cross-section is an isosceles triangle of base (upwards) 0·6 m. Water runs into the trough at the rate of 0·02 m³ s⁻¹. Find the rate at which the water level is rising when the height of the water is 0·5 m.

24 The length of a glass rod when the temperature is $T°C$ is $(1+\alpha T+\beta T^2)$ m, where $\alpha = 2 \times 10^{-6}$ and $\beta = 10^{-7}$. Find the rate of change of the length with respect to temperature at $100°C$.

25 The radius of a circular metal disc is increasing at the rate of $p\%$ per second. If its volume is constant, show that its thickness is decreasing at $2p\%$ per second.

26 If $y = (\arcsin x)^2$, prove that

$$(1 - x^2)\frac{d^2 y}{dx^2} - x\frac{dy}{dx} - 2 = 0.$$

27 Find the possible values of the constant a in order that the curve $y = ax - x^2$ may cut the line $y = 3x$ at right angles.

28 Find the values of a and b in order that the curves $y = \sin\left(x + \dfrac{\pi}{3}\right)$ and

$y = \dfrac{a + bx}{1 + x}$ may touch when $x = 0$.

29 If $y = \left(\dfrac{1 - x^2}{1 + x^2}\right)^n$ show that $(1 - x^4)\dfrac{dy}{dx} + 4nxy = 0$.

30 Find the values of the constants a and b so that if $y = a\sin x + b\sin 3x$, then $\dfrac{dy}{dx} = 1$ and $\dfrac{d^3 y}{dx^3} = 0$ when $x = 0$.

31 A vertical pole of height h stands on the summit of a hemispherical mound with horizontal base whose radius is r. An observer on the level of the base, at a distance $(r + x)$ from the centre, can just see a point at a distance y down from the top of the pole: the radius of the hemisphere to the point where the line of sight touches the surface then makes an angle θ with the horizontal. Express both x and y in terms of θ, r and h. Hence show that if x is increasing at rate u, y is increasing at rate

$$\left(\frac{r + h - y}{r + x}\right)^3 u.$$

32 In the rectangle $ABCD$ the side AB is increasing at the rate of p cm s^{-1} and the side BC is increasing at the rate of q cm s^{-1}. Find the rate of increase, at the instant when $AB = x$ cm and $BC = y$ cm, of
(a) the area of $ABCD$
(b) the diagonal AC
(c) the angle BAC.

33 Show that (a) $\displaystyle\lim_{x \to 2} \frac{x^4 - 8x}{x - 2} = 24$ (b) $\displaystyle\lim_{x \to 2} \frac{x^3 - 8}{x^2 - 2x} = 6$.

34 If f is an even function of x show that $f'(x) = -f'(-x)$.

35 Show that the gradient of the curve

$$y = (\sin x + \cos x)/(\sin x - \cos x)$$

is always negative. Show also that when $x = 0$, $\dfrac{d^2 y}{dx^2} = -4$.

36 Show that $\dfrac{d}{dx}\sin(\arctan x) = (1 + x^2)^{-3/2}$.

37 The gradient of the curve $y = 3 \sin x + k \sin 3x$ is zero when $x = \pi/3$. Show that $k = 1/2$ and find the other values of x between 0 and π for which the gradient is zero.

38 The parametric equations of a curve are

$$x = \cos t + t \sin t, \quad y = \sin t - t \cos t.$$

Show that $\dfrac{d^2 y}{dx^2} = \dfrac{\sec^3 t}{t}$.

39 The coordinates of a point on an ellipse are given by $x = 2 \sin t$, $y = 3 \cos t$. Show that (a) $\dfrac{dy}{dx} = -\dfrac{3 \tan t}{2}$ (b) $\dfrac{d^2 y}{dx^2} = -\dfrac{3 \sec^3 t}{4}$.

40 At time t seconds the distance in metres from the origin of a point moving on the x-axis is given by

$$x = t(t^3 - 4t^2 + 6t + 4).$$

Show that when the velocity of the point is $12 \ \mathrm{m\,s^{-1}}$, its acceleration is $12 \ \mathrm{m\,s^{-2}}$.

7   Coordinate geometry

7.1   Lengths and angles

The length of the line segment PQ (Fig. 7.1) can be found as follows:

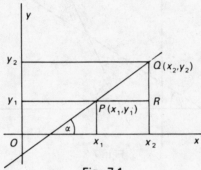

Fig. 7.1

$$PQ^2 = PR^2 + RQ^2$$
$$= (x_2 - x_1)^2 + (y_2 - y_1)^2.$$

The line PQ is at an angle α to Ox. The gradient m of the line PQ is $\tan \alpha$.

Gradient $m = \tan \alpha = \dfrac{QR}{PR} = \dfrac{y_2 - y_1}{x_2 - x_1}.$

Consider two lines l_1 and l_2 which have gradients m_1 and m_2 respectively. If l_1 and l_2 are parallel, they have the same gradient, i.e. $m_1 = m_2$. The case when l_1 is perpendicular to l_2 is shown in Fig. 7.2.

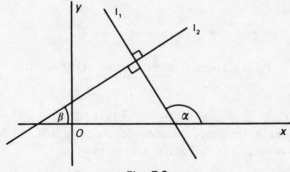

Fig. 7.2

$$m_1 = \tan \alpha \quad \text{and} \quad m_2 = \tan \beta; \, \alpha = \beta + 90°$$

$$\therefore \quad \tan \alpha = \tan (\beta + 90°) = -\cot \beta = -\frac{1}{\tan \beta}$$

$$\tan \alpha \tan \beta = -1$$

i.e.
$$m_1 m_2 = -1.$$

This is the condition for two lines with gradients m_1, m_2 to be at right angles.

In Fig. 7.3 the angle between the lines is θ.

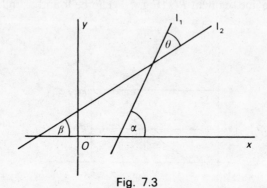

Fig. 7.3

$$\tan \theta = \tan (\alpha - \beta) = \frac{\tan \alpha - \tan \beta}{1 + \tan \alpha \tan \beta}$$

$$= \frac{m_1 - m_2}{1 + m_1 m_2}.$$

Example 1
Points P, Q and R have coordinates $(3, 4)$, $(8, 16)$ and $(-9, 9)$ respectively. Find the length and the gradient of PQ and show that the angle QPR is a right angle.

$$PQ^2 = (8-3)^2 + (16-4)^2 = 169, \text{ i.e. } PQ = 13.$$

$$\text{Gradient of } PQ = \frac{16-4}{8-3} = \frac{12}{5} (= m_1).$$

$$\text{Gradient of } PR = \frac{9-4}{-9-3} = -\frac{5}{12} (= m_2).$$

$m_1 m_2 = -1$, $\therefore$ PQ is perpendicular to PR, i.e. $\angle QPR = 90°$.

Example 2
Points A, B, C, D have coordinates $(2, -1)$, $(-2, 3)$, $(-3, -2)$, $(4, -3)$ respectively. Find the lengths of the sides of the triangle ABC and find the angle ABC. Show that the points A, B and D are collinear.

$$AB^2 = (-2-2)^2 + (3+1)^2 = 32, \text{ i.e. } AB = \sqrt{32},$$
$$BC^2 = (-3+2)^2 + (-2-3)^2 = 26, \text{ i.e. } BC = \sqrt{26},$$

$AC^2 = (-3-2)^2+(-2+1)^2 = 26$, i.e. $AC = \sqrt{26}$.

$AC = BC$ and so the triangle ABC is isosceles.

$AB^2 < AC^2+BC^2$ and therefore the triangle ABC is acute angled.

Gradient of $AB = \dfrac{3+1}{-2-2} = -1 \,(= m_1)$.

Gradient of $BC = \dfrac{-2-3}{-3+2} = 5 \,(= m_2)$.

Tan $ABC = \dfrac{-1-5}{1+(-1)(5)} = \dfrac{3}{2}$, $\therefore \angle ABC = 56\cdot3°$.

Gradient of $AD = \dfrac{-3+1}{4-2} = -1 = $ gradient of AB

$\therefore A, B$ and D are collinear.

Exercises 7.1

1   Points A, B and C have coordinates $(1, 2)$, $(4, 4)$ and $(3, -1)$ respectively. Find the lengths of the sides of the triangle ABC and show that the triangle is right-angled.

2   Find the lengths and the gradients of the sides of the quadrilateral $PQRS$, where the coordinates of the vertices are $P(3, 3)$, $Q(-1, 4)$, $R(0, -2)$, $S(4, 0)$.

3   Points A, B, C, D have coordinates $(-1, -1)$, $(0, 4)$, $(5, 5)$, $(4, 0)$ respectively. Show that $ABCD$ is a parallelogram and that the diagonal AC passes through the origin.

4   If points P, Q, R, S have coordinates $P(0, 2)$, $Q(-2, 13)$, $R(8, 8)$, $S(10, -3)$ respectively, show that $PQRS$ is a rhombus.

5   Points A, B, C, D have coordinates $A(-2, -3)$, $B(2, -1)$, $C(1, 1)$, $D(-3, -1)$ respectively. Show that $ABCD$ is a rectangle.

7.2   Division of a line segment in a given ratio

In Fig. 7.4 the points P and Q have coordinates (x_1, y_1) and (x_2, y_2) respectively

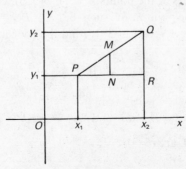

Fig. 7.4

and the mid-point of PQ is M with coordinates (X, Y).

$$PN = \tfrac{1}{2}PR = \tfrac{1}{2}(x_2 - x_1) \quad \text{and} \quad MN = \tfrac{1}{2}QR = \tfrac{1}{2}(y_2 - y_1)$$

$\therefore$
$$X = x_1 + \tfrac{1}{2}(x_2 - x_1) = \tfrac{1}{2}(x_1 + x_2)$$
and
$$Y = y_1 + \tfrac{1}{2}(y_2 - y_1) = \tfrac{1}{2}(y_1 + y_2),$$

i.e. the mid-point of PQ has coordinates $\left(\dfrac{x_1 + x_2}{2}, \dfrac{y_1 + y_2}{2}\right)$.

In Fig. 7.5 the point C divides the line joining $A(x_1, y_1)$ to $B(x_2, y_2)$ in the ratio $1:3$.

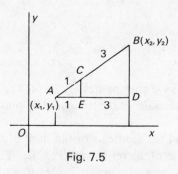

Fig. 7.5

$AC:CB = 1:3 \quad \text{and} \quad AE:ED = 1:3$
so that
$$AE = \tfrac{1}{4}AD = \tfrac{1}{4}(x_2 - x_1)$$
and also
$$EC = \tfrac{1}{4}BD = \tfrac{1}{4}(y_2 - y_1).$$

$\therefore$
$$x_3 = x_1 + \tfrac{1}{4}(x_2 - x_1) = \dfrac{3x_1 + x_2}{4}$$

Similarly
$$y_3 = y_1 + \tfrac{1}{4}(y_2 - y_1) = \dfrac{3y_1 + y_2}{4}$$

Thus C is the point $(\tfrac{1}{4}(3x_1 + x_2), \tfrac{1}{4}(3y_1 + y_2))$.

Example 1
A triangle has its vertices at the points $A(1, 7)$, $B(5, 5)$, $C(8, -1)$. M is the mid-point of AB and L is the point in BC such that $BL:LC = 1:2$. N is the point on AC produced such that $AN:NC = 3:1$. Find the coordinates of L, M and N.

The coordinates of M are $\left(\dfrac{1+5}{2}, \dfrac{7+5}{2}\right)$, i.e. $(3, 6)$.

Let L and N have coordinates (x_1, y_1) and (x_2, y_2) respectively.

$$x_1 = 5 + \tfrac{1}{3}(8 - 5) = 6 \quad \text{and} \quad y_1 = 5 + \tfrac{1}{3}(-1 - 5) = 3,$$

i.e. the coordinates of L are $(6, 3)$.

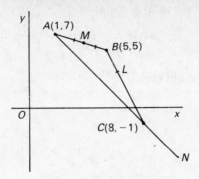

Fig. 7.6

$AN:NC = 3:1, \therefore AN:AC = 3:2$ (Fig. 7.6).

$$x_2 = 1 + \frac{3}{2}(8-1) = 11\tfrac{1}{2} \quad \text{and} \quad y_2 = 7 + \frac{3}{2}(-1-7) = -5,$$

i.e. the coordinates of N are $(11\tfrac{1}{2}, -5)$.

Example 2
Find the coordinates of the centroid of the triangle ABC, where the coordinates of the vertices are $A(x_1, y_1)$, $B(x_2, y_2)$ and $C(x_3, y_3)$.

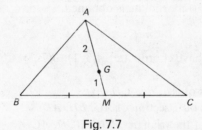

Fig. 7.7

In Fig. 7.7, M is the mid-point of BC and $G(\bar{x}, \bar{y})$ is the centroid of the triangle ABC. The coordinates of M are $(\tfrac{1}{2}(x_2 + x_3), \tfrac{1}{2}(y_2 + y_3))$.

G is the centroid of $\triangle ABC$ and so $AG = \tfrac{2}{3}AM$.

$$\therefore \qquad \bar{x} = x_1 + \tfrac{2}{3}[\tfrac{1}{2}(x_2 + x_3) - x_1] = \tfrac{1}{3}(x_1 + x_2 + x_3)$$
$$\text{and} \qquad \bar{y} = y_1 + \tfrac{2}{3}[\tfrac{1}{2}(y_2 + y_3) - y_1] = \tfrac{1}{3}(y_1 + y_2 + y_3),$$

i.e. the centroid G has coordinates $(\tfrac{1}{3}(x_1 + x_2 + x_3), \tfrac{1}{3}(y_1 + y_2 + y_3))$.

Example 3
Show that the coordinates of the point which divides the line joining $P(x_1, y_1)$ to

$Q(x_2, y_2)$ internally in the ratio $\lambda : \mu$ are

$$\left(\frac{\mu x_1 + \lambda x_2}{\lambda + \mu}, \frac{\mu y_1 + \lambda y_2}{\lambda + \mu} \right).$$

Let R be the required point (Fig. 7.8).

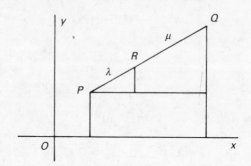

Fig. 7.8

The x-coordinate of $R = x_1 + \dfrac{\lambda}{\lambda + \mu}(x_2 - x_1)$

$$= \frac{\mu x_1 + \lambda x_2}{\lambda + \mu}.$$

Similarly the required y-coordinate is obtained.

Exercises 7.2
1   Points A, B, C, D have coordinates $A(3, 1)$, $B(7, 3)$, $C(6, 7)$, $D(1, -6)$. Find the coordinates of
   (a)   the mid-points of AB, BC, CD, AD
   (b)   the two points of trisection of AC, BD, AD
   (c)   the centroids of the triangles ABC, BCD, ACD.
2   Points A and B have coordinates $(1, 3)$ and $(-3, -5)$ respectively. The point C divides AB externally in the ratio $2:1$ and the point D divides AB externally in the ratio $1:2$. Find the coordinates of C and D.

7.3   Area of a triangle

In Fig. 7.9 the coordinates of the vertices of the triangle ABC are $A(x_1, y_1)$, $B(x_2, y_2)$ and $C(x_3, y_3)$.
Area of the triangle ABC
= area of trapezium $ACNL$ + area of trapezium $CBMN$ − area of trapezium
    $ABML$
= $\frac{1}{2}(x_3 - x_1)(y_3 + y_1) + \frac{1}{2}(x_2 - x_3)(y_2 + y_3) - \frac{1}{2}(x_2 - x_1)(y_2 + y_1)$
= $\frac{1}{2}[(x_1 y_2 - x_2 y_1) + (x_2 y_3 - x_3 y_2) + (x_3 y_1 - x_1 y_3)].$

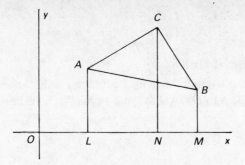

Fig. 7.9

It is difficult to remember this formula. It may help to write the coordinates as shown below, repeating the first pair.

The numerical value of the area of the triangle ABC

$= \frac{1}{2}|[(\text{sum of products marked} \searrow) - (\text{sum of products marked} \dashrightarrow)]|.$

(The modulus sign is included because the result may be positive or negative depending on the order in which the coordinates are taken.)

Putting $x_3 = 0$ and $y_3 = 0$ in the formula for the area of the triangle ABC gives

$$\text{area of triangle } ABO = |\tfrac{1}{2}(x_1 y_2 - x_2 y_1)|.$$

Example
Find the area of the triangle with vertices $(1, 2)$, $(-3, 4)$ and $(-1, -5)$.

$\text{Area} = \frac{1}{2}|[(4 + 15 - 2) - (-6 - 4 - 5)]|$
$= \frac{1}{2}|(17 + 15)|$
$= 16$ square units.

Exercises 7.3
Find the areas of the triangles with vertices at the points
1   $(0, 2)$, $(3, 5)$, $(-1, 7)$.
2   $(3, -1)$, $(4, 0)$, $(0, 6)$.

3 (0, 0),   (−2, 4),   (−5, −7).
4 (−2, 3),   (4, −7),   (3, 8).

7.4 The straight line

Figure 7.10 shows a straight line which cuts the y-axis at the point $A(0, c)$ and which is at an angle α to Ox, where $\tan \alpha = m$. $P(x, y)$ is any point on the line.

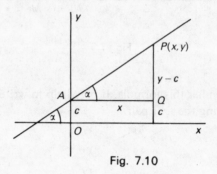

Fig. 7.10

From the triangle APQ $\qquad \tan \alpha = \dfrac{y-c}{x} = m$

∴ $\qquad\qquad\qquad\qquad\qquad y = mx + c,$

i.e. the coordinates of each point on the line satisfy this equation. Thus the set of points lying on this line is the set

$$\{(x, y) : y = mx + c\}.$$

Here the coordinates used are cartesian coordinates (after Descartes) and $y = mx + c$ is said to be the cartesian equation of the line.

The cartesian equation of any line is the equation which is satisfied by the cartesian coordinates of every point on the line.

$y = mx + c$ is the equation of a straight line which cuts the y-axis where $y = c$ and which has gradient m, i.e. is at an angle arctan (m) to Ox.

If $c = 0$, the line cuts the y-axis at the origin and so $y = mx$ is the equation of a straight line through the origin.

To find the equation of a particular straight line the values of m and c need to be found and this requires two independent pieces of information, e.g. two points on the line or the gradient and one point on the line.

Example 1
Find the equation of the line with gradient 3 which passes through the point (4, 5).

Let the equation of the line be $y = mx + c$. Gradient $= 3$ and so $m = 3$ and the line has equation $y = 3x + c$. The point (4, 5) is on the line, ∴ $5 = 3(4) + c$, i.e. $c = -7$. Thus the equation of the line is $y = 3x - 7$.

Example 2

Find the equation of the line through the points $(-1, 2)$ and $(4, 8)$.

Let the equation of the line be $y = mx + c$.

$(-1, 2)$ is on the line, $\therefore$	$2 = -m + c$	... (1)
$(4, 8)$ is on the line, $\therefore$	$8 = 4m + c$	... (2)

Solving equations (1) and (2) we get $m = 6/5$ and $c = 16/5$.

Thus the equation of the line is $y = (6/5)x + 16/5$ or $5y = 6x + 16$.

The general case will now be considered.

The equation of a line with gradient m_1 and passing through the point $A(x_1, y_1)$

Let $P(x, y)$ be any point on the line.

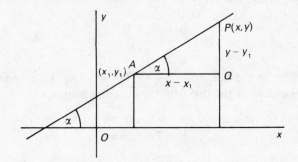

Fig. 7.11

From $\triangle APQ$ (Fig. 7.11)

$$\tan \alpha = m_1 = \frac{y - y_1}{x - x_1},$$

i.e.

$$y - y_1 = m_1 (x - x_1).$$

This relationship is satisfied by the cartesian coordinates of each and every point on the line and so is the cartesian equation of the required straight line.

This result is important and will be used frequently, e.g. in finding the equation of a tangent to a curve.

The equation of a line through the points $A(x_1, y_1)$ and $B(x_2, y_2)$

From section 7.1 the gradient of the line, $m_1, = \dfrac{y_2 - y_1}{x_2 - x_1}$

Thus the equation of the line is $y - y_1 = \left(\dfrac{y_2 - y_1}{x_2 - x_1}\right) (x - x_1).$

These results will now will now be used to deal with examples 1 and 2 on pages 170–171.

Example 1
Find the equation of the line with gradient 3 which passes through the point $(4, 5)$.

$m_1 = 3$ and $(x_1, y_1) = (4, 5)$.
$\therefore$ the equation of the line is $\quad y - 5 = 3(x - 4)$
i.e. $\qquad\qquad\qquad\qquad\qquad\quad y = 3x - 7.$

Example 2
Find the equation of the line through the points $(-1, 2)$ and $(4, 8)$.

$(x_1, y_1) = (-1, 2)$ and $(x_2, y_2) = (4, 8)$.

Gradient of the line, $m_1, = \dfrac{y_2 - y_1}{x_2 - x_1} = \dfrac{8 - 2}{4 - (-1)} = \dfrac{6}{5}$

$\therefore$ the equation of the line is $\quad y - 2 = \dfrac{6}{5}(x + 1),$

i.e. $\qquad\qquad\qquad\qquad\qquad 5y = 6x + 16.$

Example 3
Find the gradient of the line $2x + 3y + 4 = 0$. Find also the coordinates of the points of intersection of the line with the coordinate axes.

The equation of the line $2x + 3y + 4 = 0$, can be written as $3y = -2x - 4$,

i.e. $\qquad\qquad\qquad\qquad\qquad y = -\dfrac{2}{3}x - \dfrac{4}{3}.$

This is in the form $y = mx + c$ and so the gradient is $-\frac{2}{3}$. The line cuts the x-axis where $y = 0$, i.e. $2x + 4 = 0$ giving $x = -2$. The line cuts the y-axis where $x = 0$,

i.e. $3y + 4 = 0$ giving $y = -\dfrac{4}{3}$. Thus the points of intersection of the line with the

x-axis and the y-axis are $(-2, 0)$ and $\left(0, -\dfrac{4}{3}\right)$ respectively.

Example 4
Find the equation of the perpendicular bisector of the line segment AB, where the points A and B have coordinates $(2, 5)$ and $(-4, 9)$ respectively.

The coordinates of M, the mid-point of AB, are $\left(\dfrac{2-4}{2}, \dfrac{5+9}{2}\right)$ i.e. $(-1, 7)$.

Gradient of $AB = m_1 = \dfrac{9 - 5}{-4 - 2} = -\dfrac{2}{3}.$

If m_2 is the gradient of the perpendicular bisector of AB, then from section 7.1,

$m_1 m_2 = -1$ giving $m_2 = \dfrac{3}{2}.$

The equation of the perpendicular bisector of AB is

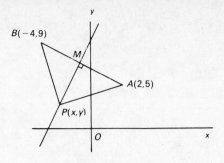

Fig. 7.12

$$y - 7 = \frac{3}{2}[x - (-1)], \text{ i.e. } 2y = 3x + 17.$$

A better method is as follows.

Let $P(x, y)$ be any point on the perpendicular bisector of AB (Fig. 7.12).

Then
$$PA^2 = PB^2$$
i.e.
$$(x - 2)^2 + (y - 5)^2 = (x + 4)^2 + (y - 9)^2$$
giving
$$2y = 3x + 17.$$

Since P is any point on the perpendicular bisector, this equation is satisfied by the coordinates of all points on the perpendicular bisector and hence is its equation.

Example 5

It is believed that the variables s and t obey a law of the form $s = a/t + b$, where a and b are constants. By drawing a suitable graph, show that the following values of s and t approximately satisfy this law and find approximate values for a and b.

t	0·25	0·5	1	2	3	4
s	6·7	3·7	2·2	1·45	1·2	1·07

$s = a(1/t) + b$ is of the form $y = mx + c$ with s for y and $1/t$ for x. Therefore if these values of s and t satisfy this law, the points $(s, 1/t)$ for the given values will lie on a straight line with gradient a and cutting the s-axis where $s = b$. Fig. 7.13 overleaf shows that the points do lie on a straight line, and so the given values of s and t do satisfy approximately a law of the form $s = a(1/t) + b$

with a = gradient of the line = $\dfrac{6·7 - 0·7}{4 - 0} = \dfrac{6}{4} = 1·5$

and b = the value of s where the line cuts the s-axis, i.e. $b = 0·7$.

Exercises 7.4

1 Find the equation of the straight line which has
   (a) gradient 2 and passes through the point (0, 1)
   (b) gradient 3 and passes through the point (2, 0)

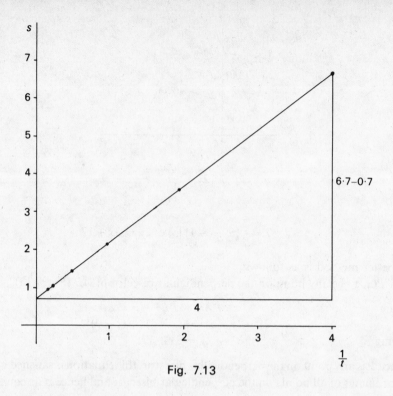

$6.7-0.7$

4

Fig. 7.13

(c) gradient -4 and passes through the point $(3, 2)$

(d) gradient -5 and passes through the point $(-1, -2)$.

2 Find the equation of the straight line through the points

(a) $(1, 2)$ and $(5, 4)$

(b) $(0, -2)$ and $(3, 2)$

(c) $(4, 0)$ and $(0, 3)$

(d) $(-1, -3)$ and $(6, -1)$.

3 Express the following equations of lines in the form $y = mx + c$. State the gradient of each line and the coordinates of the points of intersection of each line with the coordinate axes.

(a) $2x - 3y + 4 = 0$

(b) $3x - 2y - 1 = 0$

(c) $4x + y - 6 = 0$

(d) $5x + y + 5 = 0$.

4
x	1	2·1	3·7	4·5	5·2	6·7
y	5	11·8	30·4	43·5	57·1	92·8

By plotting y against x^2 show that these values of x and y satisfy approximately a law of the form $y = ax^2 + b$, where a and b are constants. From your graph find approximate values for a and b.

5

x	0·25	0·5	0·75	1·0	1·5	2·0
y	0·91	1·63	2·16	2·5	2·63	2·0

By plotting y/x against x show that these values of x and y satisfy approximately a law of the form $y = ax^2 + bx$, where a and b are constants. From your graph find approximate values for a and b.

6

s	0·2	0·4	0·6	0·8	1·0
t	2·86	1·73	1·35	1·16	1·05

By drawing a suitable graph show that these values of s and t satisfy approximately a law of the form $s(p + qt) = 1$, where p and q are constants. From your graph find approximate values for p and q.

7 Find the equation of the perpendicular bisector of the line segment AB when A and B have coordinates

(a) (0, 1)   and   (4, 7)   (d) (6, 5)   and   (−1, 2)
(b) (0, 0)   and   (−2, 8)   (e) (−3, −2) and   (−7, −3).
(c) (−2, 0) and   (6, 4)

7.5   The distance of a point from a line

The equation of a straight line can be expressed in the form

$$y = mx + c \quad \text{or} \quad px + qy + r = 0.$$

The form $y = mx + c$ gives us at a glance the gradient of the line and its intercept on the y-axis.

From the form $px + qy + r = 0$ it can be seen at once that if $p \neq 0$ the line cuts the x-axis at the point $(-r/p, 0)$. If $q \neq 0$, the line cuts the y-axis at the point $(0, -r/q)$.

The form $px + qy + r = 0$ will now be used to obtain the distance from a point to a straight line.

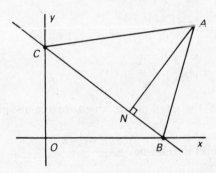

Fig. 7.14

In Fig. 7.14 the line $px + qy + r = 0$ cuts the axes at the points B and C and AN is the perpendicular from the point $A(x_1, y_1)$ to the line. The coordinates of B and

C are $\left(-\dfrac{r}{p}, 0\right)$ and $\left(0, -\dfrac{r}{q}\right)$ and by the formula in section 7.3

$$\text{the area of triangle } ABC = \tfrac{1}{2}\left[\left(\frac{r^2}{pq}\right) - \left(-\frac{ry_1}{p} - \frac{rx_1}{q}\right)\right]$$

$$= \tfrac{1}{2}\left(\frac{r}{pq}\right)\left(r + qy_1 + px_1\right).$$

Also the area of the triangle $ABC = \tfrac{1}{2}\,BC \times AN$

and since $BC^2 = \dfrac{r^2}{p^2} + \dfrac{r^2}{q^2}$ we have $BC = \left(\dfrac{r}{pq}\right)\left[p^2 + q^2\right]^{\frac{1}{2}}.$

Now equating the two expressions for the area of the triangle ABC

we get $\quad \tfrac{1}{2}\left[\left(\dfrac{r}{pq}\right)(p^2 + q^2)^{\frac{1}{2}}\right]AN = \tfrac{1}{2}\left(\dfrac{r}{pq}\right)(r + qy_1 + px_1)$

giving $\qquad\qquad AN = \left|\dfrac{px_1 + qy_1 + r}{(p^2 + q^2)^{\frac{1}{2}}}\right|.$

The modulus sign is used since we usually require the numerical value of the distance and $(px_1 + qy_1 + r)$ can produce positive or negative values.

Example 1
Find the distance to the line $y = 2x - 4$ from the points $(6, 3)$ and $(2, 5)$ and from the origin.

Putting the equation of the line in the appropriate form we get $2x - y - 4 = 0$.

Distance from $(6, 3) = \left|\dfrac{2(6) - (3) - 4}{\sqrt{[2^2 + (-1)^2]}}\right| = \dfrac{5}{\sqrt{5}} = \sqrt{5}.$

Distance from $(2, 5) = \left|\dfrac{2(2) - (5) - 4}{\sqrt{(4+1)}}\right| = \left|\dfrac{-5}{\sqrt{5}}\right| = \sqrt{5}.$

Distance from $(0, 0) = \left|\dfrac{-4}{\sqrt{5}}\right| = \dfrac{4}{\sqrt{5}}.$

Suppose we want S, the set of points which are at a distance $\sqrt{5}$ from the line $2x - y - 4 = 0$.

The distance from (h, k) to this line $= \dfrac{2h - k - 4}{\sqrt{5}}.$

If $(h, k) \in S$, then $\dfrac{2h - k - 4}{\sqrt{5}} = \sqrt{5}$

and (h, k) lies on the line $\dfrac{2x - y - 4}{\sqrt{5}} = \sqrt{5}$, i.e. $2x - y - 9 = 0$,

which is the line ℓ_1 in Fig. 7.15.

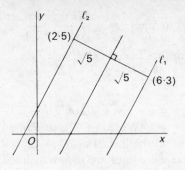

Fig. 7.15

This is only part of the set S. The set S also includes line ℓ_2.

If (h, k) is on the line ℓ_2, then $\dfrac{2h - k - 4}{\sqrt{5}} = -\sqrt{5}$, i.e. ℓ_2 is the line $2x - y + 1 = 0$.

Thus S consists of the set of points forming the lines

$$2x - y - 9 = 0 \quad \text{and} \quad 2x - y + 1 = 0$$

which are given by

$$\frac{2x - y - 4}{\sqrt{5}} = \pm\sqrt{5}.$$

Example 2

Find the equations of the bisectors of the angles between the lines $3x + 4y + 5 = 0$ (ℓ_1) and $12x - 5y + 8 = 0$ (ℓ_2).

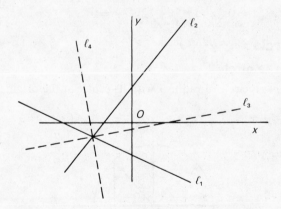

Fig. 7.16

Let (h, k) be a point on a bisector.

The distance from (h, k) to line $\ell_1 = \pm\dfrac{3h + 4k + 5}{\sqrt{(9 + 16)}}$.

The distance from (h, k) to line $\ell_2 = \pm\dfrac{12h - 5k + 8}{\sqrt{(144 + 25)}}$.

The point (h, k) is equidistant from lines ℓ_1 and ℓ_2.

$$\therefore \qquad \pm\frac{3h + 4k + 5}{5} = \pm\frac{12h - 5k + 8}{13}$$

giving $\quad 39h + 52k + 65 = 60h - 25k + 40$, i.e. $21h - 77k - 25 = 0$
or $\qquad 39h + 52k + 65 = -(60h - 25k + 40)$, i.e. $99h + 27k + 105 = 0$.
Thus the point (h, k) lies on the line $21x - 77y - 25 = 0$ or on the line $99x + 27y + 105 = 0$ and these lines are shown as lines ℓ_3 and ℓ_4 in Fig. 7.16.

The bisectors of the angles between the lines ℓ_1 and ℓ_2 therefore have
equations $\quad 21x - 77y - 25 = 0$
and $\qquad 99x + 27y + 105 = 0$, i.e. $33x + 9y + 35 = 0$.
These lines are clearly perpendicular from geometrical considerations. Also their gradients are $21/77$ (i.e. $3/11$) and $-33/9$ (i.e. $-11/3$), whose product is -1.

Exercises 7.5
Find the perpendicular distance
1 from the point $(1, 2)$ to the line $3x + 4x + 7 = 0$
2 from the point $(4, 3)$ to the line $12x - 5y + 3 = 0$
3 from the point $(0, 0)$ to the line $4x - 3y - 1 = 0$
4 from the point $(-2, 5)$ to the line $y = 4x - 1$
5 from the point $(-3, -2)$ to the line $y = 3x - 1$
6 from the origin   to the line $x + y - 1 = 0$.
7 Find the equations of the bisectors of the angles between the lines
    (a) $2x + y - 12 = 0$ and $x - 2y + 4 = 0$
    (b) $3x - 4y + 2 = 0$ and $5x + 12y - 9 = 0$.

7.6   The circle $x^2 + y^2 = r^2$

Equation of the circle
Figure 7.17 shows a circle of radius r with its centre at the origin. $P(x, y)$ is any

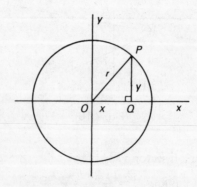

Fig. 7.17

point on this circle which is the set of points in the x–y plane which are at a distance r from the origin O. From the triangle OPQ we have $x^2 + y^2 = r^2$ so that the set of points lying on the circle is the set $\{(x, y): x^2 + y^2 = r^2\}$. Thus the cartesian equation of the circle is $x^2 + y^2 = r^2$.

The points in the x–y plane which lie inside the circle will constitute the set $\{(x, y): x^2 + y^2 < r^2\}$.

The points in the x–y plane which lie outside the circle will constitute the set $\{(x, y): x^2 + y^2 > r^2\}$.

Tangent to the circle at a given point

In Fig. 7.18 $A(x_1, y_1)$ is a point on the circle $x^2 + y^2 = r^2$.

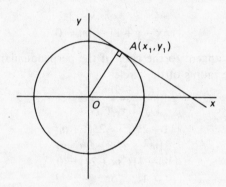

Fig. 7.18

The gradient of the radius OA is y_1/x_1,
$\therefore$ the gradient of the tangent at $A = -x_1/y_1$.
$\therefore$ the equation of the tangent at A is

$$y - y_1 = m_1(x - x_1)$$
$$= -\frac{x_1}{y_1}(x - x_1),$$

$\therefore$
$$yy_1 - y_1{}^2 = -xx_1 + x_1{}^2,$$
i.e.
$$xx_1 + yy_1 = x_1{}^2 + y_1{}^2 = r^2$$

since $A(x_1, y_1)$ is on the circle.
Thus the equation of the tangent at A is $xx_1 + yy_1 = r^2$.

Example
Find the equation of the tangent to the circle $x^2 + y^2 = 25$ at the point $(3, -4)$.

Putting $(x_1, y_1) = (3, -4)$ in the equation $xx_1 + yy_1 = 25$, the equation of the tangent is $3x - 4y = 25$.

Condition for the line $y = mx + c$ to touch the circle

The line will touch the circle if the perpendicular distance from the origin to the line is r, the radius of the circle,

i.e. if $\left| \dfrac{c}{\sqrt{(1+m^2)}} \right| = r$ from section 7.5

giving $$c^2 = r^2(1+m^2).$$

Example 1
Find the equations of the tangents from the point $(2, 11)$ to the circle $x^2 + y^2 = 25$.

Let a tangent have gradient m.
Then its equation is

$$y - 11 = m(x - 2),$$

i.e. $$mx - y + 11 - 2m = 0 \qquad \ldots (1)$$

This line will be a tangent to the circle if the perpendicular distance from the origin to it is 5, the radius of the circle.

$\therefore$ $\left| \dfrac{11 - 2m}{\sqrt{(1+m^2)}} \right| = 5$

i.e. $(11 - 2m)^2 = 25(1 + m^2)$
giving $21m^2 + 44m - 96 = 0$
$\therefore$ $(3m - 4)(7m + 24) = 0$
$\therefore$ $m = 4/3$ or $-24/7$.

Putting these values in (1), the equations of the two tangents are

$$\frac{4}{3}x - y + 11 - \frac{8}{3} = 0 \quad \text{and} \quad -\frac{24}{7}x - y + 11 + \frac{48}{7} = 0,$$

i.e. $4x - 3y + 25 = 0$ and $24x + 7y - 125 = 0$.

Example 2
Find the length of the tangent from the point $A(2, 11)$ to the circle $x^2 + y^2 = 25$.

Figure 7.19 shows the tangents AS and AT.
From $\triangle AOT$, $AO^2 = AT^2 + OT^2$
$\therefore$ $4 + 121 = AT^2 + 25$
$\therefore$ $AT^2 = 100$ and $AT = 10$.
Also $AS = AT = 10$.

Exercises 7.6

1 Find the equation of the tangent at the point $(1, -\sqrt{3})$ to the circle $x^2 + y^2 = 4$.

2 Find the equation of the tangent at the point $(-5, 12)$ to the circle $x^2 + y^2 = 169$.

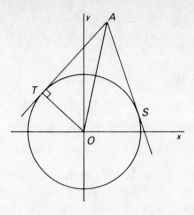

Fig. 7.19

3 Find the length of the tangents from the point $(3, -5)$ to the circle $x^2 + y^2 = 16$.
4 Find the length of the tangents from the point $(-2, -3)$ to the circle $x^2 + y^2 = 10$.
5 If $y = 2x + 3$ is a tangent to the circle $x^2 + y^2 = r^2$, find r.
6 If $y = -3x + c$ is a tangent to the circle $x^2 + y^2 = 10$, find the possible values of c.
7 If $y = mx - 5$ is a tangent to the circle $x^2 + y^2 = 9$, find the possible values of m.

7.7 The circle $(x - a)^2 + (y - b)^2 = r^2$

In Fig. 7.20 $P(x, y)$ is any point on the circle with centre (a, b) and radius r.

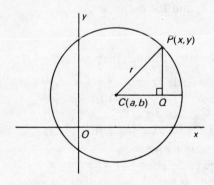

Fig. 7.20

In $\triangle CPQ$ $\qquad\qquad CQ^2 + QP^2 = CP^2,$
i.e. $\qquad\qquad (x-a)^2 + (y-b)^2 = r^2.$

This relationship is true for the cartesian coordinates of each and every point on the circle. Hence the cartesian equation of the circle, centre (a, b) and radius r, is $(x-a)^2 + (y-b)^2 = r^2$.

This can be expressed as

$$x^2 + y^2 - 2ax - 2by + a^2 + b^2 - r^2 = 0 \qquad \ldots (1)$$

Equation (1) shows that the equation of any circle may be written in the form

$$x^2 + y^2 + 2gx + 2fy + c = 0 \qquad \ldots (2)$$

Equation (2) is usually taken to be the general equation for a circle.

Note that, in the equation of the circle, the coefficients of x^2 and y^2 are equal and there is no term in xy.

By comparing equations (1) and (2) we see that

$$a = -g, \, b = -f, \, r^2 = a^2 + b^2 - c = g^2 + f^2 - c.$$

Thus the circle

$$x^2 + y^2 + 2gx + 2fy + c = 0$$

has its centre at the point $(-g, -f)$ and its radius is $\sqrt{(g^2 + f^2 - c)}$.

Example 1

Find the centre and radius of the circle $x^2 + y^2 + 4x - 6y + 7 = 0$.

Here $\quad 2g = 4$ and $2f = -6$, i.e. the centre is at the point $(-2, 3)$.

Radius $\quad = \sqrt{(g^2 + f^2 - c)} = \sqrt{(4 + 9 - 7)} = \sqrt{6}$.

Alternatively $x^2 + y^2 + 4x - 6y + 7 = 0$ can be rearranged
as $\qquad (x+2)^2 + (y-3)^2 = -7 + 4 + 9 = 6$.

$\therefore$ the centre is $(-2, 3)$ and the radius is $\sqrt{6}$.

Example 2

Find the equation of the circle through the points $P(-1, 2)$, $Q(7, -2)$ and $R(8, 5)$.

Let the equation of the circle be $x^2 + y^2 + 2gx + 2fy + c = 0$.
$(-1, 2)$ is on the circle $\therefore \quad 1 + 4 - 2g + 4f + c = 0 \qquad \ldots (1)$
$(7, -2)$ is on the circle $\therefore \quad 49 + 4 + 14g - 4f + c = 0 \qquad \ldots (2)$
$(8, 5)$ is on the circle $\therefore \quad 64 + 25 + 16g + 10f + c = 0 \qquad \ldots (3)$
Solving these three equations we get $g = -4, f = -2, c = -5$.
$\therefore$ the equation of the circle is $x^2 + y^2 - 8x - 4y - 5 = 0$.

An alternative method is to use the fact that the perpendicular bisector of a chord passes through the centre of the circle.

Using the method of example 4 on page 172 in section 7.4, the equation of the perpendicular bisector of PQ is

$$(x+1)^2 + (y-2)^2 = (x-7)^2 + (y+2)^2$$

which simplifies to $2x - y = 6$.

The centre $C(a, b)$ is on this line and so

$$2a - b = 6 \qquad \ldots (4)$$

Similarly the equation of the perpendicular bisector of QR is

$$(x-7)^2 + (y+2)^2 = (x-8)^2 + (y-5)^2$$

which simplifies to $x + 7y = 18$.

The centre $C(a, b)$ is on this line and so

$$a + 7b = 18 \qquad \ldots (5)$$

Solving equations (4) and (5) gives $a = 4$, $b = 2$, i.e. the centre is $(4, 2)$.
Radius $r = CP = \sqrt{[(-1-4)^2 + (2-2)^2]} = 5$.
$\therefore$ the equation of the circle is $(x-4)^2 + (y-2)^2 = 25$.

Example 3
Find the equation of the tangent at the point $P(4, 6)$ to the circle
$x^2 + y^2 - 2x - 4y - 20 = 0$.

The equation of the circle can be expressed as

$$(x-1)^2 + (y-2)^2 = 25$$

and its centre is at the point $A(1, 2)$.
Gradient of $AP = 4/3$.
$\therefore$ Gradient of the tangent at $P = -3/4$.
$\therefore$ The equation of the tangent at P is

$$y - 6 = -\tfrac{3}{4}(x-4)$$

i.e. $3x + 4y - 36 = 0$

The equation of a circle on a given diameter

In Fig. 7.21 overleaf, AB is a diameter of the circle and the coordinates of A and B are (x_1, y_1) and (x_2, y_2) respectively. $P(x, y)$ is any point on the circle.
Since AB is a diameter the angle $APB = 90°$.

Gradient of $AP = \dfrac{y - y_1}{x - x_1} (= m_1)$, gradient of $BP = \dfrac{y - y_2}{x - x_2} (= m_2)$.

Since AP is perpendicular to BP, $m_1 m_2 = -1$,

i.e. $\left(\dfrac{y - y_1}{x - x_1}\right)\left(\dfrac{y - y_2}{x - x_2}\right) = -1,$

giving $(x - x_1)(x - x_2) + (y - y_1)(y - y_2) = 0$.

This is true for the coordinates of all points on the circle and hence is the equation of the circle with AB as diameter.

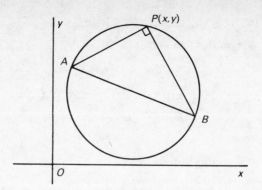

Fig. 7.21

Example 4

Find the equation of the circle on AB as diameter where the coordinates of A and B are $(1, 3)$ and $(-4, -5)$.

Putting $(x_1, y_1) = (1, 3)$ and $(x_2, y_2) = (-4, -5)$ into

$$(x - x_1)(x - x_2) + (y - y_1)(y - y_2) = 0$$

we get $\qquad (x - 1)(x + 4) + (y - 3)(y + 5) = 0,$

i.e. the equation of the circle is $x^2 + y^2 + 3x + 2y - 19 = 0$.

Exercises 7.7

1 Find the coordinates of the centre and the radius of the circle
  (a) $x^2 + y^2 - 4x - 10y + 13 = 0$
  (b) $x^2 + y^2 + 2x - 6y + 6 = 0$
  (c) $4x^2 + 4y^2 + 16x - 4y - 19 = 0$.

2 Find the equation of the tangent to the circle
  (a) $x^2 + y^2 - 4x - 2y - 20 = 0$ at the point $(6, 4)$
  (b) $x^2 + y^2 + 6x - 4y + 8 = 0$ at the point $(-5, 3)$
  (c) $(x + 1)^2 + (y + 2)^2 = 13$ at the point $(1, -5)$
  (d) $3x^2 + 3y^2 - 3x - 2y - 23 = 0$ at the point $(-2, -1)$.

3 Find the equation of the tangent to the circle
  $x^2 + y^2 - 2x - 4y - 3 = 0$ at each of the points where the circle cuts the x-axis.
  Find the angle between these two tangents.

4 Find the length of the tangents from the point
  (a) $(7, 8)$ to the circle $(x - 2)^2 + (y - 5)^2 = 16$
  (b) $(-4, 4)$ to the circle $x^2 + y^2 + 2x - 6y + 6 = 0$
  (c) $(7, 5)$ to the circle $3x^2 + 3y^2 - 3x - 2y - 23 = 0$.

5 Find the equation of the circle through the points
  (a) $(-1, -5)$, $(6, 2)$, $(-2, -2)$
  (b) $(1, 2)$, $(0, -6)$, $(4, 0)$
  (c) $(-5, -10)$, $(-6, -5)$, $(12, 7)$.

6 Find the equations of the two tangents from the origin to the circle
(a) $x^2 + y^2 - 4x - 2y + 4 = 0$
(b) $x^2 + y^2 - 8x - 4y + 10 = 0$.

7 Find the equation of the circle with AB as diameter when the coordinates of A and B are
(a) $(-2, 3)$ and $(3, -4)$
(b) $(6, 1)$ and $(-2, -5)$
(c) $(-1, 1)$ and $(0, -3)$
(d) $(8, -3)$ and $(-7, -1)$.

7.8 The parabola $y^2 = 4ax$

Sets of points forming straight lines and circles have been considered. Now the set of points, each of which is the same distance from a fixed point as it is from a fixed straight line, will be considered.

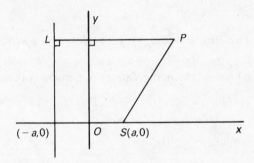

Fig. 7.22

In Fig. 7.22 the fixed point is $S(a, 0)$ and the fixed line has the equation $x = -a$ $(a > 0)$. $P(x, y)$ is any point in this set.

$$PL = PS$$

i.e.
$$x + a = \sqrt{[(x-a)^2 + y^2]}$$

or
$$(x+a)^2 = (x-a)^2 + y^2$$

which reduces to
$$y^2 = 4ax.$$

Thus this set of points can be defined as $\{(x, y): y^2 = 4ax\}$ and the cartesian equation of the curve formed by this set of points is $y^2 = 4ax$.

The fixed point $S(a, 0)$ is called the *focus*, the fixed line is called the *directrix* and the curve is a *parabola* whose *axis* is along the x-axis. Several properties of the curve $y^2 = 4ax$ are clear from its equation.

(1) The curve passes through the origin.

(2) There are no real values of y for negative values of x, i.e. no part of the curve lies to the left of the y-axis.

(3) The equation contains only an even power of y so that $y = -b$ gives the same value of x as $y = +b$ and so the curve is symmetrical about the x-axis.

(4) The y-axis is a tangent to the curve at the origin O, which is the vertex of the parabola. The curve is shown in Fig. 7.23.

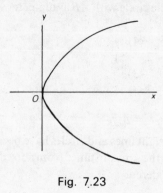

Fig. 7.23

Condition for the line $y = mx + c$ to touch the parabola $y^2 = 4ax$

Consider the line L, $y = mx + c$, and the parabola P, $y^2 = 4ax$. The points of intersection of the line and the parabola are the elements of the set $L \cap P$, where

$$L = \{(x, y): y = mx + c\} \quad \text{and} \quad P = \{(x, y): y^2 = 4ax\},$$

i.e. the set $\{(x, y): y = mx + c \text{ and } y^2 = 4ax\}$.

Solving the simultaneous equations $y = mx + c$ and $y^2 = 4ax$

gives $$(mx + c)^2 = 4ax$$
i.e. $$m^2x^2 + (2mc - 4a)x + c^2 = 0$$

This equation gives the x-coordinates of the points of intersection of the line $y = mx + c$ and the parabola $y^2 = 4ax$. If the line is a tangent to the parabola, this equation will have equal roots and then

$$(2mc - 4a)^2 - 4m^2c^2 = 0,$$
which reduces to $$mc = a \quad \text{or} \quad c = a/m.$$

Thus $y = mx + c$ is a tangent to the parabola $y^2 = 4ax$ when $mc = a$, i.e. the line $y = mx + (a/m)$ is a tangent to the parabola for all values of m.

Example 1
Find the equation of the tangent to the parabola $y^2 = 4x$ at the point $(9, -6)$.

$$\frac{d}{dx}(y^2) = \frac{d}{dy}(y^2) \times \frac{dy}{dx} = (2y)\frac{dy}{dx} \qquad \text{by the chain rule.}$$

Differentiating the equation $y^2 = 4x$ we get $2y\dfrac{dy}{dx} = 4$.

$\therefore \dfrac{dy}{dx} = \dfrac{2}{y}$ and the value of $\dfrac{dy}{dx}$ at the point $(9, -6)$ gives the gradient of the tangent at the point.

$\therefore$ at $(9, -6)$ the gradient of the tangent $= \dfrac{2}{-6} = -\tfrac{1}{3}$.

$\therefore$ the equation of the tangent at the point $(9, -6)$ is

$$y + 6 = -\tfrac{1}{3}(x - 9)$$

i.e. $\qquad\qquad\qquad x + 3y + 9 = 0.$

Example 2
Find the equations of the tangents from the point $(-3, -5)$ to the parabola $y^2 = 8x$.

Comparing $y^2 = 8x$ with $y^2 = 4ax$ we get $a = 2$ so that the line $y = mx + (2/m)$ is a tangent to the parabola $y^2 = 8x$. The tangent passes through the point $(-3, -5)$,

$$\therefore \qquad -5 = -3m + \dfrac{2}{m}, \quad \text{i.e.} \quad 3m^2 - 5m - 2 = 0,$$

$$\therefore \qquad\qquad (3m + 1)(m - 2) = 0$$

$\therefore$ the two tangents have gradients $-\tfrac{1}{3}$ and 2.
The two tangents are $y = -\tfrac{1}{3}x - 6$ and $y = 2x + 1$.

Exercises 7.8
1 Find the equation of the tangent
   (a) to the parabola $y^2 = 4x$ at the point $(4, 4)$
   (b) to the parabola $y^2 = 3x$ at the point $(12, -6)$
   (c) to the parabola $y^2 = -8x$ at the point $(-2, 4)$.
2 Find the equations of the tangents
   (a) from the point $(-4, 0)$ to the parabola $y^2 = 4x$
   (b) from the point $(-1, 1)$ to the parabola $y^2 = 3x$
   (c) from the point $(2, 1)$ to the parabola $y^2 = -12x$.
3 Sketch the parabolas whose equations are
   (a) $y^2 = 8x$ (c) $y^2 = 4(x - 2)$ (e) $y^2 = 5(3 - x)$.
   (b) $y^2 = -4x$ (d) $y^2 = 3(x + 2)$
4 Find the coordinates of the focus and the equation of the directrix
   (a) of the parabola $y^2 = 4x$ (c) of the parabola $x^2 = 8y$
   (b) of the parabola $y^2 = 12x$ (d) of the parabola $x^2 + 2y = 0$.

7.9 Parameters

In section 7.6 the circle with radius r and centre at the origin was described as the set of points $\{(x, y) : x^2 + y^2 = r^2\}$.

Since $\cos^2 \theta + \sin^2 \theta = 1$ for all values of θ, the coordinates of the point

$(r \cos \theta, r \sin \theta)$ will always satisfy the equation $x^2 + y^2 = r^2$. Thus an alternative description of this circle would be the set of points $\{(x, y) : x = r \cos \theta, y = r \sin \theta\}$, where $0 \leqslant \theta < 2\pi$. Instead of describing the circle by its cartesian equation $x^2 + y^2 = r^2$ it can be described by $x = r \cos \theta$, $y = r \sin \theta$, i.e. by expressing the coordinates of any point on the circle in terms of a third variable θ which is called a parameter.

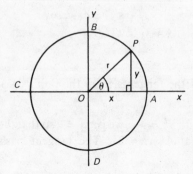

Fig. 7.24

In Fig. 7.24, as θ increases from 0 to 2π, the point $P(r \cos \theta, r \sin \theta)$ moves from A through B, C and D back to A. For each value of parameter θ there will be one point on the circle.

The parabola $y^2 = 4ax$ can also be described in parametric form, i.e. the cartesian coordinates of any point on the parabola can be expressed in terms of a parameter.

If a point has coordinates $x = at^2$, $y = 2at$, then for all values of t, $y^2 = 4ax$ and so the point lies on the parabola for all values of t.

Thus the parabola whose cartesian equation is $y^2 = 4ax$ can be expressed in parametric form as $x = at^2$, $y = 2at$, i.e. in terms of the parameter t. For each value of t there will be a corresponding point on the parabola.

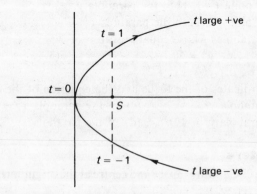

Fig. 7.25

Tangent and normal to the parabola $y^2 = 4ax$ at the point $P(ap^2, 2ap)$

In Fig. 7.26, $P(ap^2, 2ap)$ is a point on the parabola $y^2 = 4ax$, PT is the tangent at P and PN is the normal at P.

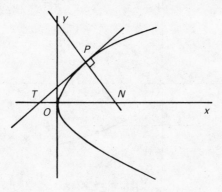

Fig. 7.26

$$x = at^2 \quad \therefore \quad \frac{dx}{dt} = 2at$$

$$y = 2at \quad \therefore \quad \frac{dy}{dt} = 2a$$

$$\therefore \quad \frac{dy}{dx} = \frac{dy}{dt} \div \frac{dx}{dt} = \frac{2a}{2at} = \frac{1}{t} \quad (t \neq 0).$$

At the point P, the parameter t takes the value p, and so

$$\frac{dy}{dx} = \frac{1}{p} = \text{gradient of the tangent at } P.$$

$\therefore$ the equation of the tangent at P is $y - 2ap = \dfrac{1}{p}(x - ap^2)$

i.e. $$py - x = ap^2.$$

The gradient of the tangent at P is $1/p$ and so the gradient of the normal at P is $-p$ since for perpendicular lines $m_1 m_2 = -1$.

$\therefore$ the equation of the normal at P is $y - 2ap = -p(x - ap^2)$

i.e. $$y + px = 2ap + ap^3.$$

The following examples will illustrate the use of parameters and will introduce some of the properties of the parabola.

Example 1
Find the point of intersection, A, of the tangents to the parabola $y^2 = 4ax$ at the points $P(ap^2, 2ap)$ and $Q(aq^2, 2aq)$. If the angle $PAQ = 90°$, show that
(i) the x-coordinate of the point A is $-a$

(ii) *PQ* passes through the focus of the parabola.

This situation is shown in Fig. 7.27.

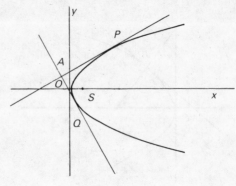

Fig. 7.27

The equation of the tangent at P is $\quad py - x = ap^2$
The equation of the tangent at Q is $\quad qy - x = aq^2$
Subtracting gives $\quad\quad\quad\quad\quad (p-q)y = a(p^2-q^2).$
Dividing by $(p-q)$, since $p \neq q$, gives $y = a(p+q)$ and then $x = apq$.
Thus the coordinates of the point A are $[apq, a(p+q)]$.

(i) When the angle $PAQ = 90°$,

$$\text{(gradient of } AP) \times (\text{gradient of } AQ) = \left(\frac{1}{p}\right) \times \left(\frac{1}{q}\right) = -1, \text{ i.e. } pq = -1.$$

Then the x-coordinate of A is $-a$ for all values of p and q. Thus when the tangents at P and Q are perpendicular, the point A always lies on the line $x = -a$, the directrix.

As P and Q move on the parabola whilst the angle PAQ remains a right angle, the set of points A forms the line $x = -a$, i.e. the locus of A is the directrix of the parabola.

(ii) Gradient of the chord $PQ = \dfrac{2ap - 2aq}{ap^2 - aq^2} = \dfrac{2}{p+q}$ since $p \neq q$.

$\therefore$ the equation of the chord PQ is $y - 2ap = \dfrac{2}{p+q}(x - ap^2)$

i.e. $\quad\quad\quad\quad\quad\quad\quad 2x = (p+q)y - 2apq.$

When the angle $PAQ = 90°$, $pq = -1$ and then when $y = 0$, $x = a$, i.e. the chord PQ passes through the point $S(a, 0)$ which is the focus of the parabola.

Example 2
The tangent at the point $P(at^2, 2at)$ to the parabola $y^2 = 4ax$ cuts the y-axis at A

and the x-axis at B. Find the locus, as t varies, of the centroid of the triangle AOB.

This situation is shown in Fig. 7.28.

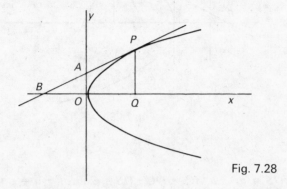

Fig. 7.28

The equation of the tangent at P is $ty - x = at^2$. This tangent cuts the y-axis at A where $x = 0$ and $y = at$, i.e. at $A(0, at)$. This tangent cuts the x-axis at B where $y = 0$ and $x = -at^2$, i.e. at $B(-at^2, 0)$. Note that $BO = OQ$ and hence that $BA = AP$ and $AO = \frac{1}{2}PQ$.

The coordinates of the centroid G are $[\frac{1}{3}(-at^2), \frac{1}{3}(at)]$. At G, $x = -\frac{1}{3}at^2$ and $y = \frac{1}{3}at$. This gives in parametric form the curve formed by the set of positions of G as t varies, i.e. as P takes different positions on the original parabola. To get the cartesian equation of this curve on which G lies we must find the relationship between the coordinates of G which is true for all values of t, i.e. we must eliminate t.

$y = \frac{1}{3}at$ and so we put $t = 3y/a$ in $x = -\frac{1}{3}at^2$ to get

$$x = -\frac{a}{3}\left[\frac{3y}{a}\right]^2, \text{ i.e. } y^2 = -\frac{1}{3}ax$$

which is the cartesian equation of the locus of G. Note that the locus of G is another parabola, which lies to the left of the y-axis.

Example 3
The tangent at the point $P(at^2, 2at)$ to the parabola $y^2 = 4ax$ cuts the x-axis at B and the normal at P cuts the x-axis at C. Show that $BS = SP = SC$, where S is the focus $(a, 0)$.

In Fig. 7.29, PQ is the perpendicular from P to the x-axis and PL is the perpendicular from P to the directrix, $x = -a$. From the previous example, B is the point $(-at^2, 0)$. The equation of the normal at P is $y + tx = 2at + at^3$. This normal cuts the x-axis where $y = 0$ and $x = 2a + at^2$, i.e. C is the point $(2a + at^2, 0)$. (Note that $QC = OC - OQ = 2a + at^2 - at^2 = 2a$.)

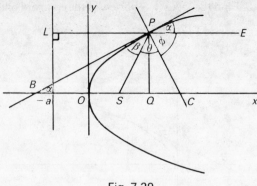

Fig. 7.29

$BS = BO + OS = at^2 + a$, $SC = OC - OS = 2a + at^2 - a = a + at^2$,
i.e. $\qquad\qquad\qquad\qquad BS = SC$.

From our definition of the parabola $SP = PL = a + at^2$

$\therefore\qquad\qquad\qquad\qquad SP = BS = SC$.

The triangle SBP is therefore isosceles giving $\alpha = \beta$. Then $\theta = \phi$.

Thus a ray of light travelling from the focus S to any point P on a mirror in the form of this parabola would be reflected along PE, i.e. parallel to the axis of the parabola. This property is utilised in the design of parabolic reflectors.

Example 4

The normal at $P(ap^2, 2ap)$ to the parabola $y^2 = 4ax$ meets the parabola again at Q. Find the coordinates of Q and the coordinates of M, the mid-point of PQ.

Let the coordinates of Q be $(aq^2, 2aq)$. The equation of the normal at P is

$$y + px = 2ap + ap^3.$$

The point Q is on this line

$\therefore\qquad\qquad 2aq + apq^2 = 2ap + ap^3$
$\therefore\qquad\qquad\qquad 2q - 2p = p(p^2 - q^2)$.
Dividing by $(p - q)$, since $p \neq q$, $\quad -2 = p(p + q)$

giving $\qquad\qquad\qquad\qquad q = -\dfrac{(p^2 + 2)}{p}$

$\therefore$ the coordinates of Q are $\left(\dfrac{a(p^2 + 2)^2}{p^2}, \dfrac{-2a(2 + p^2)}{p} \right)$.

The coordinates of M are $[\frac{1}{2}(ap^2 + aq^2), \frac{1}{2}(2ap + 2aq)]$

i.e. $\qquad\qquad\qquad [a(p^4 + 2p^2 + 2)/p^2, -2a/p]$.

Other curves may be expressed in parametric form and the following example illustrates the use of parameters with a well-known curve, the asteroid.

Example 5
A curve is given parametrically by the equations

$$x = a \cos^3 \theta, \quad y = a \sin^3 \theta, \quad 0 \leqslant \theta < 2\pi.$$

The tangent at the point P given by $\theta = \alpha$ cuts the x-axis at A and the y-axis at B. Show that, for all positions of P, the length AB is constant.

$$x = a \cos^3 \theta \quad \therefore \quad \frac{dx}{d\theta} = -3a \cos^2 \theta \sin \theta,$$

$$y = a \sin^3 \theta \quad \therefore \quad \frac{dy}{d\theta} = 3a \sin^2 \theta \cos \theta.$$

$$\therefore \quad \frac{dy}{dx} = \frac{3a \sin^2 \theta \cos \theta}{-3a \cos^2 \theta \sin \theta} = -\tan \theta$$

$$= -\tan \alpha \quad \text{at } P \text{ where } \theta = \alpha.$$

$\therefore$ the equation of the tangent at P is

$$y - a \sin^3 \alpha = -\tan \alpha (x - a \cos^3 \alpha)$$

which simplifies to $y \cos \alpha + x \sin \alpha = a \sin \alpha \cos \alpha$,

so that at A, where $y = 0$, $\quad x = a \cos \alpha$
and $\quad$ at B, where $x = 0$, $\quad y = a \sin \alpha$.
Thus the coordinates of A and B are $(a \cos \alpha, 0)$ and $(0, a \sin \alpha)$

and $\qquad\qquad\qquad AB^2 = a^2 \cos^2 \alpha + a^2 \sin^2 \alpha = a^2$.

$\therefore \;\; AB = a$ for all values of α and the length AB is constant for all positions of P.

Exercises 7.9
The following exercises relate to Fig. 7.30 overleaf in which the equation of the parabola is $y^2 = 4ax$, S is the focus, BAP is the tangent at $P(ap^2, 2ap)$ and $CPDQ$ is the normal at P. Answers to these questions should be given in terms of a and p.

1 Find the coordinates of the points A, B, C and D.
2 Find the coordinates of the point of intersection of
    (a) AD and PS $\qquad$ (b) OP and AN.
3 Find the lengths of BP, RC, AR and PD.
4 Find the areas of the triangles AOB, COD, BSP and SPD.
5 Find the area of the quadrilateral $AODP$.
6 Find the coordinates of the centroids of the triangles APC and SPD.
7 Find the angle SAP and the tangent of the angle SPD.

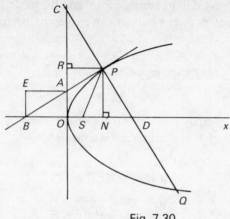

Fig. 7.30

8 Find the cartesian equation of the locus of
 (a) the vertex E of the rectangle $AOBE$
 (b) the centroid of the triangle SPD.
9 Find the equations of the tangent and normal at the point where $\theta = \alpha$ to the curve which is given parametrically by the equations
 (a) $x = a\cos\theta,$ $y = a\sin\theta,$ $0 \leqslant \theta < 2\pi$
 (b) $x = a\cos\theta,$ $y = b\sin\theta,$ $0 \leqslant \theta < 2\pi$
 (c) $x = a(\theta - \sin\theta), y = a(1 - \cos\theta), 0 \leqslant \theta < 2\pi.$
10 Find the equations of the tangent and normal at the point where $t = p$ to the curve which is given parametrically by the equations
 (a) $x = ct,$ $y = c/t,$
 (b) $x = t^2,$ $y = 2t^3.$

7.10 Polar coordinates

The position of the point P in Fig. 7.31 can be defined by its cartesian coordinates (x, y). The position of P could also be defined by the distance r and the angle θ. These are the *polar* coordinates (r, θ) of P, where r is the distance from O, called

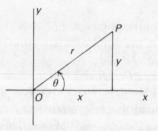

Fig. 7.31

the *pole*, and the angle θ is measured anti-clockwise from Ox to OP. The line Ox is called the *initial line*.

Just as the position of a point may be described by its polar coordinates so a set of points forming a line may be defined by the relationship satisfied by the polar coordinates of all the points in the set.

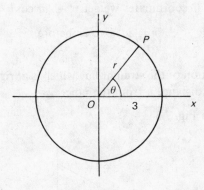

Fig. 7.32

Figure 7.32 shows the circle which has the cartesian equation $x^2 + y^2 = 9$. Clearly for all positions of P, $r = 3$. Thus the circle could be described as the set of points $\{(r, \theta) : r = 3\}$, i.e. the polar equation of this circle is $r = 3$.

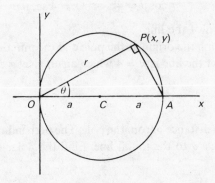

Fig. 7.33

Consider the circle shown in Fig. 7.33. This circle has its centre at the point with cartesian coordinates $(a, 0)$ and its radius is a. P is any point on this circle and has cartesian coordinates (x, y) and polar coordinates (r, θ).

Since OA is a diameter, the angle OPA is a right angle. In the triangle OPA, $r = 2a \cos \theta$ which is true for all positions of P, i.e. the polar equation of this circle is $r = 2a \cos \theta$. This equation could also be obtained from the cartesian equation of the circle, which is

$$(x - a)^2 + y^2 = a^2.$$

i.e.
$$x^2 + y^2 - 2ax = 0.$$

From Fig. 7.31, $x = r \cos \theta$, $y = r \sin \theta$.

Converting into polar coordinates we get $r^2 - 2ar \cos \theta = 0$

i.e.
$$r = 2a \cos \theta \text{ as before.}$$

Example 1

Find the polar equation of the straight line which is at right angles to the initial line and cuts it at a distance d from the pole.

The line is shown in Fig. 7.34.

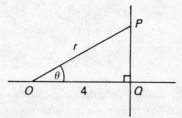

Fig. 7.34

In the triangle OPQ
$$\cos \theta = \frac{4}{r},$$

i.e.
$$r \cos \theta = 4 \quad \text{or} \quad r = 4 \sec \theta$$

is the polar equation of the line.

Alternatively, taking the origin at the pole and the initial line as the x-axis, the cartesian equation of the line is $x = 4$ which again gives $r \cos \theta = 4$ as the polar equation.

Example 2

A straight line is at a distance p from the pole. The perpendicular from the pole to the line is at an angle α to the initial line. Find the polar equation of the line.

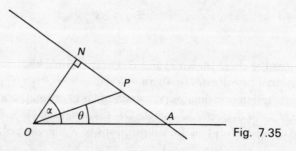

Fig. 7.35

The line is shown as APN in Fig. 7.35.

In the triangle OPN $\qquad \cos(\alpha - \theta) = \dfrac{p}{r}$,

i.e. $\qquad\qquad r = \dfrac{p}{\cos(\alpha - \theta)}$ or $r = p \sec(\alpha - \theta)$

is the polar equation of the line

Exercises 7.10

1 Find the polar equations of the lines which are at right angles to the initial line and cut it at the points $(d, 0)$ and (d, π).
2 Find the polar equations of the lines which are parallel to the initial line and pass through the points $(d, \pi/2)$ and $(d, -\pi/2)$.
3 Find the polar equations of the lines which are at an angle of $45°$ to the initial line and cut it at (a) the pole, (b) at a distance 3 from the pole.
4 Find the polar equations of the lines which are at an angle of $120°$ to the initial line and cut it at the points $(4, 0)$ and $(4, \pi)$.
5 Find the polar equation of the circle with radius a and with its centre at the point with polar coordinates (a, π).
6 Find the polar equation of the parabola $y^2 = 4ax$ taking Ox as the initial line and the pole at (a) the origin, (b) the focus of the parabola.

7.11 Curves of the form $r = f(\theta)$

The polar coordinates of a point, unlike the cartesian coordinates, are not unique. A point may be defined by polar coordinates $(4, \frac{1}{3}\pi)$ or equally well by $(4, 2\pi + \frac{1}{3}\pi)$, $(4, 4\pi + \frac{1}{3}\pi)$, ... or by $(4, \frac{1}{3}\pi - 2\pi)$, $(4, \frac{1}{3}\pi - 4\pi)$, ...

The angle θ may be positive or negative, positive angles being measured anti-clockwise from the initial line and negative angles being measured clockwise from the initial line.

The question of positive and negative values of r is less clear. Some define r as the magnitude of OP and hence allow only positive values of r. Others allow r to have both positive and negative values, positive in the direction OP and negative in the direction PO. The difference between these two schools of thought will be seen by considering the polar equation $\theta = \frac{1}{4}\pi$.

If r is taken to have only positive values (i.e. $r \in \mathbb{R}^+$), then this polar equation, $\theta = \frac{1}{4}\pi$, represents the line shown in Fig. 7.36, i.e. the line lies wholly to the right of the pole O.

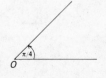

Fig. 7.36

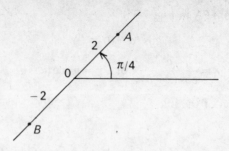

Fig. 7.37

If however r may take both positive and negative values (i.e. $r \in \mathbb{R}$), then the polar equation, $\theta = \frac{1}{4}\pi$, represents the line shown in Fig. 7.37. The meaning of positive and negative values of r is illustrated by the points A and B. The point A has polar coordinates $(2, \frac{1}{4}\pi)$ and the point B has polar coordinates $(-2, \frac{1}{4}\pi)$. Thus when the convention is that r may take both positive and negative values, the line $\theta = \frac{1}{4}\pi$ is the complete line of which BA in Fig. 7.37 is a segment, i.e. with this convention the line can lie both to the left and right of the pole.

The sketching of curves with polar equations is illustrated in the following examples. When sketching such curves the convention with regard to values of r should be stated to avoid confusion.

Example 1
Sketch the curve with polar equation $r = 3(1 + \cos\theta)$ for $0 \leqslant \theta < 2\pi$.

The values of r and θ are shown in the following table.

θ	0	$\pi/6$	$\pi/3$	$\pi/2$	$2\pi/3$	$5\pi/6$	π	$7\pi/6$	$4\pi/3$	$3\pi/2$	$5\pi/3$	$11\pi/6$	2π
$\cos\theta$	1	0·87	0·5	0	−0·5	−0·87	−1	−0·87	−0·5	0	0·5	0·87	1
r	6	5·61	4·5	3	1·5	0·39	0	0·39	1·5	3	4·5	5·61	6

The curve is shown in Fig. 7.38.

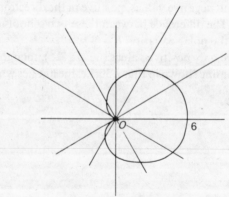

Fig. 7.38

From the table it will be seen that $r \geqslant 0$ for all values of θ and consequently in this case both conventions about the values of r produce the same curve.

Example 2

Sketch the curve whose polar equation is $r = 4 \sin 2\theta$ for $0 \leqslant \theta < \pi$.

The values of r and θ are shown in the following table.

θ	0	$\frac{\pi}{12}$	$\frac{\pi}{6}$	$\frac{\pi}{4}$	$\frac{\pi}{3}$	$\frac{5\pi}{12}$	$\frac{\pi}{2}$	$\frac{7\pi}{12}$	$\frac{2\pi}{3}$	$\frac{3\pi}{4}$	$\frac{5\pi}{6}$	$\frac{11\pi}{12}$	π
2θ	0	$\frac{\pi}{6}$	$\frac{\pi}{3}$	$\frac{\pi}{2}$	$\frac{2\pi}{3}$	$\frac{5\pi}{6}$	π	$\frac{7\pi}{6}$	$\frac{4\pi}{3}$	$\frac{3\pi}{2}$	$\frac{5\pi}{3}$	$\frac{11\pi}{6}$	2π
$\sin 2\theta$	0	0·5	0·87	1	0·87	0·5	0	−0·5	−0·87	−1	−0·87	−0·5	0
r	0	2	3·48	4	3·48	2	0	−2	−3·48	−4	−3·48	−2	0

Using the convention that r may take both positive and negative values, i.e. $r \in \mathbb{R}$, the curve is shown in Fig. 7.39.

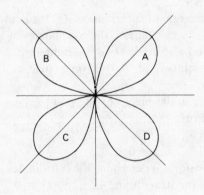

Fig. 7.39

The 'petal' marked A is given by values of θ between 0 and $\frac{1}{2}\pi$.
The 'petal' marked D is given by values of θ between $\frac{1}{2}\pi$ and π.
The 'petal' marked C is given by values of θ between π and $1\frac{1}{2}\pi$.
The 'petal' marked B is given by values of θ between $1\frac{1}{2}\pi$ and 2π.
Note that when θ is between $\frac{1}{2}\pi$ and π, the value of r is negative also that when θ is between $1\frac{1}{2}\pi$ and 2π, the value of r is negative.

Thus using the convention that r may not take negative values, i.e. $r \geqslant 0$, there would be no part of the curve in these intervals and the curve would then consist only of 'petals' A and C.

Exercises 7.11

1 Plot the points with polar coordinates, given that $r \in \mathbb{R}$,

(a) $(6, 30°)$ (f) $(-3, -135°)$ (k) $(-1, \frac{3}{4}\pi)$

(b) $(2, 150°)$ (g) $(3, \frac{1}{4}\pi)$ (l) $(-2, -\frac{1}{3}\pi)$

(c) $(3, 225°)$ (h) $(5, \frac{2}{3}\pi)$ (m) $(-3, -\frac{1}{2}\pi)$

(d) $(-1, 45°)$ (i) $(4, 1\frac{3}{4}\pi)$ (n) $(-4, -\frac{1}{4}\pi)$.

(e) $(-2, 240°)$ (j) $(3, -\frac{3}{4}\pi)$

2 Sketch, for $0 \leqslant \theta < 2\pi$, the curve with polar equation $r = \sin \theta$, given that $r \in \mathbb{R}$.

3 Sketch, for $0 \leqslant \theta < \pi$, the curve with polar equation $r = 3 \cos 2\theta$, given that $r \geqslant 0$.

4 Sketch, for $0 \leqslant \theta < 2\pi$, the curve with polar equation $r = 4 \sec (\frac{1}{4}\pi - \theta)$, given that $r \in \mathbb{R}$.

5 Sketch, for $0 \leqslant \theta < 2\pi$, the curve with polar equation $r = 1 + \cos \theta$.

6 Sketch, for $0 \leqslant \theta < 2\pi$, the curve with polar equation $r = 1 - \cos \theta$.

7 Sketch, for $0 \leqslant \theta \leqslant 4\pi$, the curve with polar equation $r = 2\theta$.

Miscellaneous exercises 7

1 The point O is the origin of coordinates, A is the point $(-5, 5)$ and C is the point $(7, 1)$. Find, by calculation, the coordinates of the point B such that both $AB = BC$ and also angle OAB is a right angle.

Prove that the points O, A, B, C lie on a circle, and find its equation.

[C]

2 The point (x_2, y_2) is the image of the point (x_1, y_1) in the straight line $ax + by + c = 0$. Prove that

(i) $a(x_1 + x_2) + b(y_1 + y_2) + 2c = 0$

(ii) $a(y_1 - y_2) - b(x_1 - x_2) = 0$.

Use these expressions to determine the coordinates of B, the image of the point $A(3, -1)$ in the straight line $4x - 2y + 11 = 0$.

State the area of the square having AB as a diagonal and determine the coordinates of the other vertices. [AEB]

3 The following values of x and y are believed to obey a law of the form

$y = \dfrac{a}{bx + c}$ where a, b and c are constants. Show that they do approximately

obey this law and hence estimate the value of the ratios $a : b : c$.

x	0	1	2	3	4
y	1·00	0·67	0·50	0·40	0·33

[AEB]

4 The following values of x and y are believed to obey a law of the form $ax^2 - by^2 = 4$, where a and b are constants. Show graphically that they do approximately obey this law and estimate the values of a and b.

x	0·50	1·00	1·50	2·00	2·50
y	2·80	2·70	2·54	2·30	1·96

[AEB]

5   The vertices of triangle ABC are $A(-16, 0)$, $B(9, 0)$ and $C(0, 12)$. Prove that the equation of the internal bisector of the angle A of the triangle is $x - 3y + 16 = 0$.

Find the equation of the internal bisector of the angle B of the triangle.

Hence, or otherwise, find the equation of a circle which touches all three sides of the triangle.   [C]

6   Find the equation of the circle, radius a and centre $C(a, b)$. Given that $b > a$, obtain the equations of the tangents from the origin O to the circle.

Calculate the coordinates of the points of contact A and B of these tangents and the circle.

By using the fact that the quadrilateral $OACB$ is cyclic, or otherwise, show that

$$\frac{\pi}{2} + \tan^{-1}\left(\frac{b^2 - a^2}{2ab}\right) = 2\tan^{-1}\left(\frac{b}{a}\right).$$ [AEB]

7   Find the equation of the circle S which passes through $A(0, 4)$ and $B(8, 0)$ and has its centre on the x-axis. If the point C lies on the circumference of S, find the greatest possible area of the triangle ABC.   [L]

8   Obtain the equation of the straight line through the points $A(1, -2)$, and $B(6, 8)$, and find the coordinates of the points P and Q on this line which divide AB internally and externally in the ratio $3:2$.

Obtain the equation of the locus of the point R which moves in such a way that $AR:BR = 3:2$, and show that this locus is a circle, with its centre at the mid-point of PQ.

Find the coordinates of the centre of the circle through P and Q such that the tangents to this circle from the origin are of length 12.   [L]

9   Prove that the point $B(1, 0)$ is the mirror-image of the point $A(5, 6)$ in the line $2x + 3y = 15$.

Find the equation of
(a) the circle on AB as diameter
(b) the circle which passes through A and B and touches the x-axis.   [L]

10   Find the equation of the circle which has its centre at the point of intersection of the common tangents to the circles $x^2 + y^2 = 9$, $x^2 + y^2 - 8x + 12 = 0$, and which passes through the points of intersection of the given circles.   [L]

11   Two circles, which touch both the x-axis and the line $3x - 4y + 3 = 0$, have their centres on the line $x + y = 3$. Show that the equation of one of these circles is

$$x^2 + y^2 - 4x - 2y + 4 = 0$$

and find the equation of the other.

Find also the equation of the second tangent from the origin to the circle whose equation is given above.   [L]

12 Show that the points $(-1, 0)$ and $(1, 0)$ are on the same side of the line $y = x - 3$.

    Find the equations of the two circles each passing through the points $(-1, 0)$, $(1, 0)$ and touching the line $y = x - 3$.   [L]

13 A square is inscribed in the circle $x^2 + y^2 - 2x + 4y + 1 = 0$. Find the area of the square.

    Also find the equations of the tangents to the circle from the origin.

    [L]

14 Show that, if the line $y = mx + c$ is a tangent to the circle $x^2 + y^2 = a^2$, then $c^2 = a^2(1 + m^2)$.

    Find the equations of the tangents from the point $(2, 6)$ to the circle $x^2 + y^2 = 4$ and find the coordinates of the points where these tangents touch the circle.   [L]

15 If P, Q are the points (x_1, y_1), (x_2, y_2), show that the equation of the circle on PQ as diameter is

$$(x - x_1)(x - x_2) + (y - y_1)(y - y_2) = 0.$$

If the tangents from the origin O touch the circle $x^2 + y^2 - 8x - 4y + 10 = 0$ at A and B, find the equation of the circle OAB and the equation of the line AB.   [L]

16 A circle whose centre is in the first quadrant touches the x-axis at $(3, 0)$ and also touches the line $4y = 3x + 36$. Find the equation of the circle and the coordinates of its point of contact with this line.

    A second circle touches the x-axis, the line $4y = 3x + 36$, and the first circle. Find all possible values of the radius of the second circle.   [L]

17 A circle with centre P and radius r touches externally both the circles $x^2 + y^2 = 4$ and $x^2 + y^2 - 6x + 8 = 0$. Prove that the x-coordinate of P is $\frac{1}{3}r + 2$, and that P lies on the curve

$$y^2 = 8(x - 1)(x - 2).$$

    [L]

18 Find the equation of the tangent to the parabola $y^2 = 4ax$ at the point $P(at^2, 2at)$.

    The line through the origin, O, parallel to this tangent, meets the parabola again at Q. Show that the line through P, parallel to the axis of the parabola, passes through the mid-point of OQ.

    Show also that, if the tangent and normal at P meet the x-axis at T and N respectively, the area of the triangle TPN is $2a^2t(1 + t^2)$.   [L]

19 The tangents to the parabola $y^2 = 4ax$ at the points $P(ap^2, 2ap)$ and $Q(aq^2, 2aq)$ meet at R. Find the coordinates of the point R. Show the area of the triangle PQR is

$$|\tfrac{1}{2}a^2(p - q)^3|.$$

    If the points P and Q move so that the area of the triangle PQR is $4a^2$, find the equation of the locus of R.   [L]

20 Points $P(ap^2, 2ap)$ and $Q(aq^2, 2aq)$ lie on the parabola $y^2 = 4ax$ and O is the origin. Show that if the angle POQ is a right angle then $pq = -4$. If P and Q vary so that this condition is satisfied, find the equation of the locus of the mid-point of PQ. State the nature of the locus. [L]

21 Prove that the equation of the chord joining the points $P(ap^2, 2ap)$ and $Q(aq^2, 2aq)$ of the parabola $y^2 = 4ax$ is

$$2x - (p+q)y + 2apq = 0.$$

S is the focus of the parabola and M is the mid-point of PQ. The line through S perpendicular to PQ meets the directrix at R. Prove that

$$2RM = SP + SQ. \qquad\qquad [L]$$

22 Obtain the equation of the chord joining the points $P(at_1^2, 2at_1)$, $Q(at_2^2, 2at_2)$ on the parabola $y^2 = 4ax$ and find the coordinates of the point of intersection of the tangents at P and Q.

Find the locus of this point of intersection
(a) if the chord PQ always passes through the point (a, a)
(b) if the chord PQ always touches the parabola $y^2 = 2ax$. [L]

23 The parabolas $x^2 = 4ay$ and $y^2 = 4ax$ meet at the origin and at the point P. The tangent to $x^2 = 4ay$ at P meets $y^2 = 4ax$ again at A, and the tangent to $y^2 = 4ax$ at P meets $x^2 = 4ay$ again at B. Prove that the angle APB is arctan $(3/4)$, and the AB is a common tangent to the two parabolas. [L]

24 Find the coordinates of the vertex and the focus of each of the two parabolas $y^2 + 4ax = 8a^2$, $y^2 - 4ax = 4a^2$.

Find the coordinates of the points in which these parabolas intersect, and find the acute angle between the tangents to these parabolas at a point of intersection. [L]

25 If the normal at $P(ap^2, 2ap)$ to the parabola $y^2 = 4ax$ meets the curve again at $Q(aq^2, 2aq)$, prove that $p^2 + pq + 2 = 0$.

Prove that the equation of the locus of the point of intersection of the tangents to the parabola at P and Q is

$$y^2(x + 2a) + 4a^3 = 0. \qquad\qquad [L]$$

26 Show that the equation of the normal to the parabola $y^2 = 4ax$ at the point $P(at^2, 2at)$ is $y + tx = 2at + at^3$.

The normal at P meets the x-axis at G and the mid-point of PG is N.
(i) Find the equation of the locus of N as P moves on the parabola.
(ii) The focus of the parabola is S. Prove that SN is perpendicular to PG.
(iii) If the triangle SPG is equilateral, find the coordinates of P. [AEB]

27 Find the equation of the normal to the parabola $y^2 = 4x$ at the point $P(t^2, 2t)$. If the normal meets the parabola again at Q show that $PQ = [4(1 + t^2)^{3/2}]/t^2$.

If PQ subtends a right angle at the focus S of the parabola, show that $t^6 - 2t^4 - 7t^2 - 4 = 0$. Verify that $t^2 = 4$ satisfies this equation and hence show that, if the perpendicular from S to PQ meets PQ at R, then $PQ = 5PR$. [AEB]

28 Sketch the graph of the parabola $x^2 = 4ay$ for $-2a \leqslant x \leqslant 2a$. State the coordinates of the focus S and the equation of the directrix.

Obtain the equation of the tangent to the curve at a point $P(2at, at^2)$.

The tangent at P meets the x-axis and the y-axis at Q and R respectively. Calculate the coordinates of Q and R and obtain the (x, y) equation of the locus of T, the mid-point of QR.

Show that the ratio of the distance of S from the tangent at P to the distance of T from the origin is $2/t$. [AEB]

29 The tangents at $P(ap^2, 2ap)$ and $Q(aq^2, 2ap)$ to the parabola $y^2 = 4ax$, where $p > q$, meet at T. Find the coordinates of T in terms of a, p and q, and prove (in any order) that

(i) the area of triangle PTQ is, $\frac{1}{2}a^2(p-q)^3$

(ii) $\sin PTQ = \dfrac{p - q}{\sqrt{[(1+p^2)(1+q^2)]}}$. [C]

30 A variable line $y = mx + c$ cuts the fixed parabola $y^2 = 4ax$ in two points P and Q. Show that the coordinates of M, the mid-point of PQ, are

$$\left(\frac{2a - mc}{m^2}, \frac{2a}{m} \right).$$

Find one equation satisfied by the coordinates of M in each of the following cases:

(a) if the line has fixed gradient m

(b) if the line passes through the fixed point $(0, -a)$. [L]

31 A point P on the parabola $(x-a)^2 = 4ay$ has coordinates $x = a + 2at$, $y = at^2$. Find the equations of the tangent and the normal to the parabola at P.

If the tangent and normal cut the x-axis at the points T and N respectively, prove that

$$PT^2/TN = at.$$

Find the coordinates of the point Q in which the normal at P intersects the parabola again. [L]

32 The points $P(2p^2, 4p)$ and $Q(2q^2, 4q)$ lie on the parabola $y^2 = 8x$ whose vertex is O. Find the equation of the chord PQ and, by putting $q = p$ or otherwise, find the equation of the tangent at P.

Find the coordinates of R, the point of intersection of the tangents at P and Q. Find the length of the perpendicular from O to the line PQ and hence find the equation of the locus of R when the ratio of the area of the triangle RPQ to that of the triangle OPQ is 4:1. [AEB]

33 Prove that the point $(a\lambda^2, 2a\lambda)$ lies on the parabola $y^2 = 4ax$ and that the tangents to the parabola at $(a\lambda^2, 2a\lambda)$ and $(a\mu^2, 2a\mu)$ meet at the point $(a\lambda\mu, a(\lambda+\mu))$.

Find the coordinates of the centroid of the triangle with vetices the above three points, and prove that if the two tangents meet on the directrix then the locus of the centroid is a parabola. [AEB]

34 A curve C_1 has equation $2y^2 = x$; a curve C_2 is given by the parametric equations $x = 4t$, $y = 4/t$. Sketch C_1 and C_2 on the same diagram and calculate the coordinates of their point of intersection, P.

The tangent to C_1 at P crosses the x-axis at T and meets C_2 again at Q. Show that T is a point of trisection of PQ. [L]

35 (i) Sketch the curve given by the parametric equations

$$x = t+1, \quad y = 4-t^2.$$

Find the equation of the normal to this curve at the point P at which $t = 1$. If this normal meets the curve again at Q and the x-axis at R, show that $PQ:QR = 5:7$.

(ii) Sketch the curve with polar equation $r = 3+2\cos\theta$, marking clearly the polar coordinates of the points A and B on the curve at which r has its greatest and least values respectively. [L]

36 If $x = \cos t$ and $y = \sec t$, express $\dfrac{dy}{dx}$ in terms of t.

(a) Sketch the curves given by $(x, y) = (\cos t, \sec t)$ in the ranges $0 < t < \frac{1}{2}\pi$ and $\frac{1}{2}\pi < t < \pi$, indicating clearly the variation of t along the curves.

(b) If P is the point of the curve with parameter t, where $0 < t < \frac{1}{2}\pi$, find the area of the trapezium bounded by the line $x = \cos t$, the tangent at P, and the coordinate axes. [L]

37 Sketch the curve with parametric equations $x = t^2$, $y = t^3$. Find the equation of the tangent to the curve at the point where $t = \sqrt{3}$.

If this tangent meets the x-axis at A, show that area of the finite region bounded by the curve and the line through A parallel to the y-axis is $4/5$. [L]

38 The parametric equations of a curve are

$$x = 3(2\theta - \sin 2\theta),$$
$$y = 3(1 - \cos 2\theta).$$

The tangent and the normal to the curve at the point P where $\theta = \pi/4$ meet the y-axis at L and M respectively. Show that the area of triangle PLM is $\frac{9}{4}(\pi-2)^2$. [AEB]

39 Plot the points on the curve given by the equations $x = 3\cos t$, $y = \cos 2t$, for the values $0, \pi/6, \pi/3, \ldots, \pi$ of t, and sketch the curve for these values of t.

Prove that the distance of any point of the curve from the point $(0, \frac{1}{8})$ is the same as its distance from the line $y = -\frac{17}{8}$. Obtain the parameters of the points on the curve where the tangents to it pass through the point $(0, -\frac{17}{8})$. [AEB]

40 Sketch on the same diagram the curve C_1 with polar equation $r = 2\cos\theta$ and the curve C_2 with polar equation $r = 2$.

If P is the point where the curves meet, Q is the point $(2, \pi/2)$ on C_2 and O is the pole, show that the centre of the circle OPQ lies on C_1. [L]

8 Real and complex numbers

8.1 Number systems

The following number systems will be considered briefly in turn:

$\mathbb{N}$—the set of positive integers and zero,
$\mathbb{Z}$—the set of all integers, positive and negative, including zero,
$\mathbb{Q}$—the set of rational numbers,
$\mathbb{R}$—the set of real numbers,
$\mathbb{C}$—the set of complex numbers.

It will be observed that each successive set in this list contains all of the preceding sets as subsets, so that, in set notation,

$$\mathbb{N} \subset \mathbb{Z} \subset \mathbb{Q} \subset \mathbb{R} \subset \mathbb{C},$$

and it will be noted that, on moving upward in the hierarchy of systems, equations that were insoluble in one system become soluble in some higher system.

8.2 The set $\mathbb{N}$ (positive integers and zero)

The set $\mathbb{N} = \{0, 1, 2, 3, \ldots\}$ has the following properties:

Property 1 The set $\mathbb{N}$ is closed under the binary operations of addition and multiplication, i.e. adding or multiplying any two members of $\mathbb{N}$ produces another member of $\mathbb{N}$. In set language this becomes:

If $a, b \in \mathbb{N}$ then $a+b \in \mathbb{N}$ and $ab \in \mathbb{N}$.

Property 2 Addition and multiplication are associative in $\mathbb{N}$:

$(a+b)+c = a+(b+c)$ for all $a, b, c \in \mathbb{N}$,
$(ab)c = a(bc)$ for all $a, b, c \in \mathbb{N}$.

Property 3 Addition and multiplication are commutative in $\mathbb{N}$:

$a+b = b+a$ for all $a, b \in \mathbb{N}$,
$ab = ba$ for all $a, b \in \mathbb{N}$.

Property 4 The 'cancellation property' holds for addition and multiplication in $\mathbb{N}$:

$a+b = a+c \Leftrightarrow b = c$ for all $a, b, c \in \mathbb{N}$,
$ab = ac \Leftrightarrow b = c$ for all $a, b, c \in \mathbb{N}$, except when $a = 0$.

Property 5 The set $\mathbb{N}$ is ordered: given any n elements of $\mathbb{N}$ they can be arranged so that $a_1 \leqslant a_2 \leqslant a_3 \leqslant \ldots \leqslant a_n$.

In $\mathbb{N}$ the identity element for multiplication is the element 1, and the identity element for addition is 0. The set $\mathbb{N}$ is not closed with respect to subtraction; a member x of $\mathbb{N}$ cannot be found which satisfies the equation $x + 5 = 3$. Also the set $\mathbb{N}$ is not closed with respect to division; a member x of $\mathbb{N}$ cannot be found which satisfies the equation $5x = 3$.

8.3 The integers, the set $\mathbb{Z} = \{0, \pm 1, \pm 2, \ldots\}$

An extension of the set $\mathbb{N}$ needs to be found so that an equation such as $x + a = b$, where $a > b$, is soluble in this extended set.

In the set $\mathbb{Z}$ the negative number $-a$, where $a \in \mathbb{N}$, is defined to be the additive inverse of a, so that

$$a + (-a) = (-a) + a = 0.$$

The five properties of $\mathbb{N}$ listed in section 8.2 still hold in $\mathbb{Z}$. The set $\mathbb{Z}$ is closed with respect to subtraction, but it is not closed with respect to division.

So, in $\mathbb{Z}$, the equation $x + 5 = 3$ can be solved, but the equation $5x = 3$ cannot.

8.4 The rational numbers, the set $\mathbb{Q} = \{m/n : m \in \mathbb{Z}, n \in \mathbb{Z}, n \neq 0\}$

Again, an extension of the set $\mathbb{Z}$ needs to be made to enable a solution of an equation of the form $ax = b$, where a and b have no common factor, to be found in the extended set.

A rational number may be exhibited in different forms, e.g. $2/5 = 4/10 = 6/15$. Where feasible, rational numbers will be exhibited in the form m/n where m and n have no common factor, and $m/1$ will be written as m. Thus rational numbers with unit denominators may be regarded as integers. Addition and multiplication for rational numbers are defined by the following rules:

$$\frac{a}{b} + \frac{c}{d} = \frac{ad + bc}{bd},$$

$$\frac{a}{b} \times \frac{c}{d} = \frac{ac}{bd}.$$

It follows that

$$\frac{a}{1} + \frac{c}{1} = \frac{a + c}{1} = a + c,$$

and

$$\frac{a}{1} \times \frac{c}{1} = \frac{ac}{1} = ac.$$

Note that the rules do not conflict with the rules for integers.

Rational numbers possess all the properties enumerated for integers and, in addition, have the property that, between any two given rational numbers, another rational number can always be found.

In $\mathbb{Q}$ there is both an additive identity element (0) and a multiplicative identity element (1), and the rational number m/n has the multiplicative inverse n/m. So, in $\mathbb{Q}$, both the equations $3x + 5 = 0$ and $3x = 5$ can be solved.

8.5 The real numbers, the set $\mathbb{R}$

But, as the Greek mathematicians found, there are still difficulties:

Problem: Find a member of $\mathbb{Q}$ which gives the length of the diagonal of a square of unit side (i.e. find a rational number whose square is 2, or, find integers a and b so that $b^2/a^2 = 2$).

It can be shown that 2 is *not* the square of any rational number.

Suppose $2 = b^2/a^2$ where b and a have no common factor.

Then $b^2 = 2a^2$.

So b^2 must be an even integer.

So b must be an even integer.

So $b = 2c$ where c is an integer.

So $4c^2 = 2a^2$ i.e. $a^2 = 2c^2$ and thus a is an even integer.

Thus a and b are both even integers, which is contrary to the assumption that a and b have no common factor. So the supposition that $2 = b^2/a^2$ must have been false, i.e. no rational number exists which has the desired property.

(This proof is an example of 'proof by contradiction'–a method often used in more advanced work.)

The Greeks called a number such as x, where $x^2 = 2$, 'incommensurable'. We call it an 'irrational' number. The Greeks found for this number a sequence of successive approximations, but they declined to regard the limit of this sequence as a number (see below). The limit of this sequence is denoted by the symbol $\sqrt{2}$, so that $\sqrt{2} \times \sqrt{2} = 2$. Thus, though $\sqrt{2}$ cannot be explicitly determined, we agree to accept it as a 'real' number. Generally, if x is any positive rational number which is not a perfect square, the irrational number $\sqrt{x}$ satisfies the definition

$$\sqrt{x} \times \sqrt{x} = x \quad (x > 0).$$

Consider the sequence of rational numbers whose first four members are $3/2$, $7/5$, $17/12$, $41/29$. It will be observed that $3 + 2 = 5$, $7 + 5 = 12$, $17 + 12 = 29$, giving the pattern of the denominators, and also that $2 + 5 = 7$, $5 + 12 = 17$, $12 + 29 = 41$, giving the pattern of numerators. So, following this pattern, the next two fractions are $99/70$, $239/169$, and the sequence may be continued as far as we wish. Note that the squares of successive terms of the sequence tend to 2.

$$3^2/2^2 = 9/4 = 2 + 1/4$$
$$7^2/5^2 = 49/25 = 2 - 1/25$$
$$17^2/12^2 = 289/144 = 2 + 1/144$$
$$41^2/29^2 = 1681/841 = 2 - 1/841$$
$$99^2/70^2 = 9801/4900 = 2 + 1/4900.$$

Thus from the set of rational numbers a sequence has been formed which can be shown to have a limiting value of $\sqrt{2}$.

So the set $\mathbb{Q}$ can be used (the details are too sophisticated for a treatment at this level) to generate another set, the set of *real numbers*, $\mathbb{R}$. This set $\mathbb{R}$ is the set in which, intuitively, most of the work at this level is carried out. The set $\mathbb{R}$ contains all numbers of the form $\sqrt{x}$, where $x > 0$, and includes the numbers e and π.

All numbers which are in the set $\mathbb{R}$ but not in the set $\mathbb{Q}$ are known as *irrational numbers*. All the properties listed in $\mathbb{Q}$ hold in $\mathbb{R}$, i.e. $\mathbb{R}$ is closed under addition and multiplication, both of which operations are commutative and associative; also $\mathbb{R}$ contains an identity element for addition (0) and for multiplication (1). Every element of $\mathbb{R}$ has an inverse for addition, and every element other than 0 has an inverse for multiplication. Also the elements of $\mathbb{R}$ can be ordered. All equations of the forms $x + a = b$, $ax = b$, $x^2 = c$, $(c \geqslant 0)$ where $a, b, c \in \mathbb{R}$ can be solved in $\mathbb{R}$.

But there are still difficulties! Suppose that in the equation $x^2 = c$ we have $c < 0$, e.g. $x^2 = -8$. Less than 500 years ago, mathematicians would have said that this equation had no solution, or that it gave an 'imaginary value' for x, just as the Greeks would have spoken similarly of the equation $x^2 = 2$.

Exercises 8.5

1 To which of the sets $\mathbb{N}$, $\mathbb{Z}$, $\mathbb{Q}$, $\mathbb{R}$ do the following numbers belong?

(a) $3/2$ (c) -13 (e) $\log_{10} 2$ (g) 2.2 (i) $\log_4 2$.

(b) $\sqrt{7}$ (d) π (f) 0 (h) $2.\dot{2}$

2 To which of the sets $\mathbb{N}$, $\mathbb{Z}$, $\mathbb{Q}$, $\mathbb{R}$ do the roots of the following equations belong?

(a) $x^2 - 3x + 2 = 0$ (d) $x^2 + 7x + 12 = 0$

(b) $4x^2 - 4x + 1 = 0$ (e) $12x^2 + 7x + 1 = 0$

(c) $2x^2 + x + 5 = 0$ (f) $\tan x = 1$.

3 Show (a) that $\sqrt{3}$ is irrational (b) that $\sqrt[3]{2}$ is irrational.

4 Show that $(\sqrt{3} + \sqrt{5})$ is irrational.

5 Show that $\log_{10} 2$ is irrational, i.e. that no rational number m/n can be found such that $10^{m/n} = 2$.

6 Show that if $a + b\sqrt{2} = c + d\sqrt{2}$, where $a, b, c, d \in \mathbb{Z}$, then $a = c$ and $b = d$.

7 Show that if a, b, c, d are positive integers such that $a/b < c/d$, then

$$a/b < (a+c)/(b+d) < c/d.$$

Deduce that between any two rational numbers another rational number can be found.

8 Show that a necessary and sufficient condition for $(\sqrt{2} - 1)$ to be a root of the cubic equation $ax^3 - bx + c = 0$, where a, b and c are rational numbers, is that $a:b:c = 1:5:2$.   [JMB]

8.6 The complex numbers, the set $\mathbb{C}$

In this section a way must be found to extend the number system so that equations such as $x^2 + 5 = 0$, $3x^2 + 2x + 1 = 0$ are soluble in the extended system.

All members of $\mathbb{R}$ may be represented as points on a directed line (often taken to be the x-axis), the number 0 conventionally being placed at the origin, other points being placed so that if $x_1 > x_2$ then the point corresponding to the number x_1 lies to the right of the number corresponding to x_2.

The system can be extended by using the concept of an 'ordered pair' of real numbers, denoted by (a, b) where $a \in \mathbb{R}, b \in \mathbb{R}$. The order is important; an ordered pair is more than just a set of two real numbers.

Formal definition of the set $\mathbb{C}$

The set $\mathbb{C}$ of complex numbers is the set of ordered pairs of real numbers such that:

(i) $(a, b) = (c, d) \Leftrightarrow a = c$ and $b = d$,

(ii) $(a, b) + (c, d) = (a + c, b + d)$,

(iii) $(a, b) \times (c, d) = (ac - bd, ad + bc)$.

Just as real numbers correspond to points on the x-axis, so an ordered pair (x, y) corresponds to a point in the x–y plane having coordinates x and y.

For points on the x-axis, the above rules give:

(i) $(a, 0) = (c, 0) \Leftrightarrow a = c$,

(ii) $(a, 0) + (c, 0) = (a + c, 0)$,

(iii) $(a, 0) \times (c, 0) = (ac, 0)$.

So, if the ordered pair $(a, 0)$ is regarded as being equivalent to the real number a, the rules for combining complex numbers do not conflict with the corresponding rules for real numbers.

Now consider the ordered pairs corresponding to points on the y-axis. This yields:

(i) $(0, b) = (0, d) \Leftrightarrow b = d$,

(ii) $(0, b) + (0, d) = (0, b + d)$,

(iii) $(0, b) \times (0, d) = (-bd, 0)$.

This last result is of great importance. In particular, if $b = 1$ and $d = 1$, then

$$(0, 1) \times (0, 1) = (-1, 0).$$

But $(-1, 0)$ corresponds to the real number -1. So a particular ordered pair, when squared, gives a result of -1.

Notation

The order pair $(0, 1)$ is denoted by the symbol i (or sometimes, in engineering books, by j). Thus $i \times i = -1$.

The rule for multiplication gives

$$(0, 1) \times (b, 0) = (0, b),$$

i.e. $$i \times b = (0, b)$$

Similarly,

$$(1, 0) \times (a, 0) = (a, 0),$$

i.e. $$1 \times a = (a, 0).$$

So the ordered pair (a, b) can be expressed as follows:

$$(a, b) = (a, 0) + (0, b),$$
i.e. $$(a, b) = a + ib.$$

From now on, for simplicity, the complex number corresponding to the ordered pair (a, b) will be denoted by $a + ib$.

This notation gives a less formal definition of the set $\mathbb{C}$.

$$\mathbb{C} = \{a + ib; a \in \mathbb{R}, b \in \mathbb{R}\} \quad \text{with the rules}$$

(i) $a + ib = c + id \Leftrightarrow a = c$, and $b = d$,
(ii) $(a + ib) + (c + id) = (a + c) + i(b + d)$,
(iii) $(a + ib) \times (c + id) = (ac - bd) + i(ad + bc)$.
Clearly $(0 + i0) + (c + id) = (c + id)$,
and $(1 + i0) \times (c + id) = (c + id)$.
In future, $1 + i0$ will be written as 1, and $0 + i0$ as 0. So the elements 0 and 1 of the set are the identity elements for addition and multiplication.

The additive inverse of $(a + ib)$ is clearly $(-a + i(-b))$.
Noting that $(0 + i1) + (0 + i1) = 0 + i2$, i.e. $2(i1) = i2$, and, in general, that $m(in) = imn$ $(m, n \in \mathbb{R})$, this inverse can be written in the form $-a - ib$.

For the multiplicative inverse of $(a + ib)$, the following equation must be satisfied:

$$(a + ib)(x + iy) = 1 + i0,$$
i.e. $$(ax - by) + i(ay + bx) = 1 + i0.$$
So $$ax - by = 1,$$
$$bx + ay = 0.$$

Solving for x and y gives $x = a/(a^2 + b^2)$, $y = -b/(a^2 + b^2)$. So the multiplicative inverse of $a + ib$ is $(a - ib)/(a^2 + b^2)$. Thus we have

$$(a + ib) - (c + id) = (a - c) + i(b - d),$$
$$(a + ib) \div (c + id) = (a + ib) \times (c - id)/(c^2 + d^2).$$

So $$\frac{(a + ib)}{(c + id)} = \left(\frac{ac + bd}{c^2 + d^2}\right) + i\left(\frac{bc - ad}{c^2 + d^2}\right).$$

Thus the operations of addition, subtraction, multiplication and division (other than by zero) are closed in $\mathbb{C}$.

Furthermore, it can be shown that any polynomial equation of degree n

$$z^n + a_1 z^{n-1} + a_2 z^{n-2} + \ldots + a_{n-1} z + a_n = 0.$$

whose coefficients are elements of $\mathbb{C}$ will have n roots in $\mathbb{C}$, which is not true in any of the systems hitherto discussed.

The set $\mathbb{C}$ does not, however, possess one property common to the sets $\mathbb{N}$, $\mathbb{Z}$, $\mathbb{Q}$, $\mathbb{R}$; the elements of $\mathbb{C}$ cannot be ordered, i.e. it is meaningless to say that one complex number is 'less than' another complex number, i.e. we cannot write $a + ib < c + id$, even when $a < c$ and $b < d$.

Nomenclature

The complex number $x+iy$ is often denoted by the single symobol z, i.e.

$$z = x+iy, \quad (z \in \mathbb{C}; x, y \in \mathbb{R}).$$

(Some writers use $z = x+yi$.)

x is known as the 'real part' of z, i.e. $x = \text{Re}(z)$,

y is known as the 'imaginary part' of z, i.e. $y = \text{Im}(z)$.

(This terminology is logically indefensible, since x and y are both real numbers, but, unfortunately, it is still current.)

The 'practical algebra' of complex numbers

The rules given for addition and multiplication indicate that, if complex numbers are exhibited in the form $a+ib$, they may be handled according to the laws of algebra for real numbers, provided that wherever i^2 occurs it is replaced by -1. In numerical work it is customary, but not essential, for the i symbol to follow the number, i.e. we write $3+5i$ instead of $3+i5$.

Thus
$$\begin{aligned}(3+5i)(2+6i) &= 3(2+6i)+5i(2+6i)\\ &= 6+18i+10i-30\\ &= -24+28i\end{aligned}$$

A procedure for division of one complex number by another complex number, based on replacement of i^2 by -1, is given in the next paragraph.

Conjugate complex numbers

The complex numbers $a+ib$, $a-ib$ are said to be complex conjugates. The complex conjugate of z is denoted by z^* (or by $\bar{z}$ in older books).

Some properties of complex conjugates are:

$z+z^* = (x+iy)+(x-iy) = 2x$, (a real number),

$z-z^* = (x+iy)-(x-iy) = i2y$, (a complex number with real part zero).

Thus
$$\text{Re}(z) = (z+z^*)/2$$
$$\text{Im}(z) = (z-z^*)/(2i)$$

$zz^* = (x+iy)(x-iy) = x^2 + y^2$, (a positive real number).

It can be shown that the complex roots of a polynomial equation with real coefficients always occur in conjugate pairs. (See Miscellaneous exercises 8, No. 16.)

Other properties of complex conjugates appear in the exercises at the end of this chapter.

Complex conjugates are useful when dividing by a complex number:

$$\frac{-24+28i}{3+5i} = \frac{(-24+28i)(3-5i)}{(3+5i)(3-5i)} = \frac{68+204i}{9+25} = \frac{34(2+6i)}{34} = 2+6i$$

Note that $(3-5i)$ is the conjugate of $(3+5i)$.

Exercises 8.6

Express as a single complex number

1 $(3+4i)+(5+6i)$
2 $(2+7i)-(3+2i)$
3 $(5+1i) \times (2+6i)$ (Note: 1i is usually written as i)
4 $(5+3i) \times (2+6i)$
5 $(5+3i) \div (2+6i)$
6 $(4+32i) \div (5+i)$
7 $(4+32i) \div (2+6i)$
8 $(5+3i)^2$
9 $(1+\sqrt{2}i)^2$
10 $(-3+4i)^2$
11 $(-4+2i)+(3-7i)$
12 $(-7-6i)-(2-4i)$
13 $(3+i)(3-i)$
14 $(1+i)^2$
15 $(1+i)/(1-i)$
16 $(1-i)/(1+i)$ (The conjugate of the previous expression)
17 $(-7+24i)/(3+4i)$
18 $(1+3i)^2$
19 $(1+\sqrt{3}i)^2$
20 $(1+\sqrt{3}i)^3$
21 $1/(1+i)+1/(1-i)$
22 $1/((1+i)(2-3i))$
23 $(2+3i)/(1+2i)-(2-3i)/(1-2i)$
24 Given that $z_1 = x_1+iy_1$, $z_2 = x_2+iy_2$, show that
  (i) $z_1{}^*+z_2{}^* = (z_1+z_2)^*$
  (ii) $z_1{}^*\dot{z}_2{}^* = (z_1z_2)^*$
These two results show that the operation of forming the conjugate is commutative with both the operations of addition and multiplication.
25 Show that the equation $z^2 - 4z + 13 = 0$ has roots $2+3i$ and $2-3i$
26 Solve the equation $z^2 - 6z + 34 = 0$.
27 Show that 3 is a root of the equation $z^3 - 7z^2 + 17x - 15 = 0$ and find the other two roots of the equation.

8.7 Geometrical representation of complex numbers (the Argand diagram)

The complex number $x+iy$ is uniquely specified by the real numbers x, y in that order. Thus the ordered pair (x, y) corresponds to the complex number $x+iy$, and conversely. But the ordered pair (x, y) corresponds to a point with co-ordinates x, y in the x–y plane, and conversely. So, *to every complex number z (= $x+iy$), there corresponds uniquely a point P with coordinates x, y, and conversely.*

In Fig. 8.1 the complex number z, where $z = x + iy$, corresponds to the point $P(x, y)$.

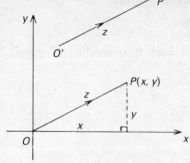

Fig. 8.1

A complex number z may be represented geometrically in three ways:
(a) by the point $P(x, y)$,
(b) by the vector $\overrightarrow{OP}$ (Fig. 8.1),
(c) by any vector $\overrightarrow{O'P'}$ which is equal to the vector $\overrightarrow{OP}$.
Any such representation of complex numbers constitutes an Argand diagram.

Addition and subtraction of complex numbers on an Argand diagram

Let
$$z_1 = x_1 + iy_1 \quad \text{and} \quad z_2 = x_2 + iy_2.$$
Then
$$z_1 + z_2 = (x_1 + x_2) + i(y_1 + y_2).$$

The points P_1 and P_2 represent z_1 and z_2 respectively.
Then $z_1 + z_2$ is represented by the point P with coordinates $(x_1 + x_2, y_1 + y_2)$.

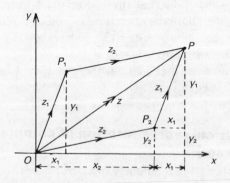

Fig. 8.2

Figure 8.2 shows that OP_2PP_1 is a parallelogram,
$$\overrightarrow{OP} = \overrightarrow{OP_1} + \overrightarrow{P_1P} = \overrightarrow{OP_2} + \overrightarrow{P_2P}.$$

So if $\overrightarrow{OP}$ represents z, then $z = z_1 + z_2$, i.e. addition of complex numbers corresponds to the addition of the related vectors.

The subtraction of complex numbers can be represented in a similar way.

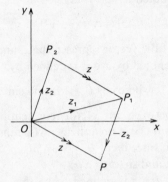

Fig. 8.3

In Fig. 8.3 $\overrightarrow{P_1P} = -\overrightarrow{OP_2} = -z_2$,

so
$$z_1 - z_2 = \overrightarrow{OP_1} + \overrightarrow{P_1P}$$
$$= \overrightarrow{OP} \qquad \text{(law of vector addition)}$$

Note that $\overrightarrow{P_2P_1} = \overrightarrow{OP}$.

So
$$\overrightarrow{P_2P_1} = z_1 - z_2,$$
(as is evident from triangle OP_2P_1), since

$$\overrightarrow{OP_2} + \overrightarrow{P_2P_1} = \overrightarrow{OP_1} \quad \text{i.e. } z_2 + (z_1 - z_2) = z_1.$$

8.8 Modulus and argument of a complex number

Let the point P represent the complex number z where $z = x + iy$. Then, in Fig. 8.4, P has cartesian coordinates (x, y) and polar coordinates (r, θ) where

$$r^2 = x^2 + y^2, \qquad (r > 0),$$
$$\cos \theta = x/r,$$
$$\sin \theta = y/r.$$

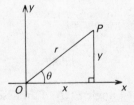

Fig. 8.4

Note that both the last two equations are necessary in order to specify θ adequately.

Definitions

Given any complex number z, where $z = x+iy$, then:

(1) The *modulus* of z, denoted by $|z|$, is defined by

$$|z| = \sqrt{(x^2+y^2)}.$$

Thus $|z|$ is the length r of the vector representing z, and $|z|^2 = zz^*$.

(2) The *argument* of z, denoted by arg z, is defined by

$$\arg z = \theta, \text{ where } \cos\theta = x/r \text{ and } \sin\theta = y/r.$$

Thus arg z is the angle between any vector representing z and the line Ox, the angle being measured from Ox in an anticlockwise sense.

Arg z has many values, since if $\cos\theta = x/r$ and $\sin\theta = y/r$ and $\phi = \theta+2n\pi$ ($n \in \mathbb{Z}$), then $\cos\phi = x/r$ and $\sin\phi = y/r$.

The *principal value* of arg z is defined to be that value of arg z satisfying $-\pi < \arg z \leqslant \pi$.

Note that arg z is (strictly speaking) the radian measure of the angle θ.

Since the position of a point is uniquely determined by its polar coordinates (or a vector by its magnitude and direction), so the complex number is uniquely specified if we are given $|z|$ and arg z. This correspondence between complex numbers and their associated vectors is of the greatest importance in the applications of complex numbers.

Note that for the complex number $0+i0$, we have $|z| = 0$ but arg z is undefined because a vector of zero length has no direction.

Example 1

If $z = -2\sqrt{3}-2i$, find $|z|$ and arg z.

Method   Represent z on an Argand diagram. (Fig. 8.5)

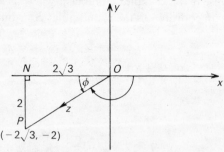

Fig. 8.5

Find the hypotenuse and base angle PON of the associated right-angled triangle. Write down $|z|$ and arg z.

Solution z is represented by $P(-2\sqrt{3}, -2)$.
So $OP^2 = (2\sqrt{3})^2 + 2^2 = 16$. Thus $OP = 4$.
$\sin \phi = \frac{1}{2}$ and thus $\phi = \pi/6$.
Thus $|z| = 4$ and $\arg z = -(\pi - \phi) = -5\pi/6 \, (+2n\pi)$.

Example 2
If $|z| = 6$, $\arg z = 2\pi/3$, express z in the form $x+iy$ (algebraic form).

Method   Represent z on an Argand diagram. (Fig. 8.6)

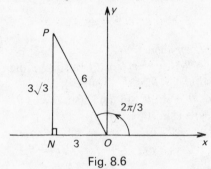

Fig. 8.6

Find the hypotenuse and base angle PON of the associated right-angled triangle.
Write down the coordinates of P.
Solution z is represented by P, with $r = 6$, $\theta = 2\pi/3$.
$ON = 6 \cos \pi/3 = 3$,
$PN = 6 \sin \pi/3 = 3\sqrt{3}$.
P is the point $(-3, 3\sqrt{3})$.
So $z = -3 + 3\sqrt{3}i$.

8.9 Trigonometric and polar forms of a complex number
If $z = x+iy$ and $|z| = r$ and $\arg z = \theta$, then from Fig. 8.4, $z = r \cos \theta + ir \sin \theta$.
Then $r(\cos \theta + i \sin \theta)$ is the trigonometric form for z and $r \angle \theta$ is the polar form
for z.
   Conversion between algebraic and trigonometric or polar forms involves the
same procedure as the determination of modulus and argument.

Example
Express in polar form $-3+4i$.

Clearly, from Fig. 8.7, $r = 5$ and $\phi = 53°$ (to the nearest degree). So $z = 5 \angle 127°$.
   Note that the trigonometric form is $5(\cos 127° + i \sin 127°)$
(this is sometimes abbreviated to 5 cis 127°.)

The polar form is very useful in the multiplication and division of complex
numbers.

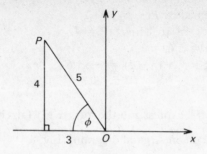

Fig. 8.7

Multiplication of complex numbers
If $z_1 = r_1(\cos \theta_1 + i \sin \theta_1)$, $z_2 = r_2(\cos \theta_2 + i \sin \theta_2)$, then
$$z_1 z_2 = [r_1(\cos \theta_1 + i \sin \theta_1)][r_2(\cos \theta_2 + i \sin \theta_2)]$$
$$= r_1 r_2[(\cos \theta_1 \cos \theta_2 - \sin \theta_1 \sin \theta_2) + i(\sin \theta_1 \cos \theta_2 + \cos \theta_1 \sin \theta_2)]$$
$$= r_1 r_2[\cos(\theta_1 + \theta_2) + i \sin (\theta_1 + \theta_2)]$$
$$= r_1 r_2 \angle (\theta_1 + \theta_2)$$
i.e. $$r_1 \angle \theta_1 \times r_2 \angle \theta_2 = r_1 r_2 \angle (\theta_1 + \theta_2).$$

This result, which depends on the addition formulae of trigonometry, is of fundamental importance.

The result can be expressed in words as follows: *To find the product of two complex numbers, multiply their moduli and add their arguments.*

Division of complex numbers
Since multiplication and division are inverse operations, it can be deduced that
$$r_1 \angle \theta_1 \div r_2 \angle \theta_2 = (r_1/r_2) \angle (\theta_1 - \theta_2),$$

i.e. to divide one complex number by another, divide the modulus of the first number by the modulus of the second and subtract the argument of the second from the argument of the first.

Geometrical interpretation of the multiplication and division of complex numbers
In Fig. 8.8 the vectors $\overrightarrow{OP_1}$, $\overrightarrow{OP_2}$ represent the complex numbers z_1, z_2 respectively, where
$$z_1 = r_1(\cos \theta_1 + i \sin \theta_1),$$
$$z_2 = r_2(\cos \theta_2 + i \sin \theta_2).$$

The vector $\overrightarrow{OP}$ represents their product, given by
$$z_1 z_2 = r_1 r_2[\cos (\theta_1 + \theta_2) + i \sin (\theta_1 + \theta_2)].$$

The effect of multiplying z_2 by z_1 is to rotate the vector $\overrightarrow{OP_2}$ through the angle θ_1 and to multiply its length by r_1. Now if N is the point $(1, 0)$, the triangles P_1ON, POP_2 will be similar, since the angles P_1ON, POP_2 are equal and

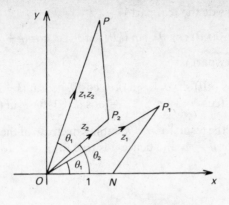

Fig. 8.8

$OP/OP_1 = OP_2/ON$. Thus if the triangle P_1ON is constructed first and then a similar triangle POP_2, the vector $\overrightarrow{OP}$ will represent the product z_1z_2.

In particular, the effect of multiplying a complex number by i is to rotate the corresponding vector anticlockwise through a right angle, since

$$i = \cos(90°) + i\sin(90°).$$

Multiplication by i^2 has the effect of rotating a vector through 180°, since

$$i^2 = -1 = \cos(180°) + i\sin(180°).$$

This should dispel any notions regarding the 'imaginary' nature of i, and reinforce the definition given in terms of ordered pairs.

Example 1
The square $ABCD$ has its centre at the origin O, and the vector $\overrightarrow{OA}$ represents the complex number $2 + 3i$. Find the complex numbers represented by $\overrightarrow{OB}, \overrightarrow{OC}, \overrightarrow{OD}$.

Multiplying $2 + 3i$ by i has the effect of rotating $\overrightarrow{OA}$ anticlockwise through a right angle. Hence $\overrightarrow{OB}$ represents $-3 + 2i$. Multiplying again by i shows that $\overrightarrow{OC}$ represents $-2 - 3i$. It follows that $\overrightarrow{OD}$ represents $3 - 2i$.

Example 2
Use the method of induction to prove de Moivre's theorem

$$(\cos\theta + i\sin\theta)^n = \cos(n\theta) + i\sin(n\theta)$$

when n is a positive integer.

The result is clearly true, when $n = 1$. Assume that it is true when $n = k$, where k is a positive integer. Then

$$(\cos\theta + i\sin\theta)^{k+1} = (\cos\theta + i\sin\theta)^k(\cos\theta + i\sin\theta)$$
$$= (\cos k\theta + i\sin k\theta)(\cos\theta + i\sin\theta).$$

On the right-hand side the real part is

$$\cos (k\theta) \cos \theta - \sin (k\theta) \sin \theta, \text{ i.e. } \cos (k+1)\theta,$$

while the imaginary part is

$$\cos (k\theta) \sin \theta + \sin (k\theta) \cos \theta, \text{ i.e. } \sin (k+1)\theta.$$
$$\text{Hence } (\cos \theta + i \sin \theta)^{k+1} = \cos (k+1)\theta + i \sin (k+1)\theta,$$

so that the truth of the result for $n = k$ implies the truth of the result for $n = k+1$. The result holds for $n = 1$, and hence it is true for $n = 2$, for $n = 3$ and for any positive integer n.

Example 3
Express $\cos 5\theta$ as a polynomial in $\cos \theta$.

$$\cos 5\theta + i \sin 5\theta = (\cos \theta + i \sin \theta)^5$$
$$= c^5 + i5c^4 s - 10c^3 s^2 - i10c^2 s^3 + 5cs^4 + is^5$$

(where $c = \cos \theta$, $s = \sin \theta$)

$$= (c^5 - 10c^3 s^2 + 5cs^4) + i(5c^4 s - 10c^2 s^3 + s^5).$$

This equation expresses the equality of two complex numbers. Equating the real parts of the two complex numbers gives

$$\cos 5\theta = c^5 - 10c^3 s^2 + 5cs^4,$$

and, on writing $s^2 = 1 - c^2$, this gives

$$\cos 5\theta = 16 \cos^5 \theta - 20 \cos^3 \theta + 5 \cos \theta.$$

By equating the imaginary parts of the two complex numbers an expression for $\sin 5\theta$ in terms of $\sin \theta$ can be deduced.

Exercises 8.9
1 Express the following in polar form, giving the argument in degrees, to the nearest degree.
(i) $5+12i$ (ii) $5-12i$ (iii) $12-5i$ (iv) $-12+5i$ (v) $-12-5i$.
2 Express the following in algebraic $(x+iy)$ form, leaving x and y in surd form where appropriate.
(i) $\sqrt{2}\angle 45°$ (ii) $2\angle 60°$ (iii) $2\angle 120°$ (iv) $2\angle 210°$ (v) $5\angle 90°$
(vi) $5\angle -90°$ (vii) $1\angle 315°$ (viii) $(2\angle 45°)^2$.
3 Simplify the following.
(i) $(\cos \theta + i \sin \theta)(\cos \theta - i \sin \theta)$
(ii) $2(\cos \pi/4 + i \sin \pi/4) \times 5(\cos \pi/3 - i \sin \pi/3)$
(iii) $2(\cos \pi/3 + i \sin \pi/3)^3$
(iv) $2(\cos 2\pi/3 + i \sin 2\pi/3)^3$
(v) $\{2(\cos 2\pi/3 - i \sin 2\pi/3)\}^3$

(vi) $8(\cos 5\pi/6 + i \sin 5\pi/6) \div 4(\cos \pi/4 - i \sin \pi/4)$.

4 Given that $z = \cos \theta + i \sin \theta$, express in trigonometric form
   (i) $1/z$ (ii) $z + 1/z$ (iii) $z^2 + 1/z^2$ (iv) $z^3 + 1/z^3$
   (v) $z - 1/z$ (vi) $z^2 - 1/z^2$.

5 Show that $(1 + \cos 2\theta) + i(\sin 2\theta) = 2 \cos \theta (\cos \theta + i \sin \theta)$.

6 Show that $(1 - \cos 2\theta) - i(\sin 2\theta) = 2 \sin \theta (\sin \theta - i \cos \theta)$.

7 Find

   (i) $\left| \dfrac{1+3i}{3+i} \right|$ (ii) $\arg \left(\dfrac{1+3i}{3+i} \right)$ (iii) $\left| \dfrac{(1+i)^6}{i^3(1+4i)^2} \right|$

   (iv) $\mathrm{Re} \left(\dfrac{(1-i)^2}{4+2i} \right)$ (v) $\mathrm{Im} \left(\dfrac{1+3i}{5-7i} \right)$.

8.10   The triangle inequality for complex numbers

It has already been stated that complex numbers cannot be ordered. Even the arguments of complex numbers cannot be ordered, since the argument of a number is not unique. The moduli, being real numbers, can be ordered.

The followed inequality, known as the triangle inequality, relates the moduli of any two complex numbers z_1 and z_2:

$$|z^1| + |z_2| \geqslant |z_1 + z_2|.$$

Proof

Let $z_1 = \overrightarrow{OP_1}$, $z_2 = \overrightarrow{OP_2}$. (Fig. 8.9)

Then $z_1 + z_2 = \overrightarrow{OP}$,

but, in triangle OP_1P, it is clear that $OP_1 + P_1P \geqslant OP$,

so $|z_1| + |z_2| \geqslant |z_1 + z_2|.$

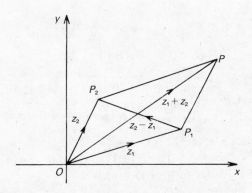

Fig. 8.9

The equality only holds when z_1 and z_2 have the same argument.
(Note that we are equating a complex number to its corresponding vector, rather than letting z_1 be represented by $\overrightarrow{OP_1}$–this procedure is acceptable, for convenience.)

Other inequalities
Clearly $OP_1 + OP_2 \geqslant P_1P_2$,

so
$$|z_1| + |z_2| \geqslant |z_2 - z_1|.$$

Also $\left. \begin{array}{l} OP_1 + P_1P_2 \geqslant OP_2, \\ \text{so } |z_1| + |z_2 - z_1| \geqslant |z_2| \end{array} \right\}$ and $\left. \begin{array}{l} OP_2 + P_2P_1 \geqslant OP_1 \\ |z_2| + |z_1 - z_2| \geqslant |z_1| \end{array} \right\}$

and combining these two results

$$|z_2 - z_1| \geqslant ||z_2| - |z_1||.$$

The following examples illustrate some of the techniques used with complex numbers.

Example 1
Show that the equation $|z+1| = 3$ represents a circle and find its centre and radius.

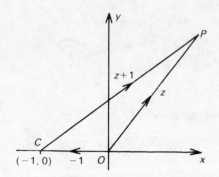

Fig. 8.10

Method 1   If $\overrightarrow{OC} = -1$ and $\overrightarrow{OP} = z$, then $\overrightarrow{CP} = (z+1)$, (Fig. 8.10),
So $|z+1| = 3 \Rightarrow CP$ is of length 3.
Hence the locus of P is a circle, centre C, of radius 3.

Method 2   If P is $x+iy$, then
$$|(x+iy)+1| = 3$$
So
$$|(x+1)+iy| = 3$$
and hence
$$(x+1)^2 + y^2 = 9$$
which is the equation of a circle, centre $(-1, 0)$ and radius 3.

Method 3   Since $|z+1|^2 = 9$,

$$(z+1)(z*+1) = 9.$$

So

$$zz* + z + z* + 1 = 9,$$
$$|z|^2 + 2\operatorname{Re}(z) + 1 = 9,$$
$$x^2 + y^2 + 2x = 8,$$
$$(x+1)^2 + y^2 = 9, \text{ as before.}$$

Example 2

Solve the equation $z^3 - 1 = 0$ and exhibit the three roots on an Argand diagram. If ω is a root not equal to 1, show that the other root may be denoted by ω^2 and prove that $1 + \omega + \omega^2 = 0$. (Fig. 8.11)

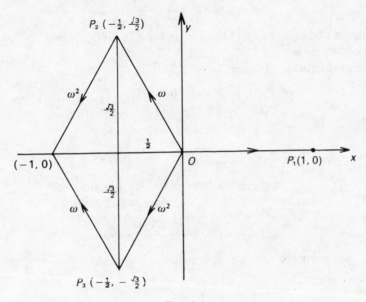

Fig. 8.11

Method 1

$$z^3 - 1 = 0$$
$$(z-1)(z^2 + z + 1) = 0$$

$$z = 1 \quad \text{or} \quad (z + \tfrac{1}{2})^2 = -\tfrac{3}{4}$$
$$z = 1 \quad \text{or} \quad z = -\tfrac{1}{2} \pm i\sqrt{3}/2.$$

So, in polar form, the roots are $1\angle 0$, $1\angle(2\pi/3)$, $1\angle(-2\pi/3)$.
If $\omega = 1\angle(2\pi/3)$, then $\omega^2 = 1\angle(4\pi/3) = 1\angle(-2\pi/3)$.
If $\omega = 1\angle(-2\pi/3)$, then $\omega^2 = 1\angle(-4\pi/3) = 1\angle(2\pi/3)$.
Since the roots of the equation $z^3 - 1 = 0$ are $1, \omega, \omega^2$, and since the coefficient of z^2 is zero, we have $1 + \omega + \omega^2 = 0$ (sum of roots).

Method 2
$$z^3 = 1 = 1\angle 0 = 1\angle 2\pi = 1\angle 4\pi,$$
so $z = 1\angle 0, 1\angle(2\pi/3), 1\angle(4\pi/3)$
since, if cubing a complex number trebles its argument, then taking a cube root must divide the argument by 3. (The remainder of the solution follows as before.)

Example 3
Show that if z_1 is a root of the equation $az^2 + bz + c = 0$, where a, b, c are real, z_1^* is also a root.

Method 1
If z_1 is a root of the equation then

$$az_1^2 + bz_1 + c = 0$$

i.e. the three numbers in $\mathbb{C}$ add to zero (Fig. 8.12). So the corresponding three vectors add to zero.
Reflecting the figure in Ox gives

$$a(z_1^2)^* + bz_1^* + c = 0$$
So $\qquad a(z_1^*)^2 + bz_1^* + c = 0 \Rightarrow z_1^*$ is a root.

Method 2
Let $z_1 = x + iy$.

Then $\qquad a(x^2 - y^2 + i2xy) + b(x + iy) + c = 0$
So $\qquad a(x^2 - y^2) + bx + c = 0 \quad$ and $\quad 2axy + by = 0$
So $\qquad a(x^2 - y^2) + bx + c - i(2axy + by) = 0$
So $\qquad a(x^2 - y^2 - i2xy) + b(x - iy) + c = 0$

showing that $x - iy$ is a root, i.e. z_1^* is root.

Example 4
Show that the equations $|z| = |z - 2|$ and $|z - i| = |z - 1|$ correspond to two straight lines in the Argand diagram, and find the value of z which satisfies both equations.

Algebraic solution
If $z = x + iy$ then $|z| = |z - 2|$

$\Rightarrow \qquad x^2 + y^2 = (x - 2)^2 + y^2 \quad$ (on squaring)
$\Rightarrow \qquad\qquad 4x = 4 \qquad\qquad$ (on reduction)
$\Rightarrow \qquad\qquad x = 1$
$\qquad\qquad |z - i| = |z - 1|$
$\Rightarrow \qquad x^2 + (y - 1)^2 = (x - 1)^2 + y^2 \quad$ (on squaring)
$\Rightarrow \qquad\qquad x = y. \qquad\qquad$ (on reduction)

So the lines meet at the point $(1, 1)$. Hence, $z = 1 + i$.

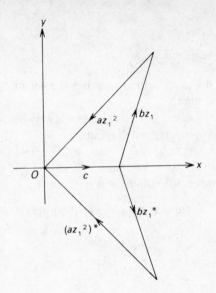

Fig. 8.12

Fig. 8.13

Geometric solution (Fig. 8.13.)

If A is the point $(2, 0)$ and if P is the point representing z, then $\overrightarrow{OP} = z$ and $\overrightarrow{AP} = z - 2 (\overrightarrow{OA} + \overrightarrow{AP} = \overrightarrow{OP})$. So $|z| = |z - 2| \Rightarrow OP = AP$, i.e. P is equidistant from O and A. Hence P lies on the perpendicular bisector of OA, i.e. on the line $x = 1$. In the same way, $|z - i| = |z - 1|$ show that P is equidistant from the points $(0, 1)$ and $(1, 0)$, i.e. P lies on the perpendicular bisector of the join of these points, and this is the line $x = y$. The two lines meet at the point $(1, 1)$. Hence, as before, $z = 1 + i$.

Example 5

Show, without reference to an Argand diagram, that

$$|z_1| + |z_2| \geqslant |z_1 + z_2|.$$

Consider
$$\begin{aligned}
|z_1 + z_2|^2 &= (z_1 + z_2)(z_1{}^* + z_2{}^*) \\
&= z_1 z_1{}^* + z_2 z_2{}^* + z_1 z_2{}^* + z_2 z_1{}^* \\
&= |z_1|^2 + |z_2|^2 + 2\,\mathrm{Re}(z_1 z_2{}^*),
\end{aligned}$$

since $(z_1 z_2{}^*)^* = z_1{}^* z_2$.

Hence
$$\begin{aligned}
|z_1 + z_2|^2 &\leqslant |z_1|^2 + |z_2|^2 + 2|z_1 z_2{}^*| \quad \text{since } \mathrm{Re}(z) \leqslant |z|. \\
&= |z_1|^2 + |z_2|^2 + 2|z_1||z_2| \quad \text{since } |z_2{}^*| = |z_2|.
\end{aligned}$$

Since both sides of the inequality are non-negative, it follows that

$$|z_1 + z_2| \leqslant |z_1| + |z_2|.$$

Real and complex numbers **225**

Miscellaneous exercises 8

1   Prove that

$$(1+\cos\theta+i\sin\theta)^n = 2^n\cos^n\tfrac{1}{2}\theta\,(\cos\tfrac{1}{2}n\theta+i\sin\tfrac{1}{2}n\theta).$$

2   Express $(1+\sqrt{3}i)$ in polar form and hence, or otherwise, find the set of values of n for which $(1+\sqrt{3}i)^n$ is a real number.

3   Given that $z = \cos\theta+i\sin\theta$, find $|z+1|$ and $\arg(z+1)$. (The results may be written down directly by consideration of an Argand diagram.)

4   Given that $z_1 = 1\angle\theta$ and $z_2 = 1\angle\phi$, where $0 < \theta < \phi < \pi/2$, express z_1+z_2 in polar form. (Hint: consider an Argand diagram.)

5   Given that $z+1/z = a$, where $a \in \mathbb{R}$, prove that either $z \in \mathbb{R}$ or $|z| = 1$.

6   Prove that, if $z \neq 0$, then $\left(\dfrac{z}{z^*}+\dfrac{z^*}{z}\right)$ is a real number lying in the interval $[-2, 2]$.

7   Show that $(\sin\theta+i\cos\theta)^n = \cos n\left(\dfrac{\pi}{2}-\theta\right)+i\sin n\left(\dfrac{\pi}{2}-\theta\right)$ where $n \in \mathbb{N}$.

8   Simplify $(\sqrt{3}+i)^{10}-(\sqrt{3}-i)^{10}$.

9   Find values for the real numbers a and b such that $(a+ib)^2 = 8+6i$.

10   Use the method of question 9 to find the exact square roots of
   (i) $7-24i$ (ii) $15-8i$ (iii) $(4+3i)/(12+5i)$ (iv) $2i$
   (v) $-2i$

11   Use de Moivre's theorem to show that, for values of θ such that $\sin\theta \neq 0$,

$$\frac{\sin 5\theta}{\sin\theta} = 16\cos^4\theta-12\cos^2\theta+1.$$

12   Factorise
   (i) x^2+9 (ii) x^2+a^2 (iii) x^2+2x+5 (iv) $x^2+2ax+a^2+b^2$.

13   By assuming $\dfrac{1}{x^2+a^2} \equiv \dfrac{P}{x-ia}+\dfrac{P}{x-ia}$, where P and Q may be complex, express $\dfrac{1}{x^2+a^2}$ in partial fractions.

14   Express $\dfrac{x}{x^2+a^2}$ in partial fractions.

15   By noting that the point corresponding to z^* in an Argand diagram is the reflection in the x-axis of the point corresponding to z, or otherwise, establish the results
   (i) $(z^*)^* = z$
   (ii) $(z_1+z_2)^* = z_1^*+z_2^*$
   (iii) $(z_1z_2)^* = z_1^*z_2^*$
   (iv) $(1/z)^* = 1/z^*$.

16   If $f(z)$ is any polynomial in z with real coefficients, use the results of question 15 (ii), (iii) to show that $[f(z)]^* = f(z^*)$, and deduce that, if ω is a root of $f(z) = 0$, then ω^* is also a root.

17 Express in the form $f(x) = 0$, where $f(x)$ is a polynomial of degree 4 having real coefficients, the equation having as two of its roots the numbers $3+i$, $1+3i$.

18 Show that $1+2i$ is a root of the equation $z^2 - 3(1+i)z + 5i = 0$ and find the other root.

19 Given that $a+ib$ is a root of equation $z^3 + pz + q = 0$, where p and q are real and b is not zero, prove that
  (i) $2a(a^2 + b^2) = q$
  (ii) $3a^2 - b^2 = -p$
  (iii) a is a root of the equation $8x^3 + 2px - q = 0$.

20 If $z = x+iy$ and $\alpha = a+ib$, show that the equation

$$zz^* - \alpha^* z - \alpha z^* + \alpha\alpha^* = r^2$$

represents a circle, and find, in terms of a, b and r, the centre and radius of the circle.

21 A regular hexagon $ABCDEF$ in the Argand diagram has its centre at the origin O and its vertex A at the point $z = 2$.
  (a) Indicate in a diagram the region within the hexagon in which both the inequalities $|z| \geqslant 1$ and $-\pi/3 \leqslant \arg z \leqslant \pi/3$ are satisfied.
  (b) Find, in the form $|z-c| = R$, the equation of the circle through the points O, B, F.
  (c) Find the values of z corresponding to the points C and E.
  (d) The hexagon is rotated anticlockwise about the origin through an angle of $45°$. Express, in the form $r(\cos\theta + i\sin\theta)$ the values of z corresponding to the new positions of the points C and E. [L]

22 If $z = \cos\theta + i\sin\theta$, show that

$$z + 1/z = 2\cos\theta, \quad z^n + 1/z^n = 2\cos n\theta, \quad (n \in \mathbb{N}).$$

Hence express $\cos^5\theta$ in the form

$$a\cos 5\theta + b\cos 3\theta + c\cos\theta,$$

where a, b and c are real numbers.

23 Express each of the complex numbers z_1, z_2, z_3 in the form $a+ib$ where

$$z_1 = (1-i)(1+2i), \quad z_2 = \frac{2+6i}{3-i}, \quad z_3 = \frac{-4i}{1-i}.$$

Show that $|z_2 - z_1| = |z_1 - z_3|$ and that, for principal values of the argument, $\arg(z_2 - z_1) - \arg(z_1 - z_3) = \pi/2$.
  If z_1, z_2, z_3 are represented by points P_1, P_2, P_3 respectively in an Argand diagram, prove that P_1 lies on the circle with $P_2 P_3$ as diameter. [L]

24 Use de Moivre's theorem to show that

$$\tan 5\theta = \frac{5\tan\theta - 10\tan^3\theta + \tan^5\theta}{1 - 10\tan^2\theta + 5\tan^4\theta}.$$

Deduce that $\tan 18°$ is a root of the equation $5x^4 - 10x^2 + 1 = 0$, and explain why $\tan 162°$ is also a root of this equation.

25 If $z = \cos\theta + i\sin\theta$ and $w = 2z - 3 - 3/z$, express $|w|^2$ in terms of $\cos\theta$, and hence show that the greatest value of $|w|^2$ is $275/8$. [L]

26 (i) If $z_1 = 1 - i$ and $z_2 = 7 + i$, find the modulus of
    (a) $z_1 - z_2$ (b) $z_1 z_2$ (c) $(z_1 - z_2)/(z_1 z_2)$.
    (ii) Sketch on an Argand diagram the locus of a point P representing the complex number z, where

$$|z - 1| = |z - 3i|,$$

and find z when $|z|$ has its least value on this locus. [L]

27 Find, in algebraic form, all the fourth roots of $28 + 96i$.

28 (i) Find, in the form $a + ib$, the complex number z satisfying equation

$$\frac{2z - 3}{1 - 4i} = \frac{3}{1 + i}.$$

(ii) Use de Moivre's theorem to prove that

$$\cos 5\theta = 16\cos^5\theta - 20\cos^3\theta + 5\cos\theta.$$

By considering the equation $\cos 5\theta = 0$, prove that

$$\cos(\pi/10)\cos(3\pi/10) = \frac{1}{4}\sqrt{5}.$$

29 Determine all pairs of values of the real numbers p, q for which $1 + i$ is a root of the equation

$$z^3 + pz^2 + qz - pq = 0.$$ [JMB]

30 The complex numbers z_1 and z_2 satisfy the equation

$$z_2{}^2 - z_1 z_2 + z_1{}^2 = 0.$$

If $z_1 = a + ib$, (a and b are real), show that a possible value for z_2 is given by

$$z_2 = \tfrac{1}{2}(a - b\sqrt{3}) + i\tfrac{1}{2}(b + a\sqrt{3}).$$

In an Argand diagram, the points P and Q represent z_1 and z_2 respectively, and O is the origin. Show that the triangle OPQ is equilateral. [JMB]

31 The roots of the equation

$$(1 + i)z^2 - i2z + 3 + i = 0$$

are α and β. Find $\alpha + \beta$ and $\alpha\beta$ in the form $p + iq$ where p and q are real.
    Find, in a form not involving α and β, the quadratic equation whose roots are $\alpha + 3\beta$, $3\alpha + \beta$. [JMB]

32 Solve, giving your answer to two places of decimals, the equation $x^4 - 2x^2 + 5 = 0$.

33 The points P and Q in the Argand diagram represent the non-zero complex

numbers z_1 and z_2 respectively. Show that a necessary and sufficient condition for the triangle OPQ to be isosceles and right-angled at O is $z_1{}^2 + z_2{}^2 = 0$.

One vertex of a square is represented by the complex number $5 + 2i$ and its centre is represented by the complex number $2 + i$. Find, in the form $a + ib$, the complex numbers representing the other three vertices.

34 The sets A, B, C of points z in the complex plane are given by

$$A = \{z : |z - 2| = 2\},$$
$$B = \{z : |z - 2| = |z|\},$$
$$C = \{z : \arg (z - 2) = \arg z\}.$$

Determine the locus of z in each case and illustrate the sets of points on one diagram.

Show also that the points of the set $A \cap (B \cup C)$ are the vertices of an equilateral triangle.

35 If P_1, P_2, P_3 are the points in the Argand diagram that correspond respectively to the numbers

$$z_1, z_2, (1 + \tfrac{1}{2}i)z_2 - \tfrac{1}{2}iz_1,$$

show that $P_1 P_2 P_3$ is a right-angled triangle. Also find in terms of z_1, z_2 the complex number that corresponds to P_4 if $P_1 P_2 P_3 P_4$ is a rectangle.

36 Show that the locus of the points in the complex plane which represent numbers such that

$$\left| \frac{z - 1}{z + 1} \right| = k,$$

where k is a constant not equal to 0 or 1, is a circle. Show that if such a locus meets the x-axis at the distinct points $(x_1, 0)$ and $(x_2, 0)$, then $x_1 x_2 = 1$.

[JMB]

9 Differentiation 2

9.1 Maximum and minimum values

If the gradient of the curve $y = f(x)$ is zero when $x = a$, the point $(a, f(a))$ is called a stationary point.

If the gradient changes sign as x increases through the value a, the point is called a turning point. When the sign changes from positive to negative, $f(a)$ will be a maximum value of y. When the sign changes from negative to positive, $f(a)$ will be a minimum value of y.

In Fig. 9.1, the sign of the gradient of the curve $y = \sin x$ changes from $+$ to $-$ as x increases through the values $-3\pi/2$ and $\pi/2$, so that P and R are maximum points. The sign of the gradient changes from $-$ to $+$ as x increases through the values $-\pi/2$ and $3\pi/2$, and so Q and S are minimum points.

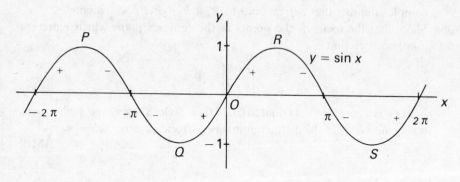

Fig. 9.1

There are three steps in the process of finding the maximum and minimum values of y.

Step 1   Differentiate y and find the values of x for which $dy/dx = 0$.

Step 2   Examine the behaviour of the sign of dy/dx as x increases through each of these values.

A change from $+$ to $-$ gives a maximum value for y.

A change from $-$ to $+$ gives a minimum value for y.

Step 3   Calculate the value of y at each turning point.

There is an alternative method which uses the second derivative of y. When

$\dfrac{d^2y}{dx^2}$ is negative at a turning point, the gradient decreases as x increases. The sign of the gradient will change from $+$ to $-$, and y will have a maximum value.

When $\dfrac{dy}{dx} = 0$ and $\dfrac{d^2y}{dx^2} < 0$ at $x = a$, y has a maximum value at $x = a$.

If $\dfrac{d^2y}{dx^2}$ is positive at a turning point, the gradient increases as x increases. The sign of the gradient will change from $-$ to $+$, and y will have a minimum value.

When $\dfrac{dy}{dx} = 0$ and $\dfrac{d^2y}{dx^2} > 0$ at $x = a$, y has a minimum value at $x = a$.

Example 1
Show that y has a minimum value of 1 at the turning point on the curve $y = x^2 - 4x + 5$. (Fig. 9.2.)

Step 1 The gradient equals $(2x - 4)$, which is zero when $x = 2$.
Step 2 As x increases through the value 2, the sign of $(2x - 4)$ changes from negative to positive. Hence y has a minimum value at $x = 2$.
Step 3 By substituting $x = 2$ in $(x^2 - 4x + 5)$ we find that this minimum value is 1.

The same result is obtained by writing $x^2 - 4x + 5$ in the form $(x - 2)^2 + 1$ and noting that $(x - 2)^2$ is never negative.

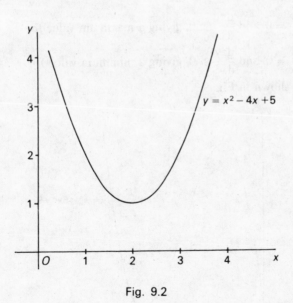

Fig. 9.2

Example 2
Find the maximum and minimum values of y on the curve $y = x^3 - 6x^2 + 9x$.

Step 1
$$\frac{dy}{dx} = 3x^2 - 12x + 9 = 3(x-1)(x-3).$$

The stationary points are given by $x = 1$ and $x = 3$.

Step 2 The behaviour of the sign of the gradient is shown in the following table.

	$x < 1$	$1 < x < 3$	$x > 3$
$\dfrac{dy}{dx}$	$+$	$-$	$+$

As x increases through the value 1, the sign of the gradient changes from $+$ to $-$, indicating a maximum value for y. As x increases through the value 3, the sign of the gradient changes from $-$ to $+$, indicating a minimum value for y.

Step 3 When $x = 1$, $y = 4$. This is the maximum value of y. When $x = 3$, $y = 0$. This is the minimum value of y. The maximum value of y is a local maximum, i.e. it is the greatest value of y when x lies in a small interval containing the point $x = 1$. The minimum value of y is a local minimum, i.e. it is the smallest value of y for values of x in a small interval containing the point $x = 3$.

Instead of using the table shown in step 2 the sign of the second derivative could be considered. We have

$$\frac{d^2y}{dx^2} = 6x - 12.$$

At $x = 1$, $\dfrac{dy}{dx} = 0$ and $\dfrac{d^2y}{dx^2} < 0$, giving a maximum value for y.

At $x = 3$, $\dfrac{dy}{dx} = 0$ and $\dfrac{d^2y}{dx^2} > 0$, giving a minimum value for y.

The curve is shown in Fig. 9.3.

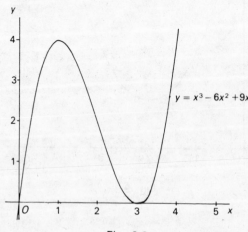

$$y = x^3 - 6x^2 + 9x$$

Fig. 9.3

Example 3
The total surface area of a rectangular block of wood is 2 square metres. Given that the block has a square base, find the maximum volume of the block.

In a problem such as this we must first construct an expression for the quantity of which the maximum or minimum value has to be found.

Let the length of the edge of the base be x metres and let the height of the block be h metres. Then the volume V m³ of the block is given by $V = x^2h$.

Next V must be expressed in terms of x alone or in terms of h alone, whichever is simpler. The surface area is given as $2\,\text{m}^2$, so that

$$2x^2 + 4xh = 2.$$

Hence
$$h = (1 - x^2)/(2x)$$
and
$$V = \tfrac{1}{2}(x - x^3).$$

Now proceed as before.
Step 1

$$\frac{dV}{dx} = \tfrac{1}{2}(1 - 3x^2).$$

This is zero when $x = 1/\sqrt{3}$. (Since x must be positive, we reject the value $-1/\sqrt{3}$.)
Step 2

$$\frac{d^2V}{dx^2} = -3x.$$

This is negative when $x = 1/\sqrt{3}$, indicating a maximum value for V.
Step 3
When $x = \dfrac{1}{\sqrt{3}}$, $V = \dfrac{1}{2}\left(\dfrac{1}{\sqrt{3}} - \dfrac{1}{3\sqrt{3}}\right) = \dfrac{1}{3\sqrt{3}}$.

Thus the maximum volume of the block is $1/(3\sqrt{3})\,\text{m}^3$.

Exercises 9.1
1 Find the coordinates of the stationary points on the following curves. State in each case whether the point is a maximum point, a minimum point or neither.
   (a) $y = x^2 - 6x + 2$ (d) $y = 2x^2 + 4x - 2$
   (b) $y = -3x^2 + 6x + 3$ (e) $y = x^3 + 2$
   (c) $y = -x^2 + 6x - 5$ (f) $y = 3x^2 + 2x + 1$.
2 Find the coordinates of any turning points on the following curves, and determine the nature of each turning point.
   (a) $y = x^4 - 4x + 3$ (d) $y = (x^3 + 16)/x$
   (b) $y = x^3 + 3x + 2$ (e) $y = (x - 1)(x + 2)^2$
   (c) $y = x^3 - x^2$ (f) $y = x^2/(x + 2)$.

3 Find the maximum and minimum values of
   (a) $x^3 + 3x^2 - 9x + 5$ (b) $2x^3 - 3x^2 - 12x + 5$ (c) $x^3 - 9x^2 + 24x$.
4 The sum of two numbers is 20. Find the maximum value of their product.
5 Find the minimum value of the sum of a positive number and its reciprocal.
6 An open box (with no lid) is to be made using one square metre of cardboard. Find the maximum volume of the box given that it has a square base.
7 A wire of length $4a$ is bent into the form of a sector of a circle. Find the maximum possible area of the sector.
8 Find the maximum value of $x^2 + y^2$ given that $x + y = 1$.
9 One side of a rectangular playground is a brick wall, and the other three sides are formed by fencing. The area of the playground is $5000\,\mathrm{m}^2$. Find the least length of fencing required.
10 Show that on the curve $y = (x^2 + 4)/x$, the maximum value of y is -4 and the minimum value of y is 4.
11 The vertices of a rectangle lie on a circle of radius a. Show that the area of the rectangle is not greater than $2a^2$.
12 A right circular cone has a fixed slant height a. Find the area of the circular base of the cone when the volume of the cone is a maximum.

9.2   Points of inflexion

When a tangent to the curve $y = f(x)$ crosses the curve at its point of contact, this point is known as a point of inflexion. At such a point, the gradient of the curve has either a maximum or a minimum value. It follows that the curve $y = f(x)$ will have a point of inflexion when $x = a$ if d^2y/dx^2 changes sign as x passes through the value a.

There are two steps in the process of finding points of inflexion.
Step 1   Find the values of x for which $d^2y/dx^2 = 0$.
Step 2   Test whether d^2y/dx^2 changes sign as x increases through each of these values of x. A change of sign will indicate a point of inflexion.

In Fig. 9.4 the line $y = x$ is the tangent to the curve $y = \sin x$ at the origin. This

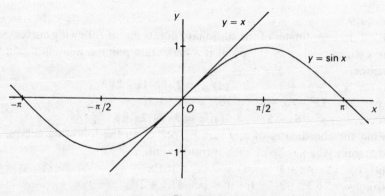

Fig. 9.4

tangent crosses the curve at the origin, which is a point of inflexion of the curve. As x moves from $-\pi/2$ to zero, the gradient of the curve increases and d^2y/dx^2 is positive. At the origin, the gradient has its maximum value 1 and $d^2y/dx^2 = 0$. Then as x moves from zero to $\pi/2$, the gradient decreases and d^2y/dx^2 is negative. When $x = \pi$ the gradient takes its minimum value -1, and the point $(\pi, 0)$ is another point of inflexion of the curve. The table below shows the changes in sign of d^2y/dx^2.

	$-\pi/2 \leqslant x < 0$	$x = 0$	$0 < x < \pi$	$x = \pi$	$\pi < x \leqslant 3\pi/2$
$\dfrac{d^2y}{dx^2}$	$+$	0	$-$	0	$+$

Example 1
Show that the point $(2, 2)$ is a point of inflexion on the curve $y = x^3 - 6x^2 + 9x$. (Fig. 9.3 on p. 232)

Step 1   By differentiating twice with respect to x we find that d^2y/dx^2 equals $6x - 12$. This is zero when $x = 2$.

Step 2   When x is less than 2, $(6x - 12)$ is negative, and when x is greater than 2, $(6x - 12)$ is positive.

This shows that the sign of d^2y/dx^2 changes from $-$ to $+$ as x increases through the value 2. When $x = 2$, $y = 2$. Therefore the curve has a point of inflexion at the point $(2, 2)$.

Example 2
Find whether the origin is (i) a turning point, (ii) a point of inflexion, on the curves (a) $y = x^3$ (b) $y = x + x^3$ (c) $y = x^4$.

(a) If $y = x^3$, $dy/dx = 3x^2$ and $d^2y/dx^2 = 6x$.

At the origin, $dy/dx = 0$, but the gradient does not change sign as x increases through the value 0. This means that the origin is a stationary point but not a turning point on the curve.

At the origin, $d^2y/dx^2 = 0$, and the sign of d^2y/dx^2 changes from $-$ to $+$ as x increases through the value 0. Thus the origin is a point of inflexion.

(b) If $y = x + x^3$, $dy/dx = 1 + 3x^2$ and $d^2y/dx^2 = 6x$.

The gradient of the curve at the origin is 1, and so the origin is not a turning point. The second derivative is zero at the origin, and its sign changes from $-$ to $+$ as x increases through the value 0. Thus the origin is a point of inflexion.

(c) If $y = x^4$, $dy/dx = 4x^3$ and $d^2y/dx^2 = 12x^2$.

At the origin, $dy/dx = 0$ and the gradient changes sign from $-$ to $+$ as x increases through the value 0. Hence the origin is a turning point at which y takes a minimum value.

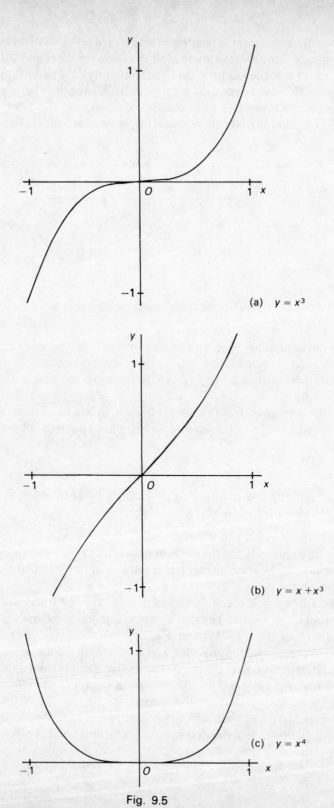

(a) $y = x^3$

(b) $y = x + x^3$

(c) $y = x^4$

Fig. 9.5

At the origin, $d^2y/dx^2 = 0$ but the sign of d^2y/dx^2 does not change as x increases through the value 0. It follows that the origin is not a point of inflexion. The behaviour of each curve near the origin is shown in Fig. 9.5.

Exercises 9.2

1 Which of the following statements are true for the curve $y = f(x)$?
   (a) If $dy/dx = 0$ when $x = a$, the curve has a turning point when $x = a$.
   (b) If the sign of dy/dx changes from $-$ to $+$ as x increases through the value a, the curve has a maximum point when $x = a$.
   (c) If $d^2y/dx^2 = 0$ when $x = a$, the curve has a point of inflexion when $x = a$.
   (d) If the curve has a point of inflexion when $x = a$, $dy/dx = 0$ when $x = a$.
2 Find whether the origin is a point of inflexion on the curves
   (a) $y = x^5$ (b) $y = x^6$ (c) $y = 3x - x^3$.
3 Show that the curve $y = \cos x$ has a point of inflexion when $x = \pi/2$ and also when $x = -\pi/2$.
4 Show that the curve $y = \tan x$ has a point of inflexion at the origin.
5 Find the coordinates of the points of inflexion on the curves
   (a) $y = (x + 1)^3$ (c) $y = x^2(x - 2)^2$
   (b) $y = x^3 + 3x^2$ (d) $y = 1/(x^2 + 3)$.
6 Find the coordinates of the point on the curve

$$y = 7 - 11x + 6x^2 - x^3$$

at which the gradient is a maximum. Sketch the curve and draw its tangent at this point.

9.3 Implicit differentiation

The equation $x^2 + y^2 = 1$ represents a circle with unit radius. This equation is said to express y implicitly in terms of x. The upper semicircle is given explicitly by the equation

$$y = \sqrt{(1 - x^2)},$$

and so the gradient at a point on the upper semicircle is given by

$$\frac{dy}{dx} = \frac{-x}{\sqrt{(1 - x^2)}} = -\frac{x}{y}.$$

The lower semicircle is given by the explicit equation

$$y = -\sqrt{(1 - x^2)},$$

which gives

$$\frac{dy}{dx} = \frac{x}{\sqrt{(1 - x^2)}} = -\frac{x}{y}.$$

The same result is obtained if we differentiate each term of the implicit equation with respect to x. From the equation $x^2 + y^2 = 1$ we find by using the chain rule

$$2x + 2y\left(\frac{dy}{dx}\right) = 0.$$

Provided that y is not zero this gives $dy/dx = -x/y$. This approach is called implicit differentiation.

Example

Find the gradient at the point $(2, 1)$ on the curve $x^2y - 2xy^2 + y^3 = 1$.

Differentiate each term with respect to x.

$$\left(2xy + x^2\frac{dy}{dx}\right) - \left(2y^2 + 4xy\frac{dy}{dx}\right) + 3y^2\frac{dy}{dx} = 0.$$

By substituting for x and y, we find that at the point $(2, 1)$, $dy/dx = 2$.

Exercises 9.3

1 Find the gradient at the point $(1, 0)$ on the curve $2x^2 + xy + \sin y = 2$.
2 Find the equations of the tangents to the curve $2x^2 + 2xy + y^2 = 8$ which are parallel to the x-axis.
3 Find the equations of the tangents to the curve $x^2 - xy + y^2 = 4$ which are parallel to the line $x + y = 0$.
4 If $x^2 - y^2 = 1$, find dy/dx and d^2y/dx^2.
5 Show that at the point $(1, 1)$ the gradient of the curve $x^3 + 3xy + y^3 = 5$ is -1.
6 Show that the stationary points on the curve $(x + y)^2 = (x + 1)(y + 1)$ are the points $(1, -1)$, $(-1/3, 5/3)$.

9.4 Curve sketching

In preparation for sketching the curve $y = f(x)$ the following questions should be considered.

(a) Is f an even function, an odd function or neither?

If f is even, so that $f(x) = f(-x)$, the curve is symmetrical about the y-axis. If f is odd, so that $f(-x) = -f(x)$, the curve is said to be symmetrical with respect to the origin. In this case, if the point (a, b) lies on the curve, so does the point $(-a, -b)$.

(b) Where does the curve meet the axes?

The curve meets the x-axis at the points given by the roots of the equation $f(x) = 0$. It meets the y-axis at the points given by $y = f(0)$. If the curve passes through the origin, it may be possible to find the behaviour of the curve near the origin by inspection.

(c) Which values of x make y large? How does y behave when x is large?

If y tends to infinity as x approaches the value a, the line $x = a$ will be a vertical asymptote. If y approaches the value b as x tends to infinity, the line $y = b$ will be a horizontal asymptote.

(d) Are there any values of x for which f is not defined? Are there any values which y cannot take?

(e) When is the gradient positive, and when is it negative?

When dy/dx is positive the curve is rising, and when dy/dx is negative the curve is falling. A change in the sign of dy/dx indicates a turning point.

It is always useful to plot any turning points and to plot the points at which the curve crosses or touches the axes. Occasionally it is helpful to plot a point which will clear up a doubt. For most curves there is available a variety of approaches, but in the examples which follow the scheme above will be adhered to.

Example 1

Sketch the curve $y = x^2/(x^2 + 3)$.

(a) Since y is an even function of x, the curve is symmetrical about the y-axis.

(b) The curve passes through the origin but does not cross the x-axis. Near the origin, y is approximately equal to $x^2/3$, since $(x^2 + 3) \approx 3$.

(c) The equation of the curve can be written in the form

$$y = 1 - \frac{3}{x^2 + 3}.$$

This shows that as x tends to infinity, y tends to 1, and that the curve always lies below the horizontal asymptote $y = 1$.

(d) The value of y is never negative and never greater than 1.

(e) The gradient is given by

$$\frac{dy}{dx} = \frac{6x}{(x^2 + 3)^2}.$$

The sign of the gradient changes from $-$ to $+$ as x increases through the value 0,

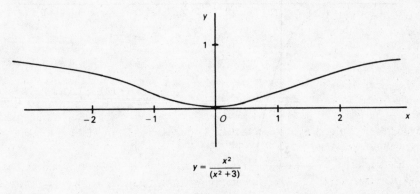

$$y = \frac{x^2}{(x^2 + 3)}$$

Fig. 9.6

indicating a minimum turning point at the origin. The second derivative is found to be

$$\frac{18(1-x^2)}{(x^2+3)^3},$$

which changes sign when $(1-x^2)$ changes sign. It follows that there is a point of inflexion when $x = -1$ and another when $x = 1$. (Fig. 9.6)

Example 2
Sketch the curve $y = (x+1)/(x-1)$.

(a) y is not an even function or an odd function of x.
(b) The curve crosses the x-axis at the point $(-1, 0)$, and it crosses the y-axis at the point $(0, -1)$.
(c) As x tends to 1, y tends to infinity, and so the line $x = 1$ is a vertical asymptote.

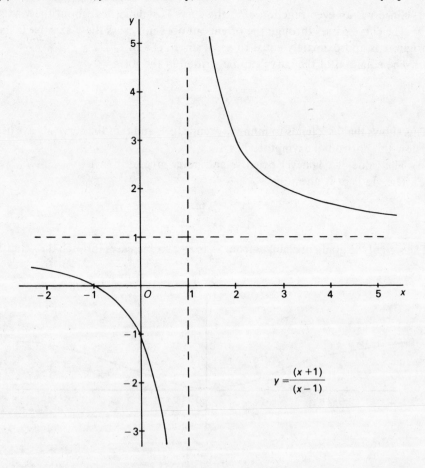

$$y = \frac{(x+1)}{(x-1)}$$

Fig. 9.7

When x is a little larger than 1, y is large and positive. When x is a little smaller than 1, y is large and negative. As x tends to infinity, y tends to 1, and so the line $y = 1$ is a horizontal asymptote. When x is large and positive, y is greater than 1 and the curve lies above the asymptote. When x is large and negative, y is less than 1 and the curve lies below the asymptote.

(d) The only restriction is that x cannot equal 1.

(e) The gradient of the curve is given by

$$\frac{dy}{dx} = \frac{-2}{(x-1)^2},$$

which is always negative. The second derivative equals $4/(x-1)^3$, which is positive when $x > 1$ and negative when $x < 1$. This means that, as x increases, the gradient decreases while $x < 1$ and increases when $x > 1$. (Fig. 9.7)

Example 3
Sketch the curve $y = x/(1-x^2)$.

(a) Since y is an odd function of x, the curve is symmetrical with respect to the origin.

(b) The curve meets the axes only at the origin, and near this point y is approximately equal to x, since x^2 will be small.

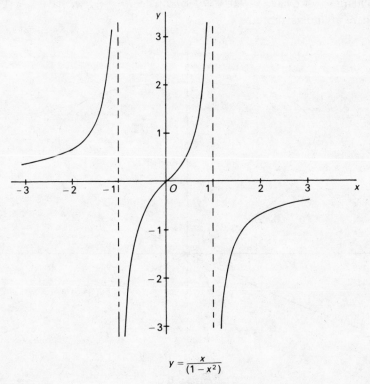

$$y = \frac{x}{(1-x^2)}$$

Fig. 9.8

(c) The line $y = 0$ is a horizontal asymptote. When x is large and positive, y is negative and the curve lies beneath the x-axis. When x is large and negative, y is positive and the curve lies above the x-axis. The lines $x = 1$ and $x = -1$ are vertical asymptotes. The behaviour of the curve near these lines can be deduced from the fact that the gradient is always positive.

(d) All values can be taken by x except 1 and -1.

(e) The first and second derivatives are found to be

$$\frac{dy}{dx} = \frac{(1 + x^2)}{(1 - x^2)^2}, \quad \frac{d^2 y}{dx^2} = \frac{6x + 2x^3}{(1 - x^2)^3}.$$

It can be seen that the gradient is always positive, and that the second derivative changes sign as x passes through the value 0. Thus the origin is a point of inflexion. (Fig. 9.8)

Example 4

Sketch the curve given by the parametric equations

$$x = 2t(1 - t^2), \quad y = 1 - t^2.$$

(a) The curve is symmetrical about the y-axis, for when t is replaced by $-t$ only the sign of x is changed.

(b) When $t = 0$ we have $x = 0$, $y = 1$, and when $t = \pm 1$ we have $x = 0$, $y = 0$.

(c) There are no asymptotes.

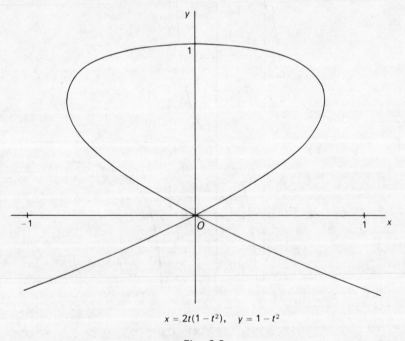

$$x = 2t(1 - t^2), \quad y = 1 - t^2$$

Fig. 9.9

(d) The value of y cannot exceed 1.

(e) $\dfrac{dy}{dx} = \dfrac{dy}{dt} \bigg/ \dfrac{dx}{dt} = \dfrac{t}{3t^2 - 1}$.

When $t = 0$, the curve has a turning point at which y takes its maximum value. At the origin, one branch has gradient $1/2$ and the other has gradient $-1/2$. As t^2 approaches the value $1/3$ the gradient tends to infinity. (Fig. 9.9)

Exercises 9.4

1   Sketch the curves
   (a) $y = x^3 - 6x^2$ (c) $y = x^5 - x^3$
   (b) $y = x^4 - 8x^2$ (d) $y = (x^2 - 1)(x^2 - 4)$.

2   Sketch the curves
   (a) $y = 1/(x+1)$ (c) $y = (x+2)/(x+4)$
   (b) $y = x/(x+2)$ (d) $y = 1/(x^2 - 4)$.

3   Sketch the curves
   (a) $y = 1/x^2$ (c) $y = (x^2 - 1)/(x^2 + 1)$
   (b) $y = x/(x^2 + 1)$ (d) $y = (x^2 + 1)/(x^2 - 1)$.

4   Sketch the curves
   (a) $y = (x-1)/(x^2 - 2x)$ (c) $y = 1/(x^2 - 2x - 3)$
   (b) $y = x^2/(x+1)$ (d) $y = 1/[x(x-1)]$.

5   Sketch the curves given by the parametric equations
   (a) $x = t^2$, $y = t^3$
   (b) $x = 2t + 1$, $y = t^2$
   (c) $x = 2\cos t$, $y = \sin t$, where $0 \leqslant t < 2\pi$
   (d) $x = \sin t$, $y = \sin^2 t$, where $-\pi/2 \leqslant t \leqslant \pi/2$.

9.5   Small increments

When $y = f(x)$, the rate of change of y with respect to x is denoted by $f'(x)$ and is defined by

$$f'(x) = \lim_{\delta x \to 0} \frac{f(x + \delta x) - f(x)}{\delta x}.$$

By taking δx sufficiently small, the difference

$$\frac{f(x + \delta x) - f(x)}{\delta x} - f'(x)$$

can be made as small as we please, and then

$$f(x + \delta x) - f(x) \approx f'(x)\,\delta x.$$

Now $f(x + \delta x) - f(x)$ is the change in the value of y, denoted by δy. This gives

$$\delta y \approx f'(x)\delta x \quad \text{or} \quad \delta y \approx \frac{dy}{dx}\,\delta x.$$

Example 1
Estimate the value of the cube root of 8·3.

Let $y = x^{1/3}$, so that $dy/dx = \frac{1}{3}x^{-2/3}$.
When $x = 8$, $y = 2$ and $dy/dx = \frac{1}{3} \times \frac{1}{4} = 1/12$.
Since $\delta x = 0\cdot3$, δy will be approximately $(0\cdot3)/12$, i.e. $\delta y \approx 0\cdot025$, and so the cube root of 8·3 is 2·025 approximately.

Example 2
A point moves along the x-axis in such a way that x m, its distance from the origin at time t s, is equal to $(4t + 3t^2)$ m. Find approximately the distance travelled while t increases from 3 to 3·1.

The increment δt in the value of t is 0·1. When $t = 3$, $dx/dt = 4 + 6t = 22$.
The change δx is given approximately by

$$\delta x \approx \frac{dx}{dt}\delta t,$$

which gives $\delta x \approx (22)(0\cdot1) = 2\cdot2$.
The exact change in x is easily shown to be 2·23. (The approximation is based on the assumption that the velocity remains constant while t increases from 3 to 3·1.)

Example 3
Find approximately the change in the value of $2x \cos 3x$ when x increases from 0 to 0·1.

Let $y = 2x \cos 3x$. Then $\delta y \approx (2 \cos 3x - 6x \sin 3x)\delta x$. When $x = 0$ and $\delta x = 0\cdot1$, $\delta y \approx 0\cdot2$.

Exercises 9.5
1 Calculate δy approximately when
   (a) $y = \sin x$, $x = \pi/3$ and $\delta x = 0\cdot2$
   (b) $y = \tan x$, $x = \pi/4$ and $\delta x = 0\cdot1$
   (c) $y = x^4$, $x = 2$ and $\delta x = 0\cdot01$.
2 Given that $y = x^4 + 4x^2 + 3$, find δy approximately given that
   (a) $x = 2$ and $\delta x = 0\cdot005$
   (b) $x = 5$ and $\delta x = 0\cdot001$.
3 The length of the side of a square is 4 m. Find approximately the increase in this length required to increase the area of the square by one per cent.
4 In the triangle ABC, $b = 4\cdot3$ m, $c = 5\cdot2$ m and $\angle A = 60°$. If the angle A is increased by one per cent, find the percentage increase in the area of the triangle.
5 The radius of a sphere increases from 6 m to 6·01 m. Show that the increase in the volume of the sphere is approximately $(1/200)$th of the original volume.
6 Find approximately the change in the value of y when x changes to

$(x + \delta x)$, given that y equals

(a) x^3 (b) $1/x$ (c) $\cos x$ (d) $\arctan x$.

7   If x increases from 2 to 2·1, show that the decrease in the value of $x/(x^2 + 1)$ is approximately 0·012.

8   If x increases from 0·75 to 0·76 show that the value of $1/\sqrt{(x^2 + 1)}$ decreases by approximately 0·003 84.

9   Taking one degree to equal 0·0175 radians, show that $\sin(30·2°) \approx 0·503$ and $\cos(30·2°) \approx 0·864$.

10   In the triangle ABC, the length of the side b is given a small increase δb, while the side c and the angle A are unchanged. By expressing a in terms of b, c and A show that $\delta a \approx (\cos C)\,\delta b$.

Miscellaneous examples 9

1   A solid is formed by a cylinder and a hemisphere which have a common circular base of radius r, the height of the cylinder being h. Prove that the solid has a minimum surface area for a given volume when $r = h$.

The surface area S of the solid is given by

$$S = (2\pi rh + \pi r^2) + 2\pi r^2 = 2\pi rh + 3\pi r^2,$$

and the volume V is given by

$$V = \pi r^2 h + \tfrac{2}{3}\pi r^3.$$

The first step is to express S in terms of V and r.

Since
$$\pi rh = \frac{V}{r} - \frac{2}{3}\pi r^2,$$

$$S = \frac{2V}{r} - \frac{4}{3}\pi r^2 + 3\pi r^2 = \frac{2V}{r} + \frac{5\pi r^2}{3}.$$

We differentiate with respect to r, bearing in mind that V is constant.

$$\frac{dS}{dr} = -\frac{2V}{r^2} + \frac{10}{3}\pi r.$$

When $dS/dr = 0$, we have $V = (5/3)\pi r^3$,

i.e.
$$\pi r^2 h + \frac{2}{3}\pi r^3 = \frac{5}{3}\pi r^3.$$

This gives
$$r = h.$$

Now
$$\frac{d^2 S}{dr^2} = \frac{4V}{r^3} + \frac{10}{3}\pi.$$

When V equals $5\pi r^3/3$, $d^2 S/dr^2$ is positive, and so the surface area S has a minimum value when $r = h$.

2   The radius r and the height h of a circular cylinder vary in such a way that the volume V of the cylinder remains constant. Find the rate of change of h with respect to r at the instant when h equals r.

We differentiate the equation $V = \pi r^2 h$ implicitly with respect to r. Since V is constant, $dV/dr = 0$ and we obtain the equation

$$\pi(2r)h + \pi r^2 \frac{dh}{dr} = 0.$$

Therefore when h equals r, $dh/dr = -2$.

Alternatively, we can express h in terms of r, and then differentiate with respect to r.

$$h = \frac{V}{\pi r^2},$$

$$\frac{dh}{dr} = \frac{-2V}{\pi r^3}.$$

When $h = r$, V is equal to πr^3 and so $dh/dr = -2$.

Miscellaneous exercises 9

1 A circular cylinder of height x is inscribed in a right circular cone of height h and radius of base a, with the axis of the cylinder coinciding with that of the cone. Show that the volume of the cylinder is

$$\pi a^2 x (h-x)^2 / h^2.$$

Prove that as x varies the volume of the cylinder is not greater than $\frac{4}{27} \pi a^2 h$.

2 A right pyramid stands on a square base. Prove that, for a given volume, the total surface area of the triangular faces is least when the side of the base is $\sqrt{2}$ times the height.

3 A cylindrical tin canister without a lid is made of sheet metal. If S is the area of the sheet used, without waste, V the volume of the canister and r the radius of the cross-section, prove that

$$2V = Sr - \pi r^3.$$

If S is given, prove that the volume of the canister is greatest when the ratio of the height to the diameter is $1:2$.

4 The plane ends of a right circular cylinder of height h and radius r are scooped out to form hollow surfaces of radius r. If the volume, V, remaining is given, find the value of r/h in order that the total surface area S, may be a minimum.

5 Three boards, each of length l and width a, are used to construct the horizontal bottom and sloping sides of an open water trough of length l. At what angle to the vertical should the sides be placed in order that the trough may hold the greatest possible volume? Show that this volume is $\dfrac{3\sqrt{3}}{4} a^2 l$.

6 A rectangular box is made from material of negligible thickness, and has a square base of side a, a depth h and no lid. Determine the ratio $a:h$ in the following two cases:

(i) when the box has a given volume and its external surface area is a minimum;

(ii) when the external surface area is given and the volume is a maximum.

7 Prove that the curve $y = 4x^5 + kx^3$ has two turning points when $k < 0$ and none when $k \geqslant 0$. Find the turning points when $k = -5/3$, and distinguish between them. Sketch the curve for this value of k.

8 Show that for any integer k the point $(k\pi, k\pi)$ is a point of inflexion on the curve $y = x + \sin x$.

Show also that the tangent at any point of inflexion is parallel either to the ·line $y = 0$ or to the line $y = 2x$. Sketch the curve.

9 The curve $y = x^4 + ax^3 + bx^2$ has a point of inflexion at the point $(-2, 0)$. Find the values of the constants a and b, and show that the curve has another point of inflexion at the point $(-1/2, 15/16)$. Sketch the curve.

10 Show that the gradient of the curve $y = x^3 - 3x^2 + 4x - 1$ is never less than 1. Show also that this curve crosses its tangent at the point $(1, 1)$.

11 Find the maximum and minimum values of y where

$$y = 2 \cos x + \sin 2x.$$

Sketch the graph of y in the interval $0 \leqslant x \leqslant 2\pi$, and indicate on the graph the points of inflexion. Verify that the sign of $d^2 y/dx^2$ changes at each point of inflexion. [L]

12 If $(1 + x)(2 + y) = x^2 + y^2$, find dy/dx in terms of x and y. Find the gradient of the curve $(1 + x)(2 + y) = x^2 + y^2$ at each of the two points where the curve cuts the y-axis.

13 Show that, if $x^3 + y^3 = 3axy$, then $\dfrac{dy}{dx} = \dfrac{x^2 - ay}{ax - y^2}$.

14 Show that the line $2x - y = 0$ cuts at right angles the curve

$$4x^2 - 4xy + y^2 - 4x - 8y + 10 = 0.$$

15 Show that the stationary points of the curve

$$y = 1 + 2 \cos x + \cos 2x$$

occur when $\sin x + \sin 2x = 0$. Find all the stationary points lying in the range $0 \leqslant x \leqslant 2\pi$ and determine whether each is a maximum, a minimum or a point of inflexion. Find also the points in the range $0 \leqslant x \leqslant 2\pi$ at which $y = 0$. Sketch the graph of y in this range of x. [JMB]

16 Given that $f(x) \equiv \dfrac{2x - x^2}{x^2 - 2x - 3}$,

(a) state the values of x for which $f(x) = 0$, and the values of x for which $f(x) > 0$.

(b) show that $f(x) \to -1$ as $x \to \pm \infty$

(c) sketch the graph $y = f(x)$, showing particularly where the curve crosses the x-axis and how it approaches its asymptotes. [L]

17 Find the coordinates of the point of intersection of the curve

$$y = \frac{x^3 + 1}{x}$$

and the x-axis. Find also the gradients when $x = 1$ and $x = -1$, the value of x at the stationary point, and the nature of this stationary point. Sketch this curve and also the curve

$$y = \frac{x}{x^3 + 1}.$$ [L]

18 Sketch the graph of

$$y = \frac{(x-2)^2}{(x-6)(x+1)},$$

giving the coordinates of the points of intersection with the coordinate axes and the coordinates of the stationary points. Give the equations of the asymptotes of the curve.   [L]

19 Sketch the curves given by the equations
(a) $x = 2 + 2\cos t$, $y = \sin t$
(b) $x = 2a(1 + t)$, $y = a(1 - t^2)$.

20 The parametric equations of a curve are $x = \cos t$, $y = \sin^2 t$. The tangent to the curve at a point P cuts off intercepts OA and OB respectively on the positive x and y axes. Show that the area of triangle OAB is

$$(1 + \cos^2 t)^2 / (4 \cos t)$$

and determine the coordinates of P for this area to be a minimum. State the minimum area of the triangle.   [AEB]

21 If the expression $ax^2/(x + b)$ has a stationary value of 12 when $x = 3$, find the values of the constants a and b. Find also the second stationary value of the expression and determine whether it is a maximum or a minimum.

22 Find the coordinates of the points on the curve $3x^2 + 2xy + 3y^2 = 8$ at which the gradient is -1. Sketch the curve.

23 If $xy^2 = (x-1)^2$, prove that

$$y\frac{d^2y}{dx^2} + \left(\frac{dy}{dx}\right)^2 = \frac{1}{x^3}.$$

24 Determine the equations of the tangents at the points of inflexion on the curve $y = x^2(x-4)(x-6)$, and the coordinates of the points where these tangents meet the curve again.

25 A right circular cone of height $a + x$ is inscribed in a sphere of radius a. Find, in terms of a, the value of x for which the volume of the cone is a maximum.

26 An isosceles triangle has a constant area A. Equilateral triangles are described outwards on the three sides of the triangle as bases. Prove that the minimum area of the figure formed by the four triangles is $4A$.

27 A right circular cone of semi-vertical angle θ is inscribed in a sphere of given radius a. Show that the curved surface area of the cone is $4\pi a^2 \sin\theta \cos^2\theta$.

Find the maximum volume of the cone if θ may vary, and show that the curved surface area is a maximum when the volume is a maximum.

28 An ornament consists of two equal right circular cones whose bases are separated by a sphere. The axes of the cones and the centre of the sphere are in one straight line: the distance between the vertices of the cones is $2l$ and the radii of the bases of the cones are $\sqrt{(3/2)}l$, where l is a constant. Find the radius of the sphere if the total volume is to be a minimum.   [L]

29 A man on a lake at a distance a from the nearest point A of the straight shore wishes to reach a point B on the shore, where $AB = a$. He can proceed by boat, at speed u, direct to any point of the shore and then along the shore to B at speed $2u$. Find the least time in which he can get to B.

30 The loss of heat from a closed full hot-water tank is proportional to its surface area. A cylindrical tank has flat ends. If its volume is fixed, determine the ratio of the length to the radius for the loss of heat to be a minimum.

Find also whether such a tank would retain heat more efficiently than a cubical one of equal volume.   [JMB]

31 Cans in the form of right circular cylinders are to be manufactured from sheet metal. Some are to be open at one end and closed at the other and others are to be closed at both ends. Prove that, in order to obtain a maximum volume from a given area of sheet metal, the height of the cylinder should be made equal to the radius of the base in the one case and to the diameter of the base in the other.   [L]

10 Integration

10.1 The indefinite integral

By differentiating a function its derived function is obtained. The reverse process consists of finding the function when its derived function is known.

Since the derivative of x^3 is $3x^2$, the equation

$$\frac{dy}{dx} = 3x^2$$

is satisfied when $y = x^3$. But it is satisfied when y equals $(x^3 + 1)$, or $(x^3 - 2)$, or $(x^3 + c)$, where c is any constant. We call $(x^3 + c)$ the indefinite integral of $3x^2$, and we write

$$\int 3x^2 \, dx = x^3 + c,$$

where c is an arbitrary constant.

The derivative of $(x^2 + x)$ is $(2x + 1)$, and so

$$\int (2x + 1) \, dx = x^2 + x + c.$$

If n is a rational number not equal to -1, we have shown that

$$\frac{d}{dx}(x^{n+1}) = (n+1)x^n,$$

so that the indefinite integral of x^n is given by

$$\int x^n \, dx = \frac{1}{(n+1)}x^{n+1} + c.$$

Note that we increase the index n by 1 and divide by $(n+1)$.

For $n = 4$, we have $\int x^4 \, dx = (x^5/5) + c$.

For $n = -2$, we have $\int (1/x^2) \, dx = (-1/x) + c$.

For $n = \frac{1}{2}$, we have $\int x^{1/2} \, dx = \frac{2}{3}x^{3/2} + c$.

Example 1

Find $\displaystyle\int \frac{x^4 + 1}{x^2} \, dx$.

First break the expression to be integrated, i.e. the integrand, into two parts, which can then be dealt with separately. Since

$$\frac{x^4 + 1}{x^2} = x^2 + \frac{1}{x^2},$$

$$\int \frac{x^4+1}{x^2}\,dx = \int x^2\,dx + \int \frac{1}{x^2}\,dx = (x^3/3)-(1/x)+c.$$

Example 2
Find the indefinite integrals of sin x and cos x.

Since sin x is the derivative of $-\cos x$ we have

$$\int \sin x\,dx = -\cos x + c.$$

Since cos x is the derivative of sin x we have

$$\int \cos x\,dx = \sin x + c.$$

Example 3
The gradient at any point of a curve is given by

$$\frac{dy}{dx} = 4x^3 + 4x.$$

If the curve passes through the point (0, 2), find its equation.

First we find the indefinite integral of $(4x^3 + 4x)$.

$$y = \int 4x^3\,dx + \int 4x\,dx$$
$$= x^4 + 2x^2 + c.$$

The particular value required for the constant c is found by substituting $x = 0$ and $y = 2$. This gives $c = 2$, and so the equation of the curve is

$$y = x^4 + 2x^2 + 2.$$

Example 4
At time t seconds the speed of a point moving in a straight line is $(2t + t^2)$ metres per second. Find the distance travelled in 6 seconds from rest.

Let s m denote the distance travelled in time t s. Then the speed is given by

$$\frac{ds}{dt} = 2t + t^2.$$

It follows that

$$s = t^2 + (t^3/3) + c.$$

Since $s = 0$ when $t = 0$, the value required for the constant c is zero. Then when $t = 6$, $s = 108$, so that the distance travelled is 108 metres.

Exercises 10.1
1 Write down the indefinite integrals of
   (a) x^6 (c) $1+2x$ (e) $2x-3$
   (b) $4x^3$ (d) $6x^5$ (f) $x+3x^2$.

2 Find the indefinite integrals of
(a) $3x^{1/2}$ (c) $(x^2+1)/x^2$ (e) $x^{3/4}$
(b) $1/x^2$ (d) $12/x^7$ (f) $(1-x^2)^2$.

3 Find (a) $\int x^{1/3}\,dx$ (b) $\int 5x^{3/2}\,dx$ (c) $\int x^{-1/2}\,dx$.

4 Find (a) $\int 10t^4\,dt$ (b) $\int (1-t)^2\,dt$ (c) $\int t^{2/3}\,dt$.

5 Find (a) $\int u^{-3}\,du$ (b) $\int (2u+1)^3\,du$ (c) $\int u^{1/4}\,du$.

6 Find the indefinite integrals of
(a) $3\cos x - 2\sin x$ (d) $(x^2-2x)^2$
(b) $(\cos \tfrac{1}{2}x + \sin \tfrac{1}{2}x)^2$ (e) $(x+1)/\sqrt{x}$
(c) $(x+1)(2x-1)$ (f) $(x^2+1)^3/x^2$.

7 Find the equation of the curve
(a) with gradient $(4x-3)$, passing through the point $(1, 2)$
(b) with gradient $(4x-x^3)$, passing through the point $(2, 0)$
(c) with gradient $(2x+3x^2)$, passing through the point $(2, 3)$
(d) with gradient $(\cos x + \sin x)$, passing through the point $(0, 1)$.

8 At time t s, the velocity of a point moving in a straight line is v m s^{-1} and the distance it has travelled is s m.
(a) If $v = t(1+t)$ and $s = 10$ when $t = 0$, find s when $t = 6$.
(b) If $v = (6t^2 - 4t)$ and $s = 4$ when $t = 1$, find s when $t = 3$. .
(c) If $v = (1+3t)(3+t)$ and $s = 10$ when $t = 1$, find s when $t = 2$.

10.2   Integration by substitution

It is often possible to simplify an integral by a suitable change of variable. The aim is to convert the integral into a standard type.

Let $dy/dx = f(x)$ so that $y = \int f(x)\,dx$, and let x be a function of a variable t. By the chain rule,

$$\frac{dy}{dt} = \frac{dy}{dx} \times \frac{dx}{dt} = f(x) \times \frac{dx}{dt}.$$

It follows that

$$y = \int \left(f(x) \times \frac{dx}{dt} \right) dt.$$

If $x = g(t)$, this can be expressed in the form

$$\int f(x)\,dx = \int fg(t) \times g'(t)dt.$$

Example 1
$y = \int (2x+1)^3\,dx$.

The linear substitution $(2x+1) = t$ is suitable. Since $dx/dt = \tfrac{1}{2}$, we have

$$y = \int \left(t^3 \times \frac{1}{2} \right) dt = t^4/8 + c.$$

Hence $y = (2x+1)^4/8 + c$.

Example 2

$$y = \int \frac{4}{(4x-3)^2}\, \mathrm{d}x.$$

Let $(4x-3) = t$, so that $\mathrm{d}x/\mathrm{d}t = \frac{1}{4}$. Then

$$y = \int\left(\frac{4}{t^2} \times \frac{1}{4}\right)\mathrm{d}t = \int \frac{1}{t^2}\mathrm{d}t = -\frac{1}{t}+c.$$

Hence $\quad y = c - 1/(4x-3)$.

Example 3

$y = \int 6 \sin (3x)\, \mathrm{d}x.$

If we put $3x = t$, we have $\mathrm{d}x/\mathrm{d}t = 1/3$.
Hence $y = \int (6 \sin t)(1/3)\, \mathrm{d}t = -2\cos t + c.$
This gives $y = -2 \cos (3x) + c.$

Example 4

$\int 8x(x^2+2)^3\, \mathrm{d}x.$

Put $(x^2+2) = t$, so that $2x\,(\mathrm{d}x/\mathrm{d}t) = 1$.
The integral becomes

$$\int 4t^3\, \mathrm{d}t = t^4 + c = (x^2+2)^4 + c.$$

Example 5

$$\int \frac{1}{\sqrt{(1-x^2)}}\, \mathrm{d}x.$$

Use the substitution $x = \sin \theta$, which gives $\sqrt{(1-x^2)} = \cos \theta$ and $\mathrm{d}x/\mathrm{d}\theta = \cos \theta$.
Then

$$\int \frac{1}{\sqrt{(1-x^2)}}\, \mathrm{d}x = \int \frac{\cos \theta}{\cos \theta}\, \mathrm{d}\theta = \int \mathrm{d}\theta$$

$$= \theta + c = \arcsin x + c.$$

Example 6

$$\int \frac{1}{1+x^2}\, \mathrm{d}x.$$

The substitution $x = \tan \theta$ gives $\mathrm{d}x/\mathrm{d}\theta = \sec^2 \theta$ and $(1+x^2) = \sec^2 \theta$.
Then

$$\int \frac{1}{1+x^2}\, \mathrm{d}x = \int \frac{\sec^2 \theta}{\sec^2 \theta}\, \mathrm{d}\theta = \theta + c = \arctan x + c.$$

Example 7

$$\int \frac{1}{1+\cos x}\,dx.$$

Make the substitution $t = \tan(x/2)$. Then $x = 2\arctan t$ and $dx/dt = 2/(1+t^2)$.

Since
$$\cos x = (1-t^2)/(1+t^2),$$
$$1+\cos x = 2/(1+t^2).$$

We now have

$$\int \frac{1}{(1+\cos x)}\,\frac{dx}{dt}\,dt = \int \left(\frac{1+t^2}{2}\right)\left(\frac{2}{1+t^2}\right)\,dt$$

$$= \int dt = \tan(x/2)+c.$$

Example 8

$$\int \frac{1}{x^2+2x+2}\,dx.$$

Since $(x^2+2x+2) = (x+1)^2 + 1$, we put $(x+1) = \tan\theta$.
Then $(x^2+2x+2) = \sec^2\theta$ and $dx/d\theta = \sec^2\theta$.
The integral becomes

$$\int d\theta = \theta + c = \arctan(x+1)+c.$$

Exercises 10.2

Use the substitutions given to find the following indefinite integrals.

1 $\int (2x+3)^2\,dx,$ $(2x+3) = t.$

2 $\int (3x-2)^3\,dx,$ $(3x-2) = t.$

3 $\int x(x+1)^4\,dx,$ $(x+1) = t.$

4 $\int \sqrt{(4x-3)}\,dx,$ $(4x-3) = t.$

5 $\int \cos(3\theta)\,d\theta,$ $3\theta = t.$

6 $\int x^2(1+x^3)^2\,dx,$ $(1+x^3) = t.$

7 $\int \sin^2\theta\cos\theta\,d\theta,$ $\sin\theta = t.$

8 $\int (2x+1)^4\,dx,$ $(2x+1) = t.$

9 $\int \sec^2(2x)\,dx,$ $\tan(2x) = t.$

10 $\int x\sqrt{(1-x^2)}\,dx,$ $(1-x^2) = t^2.$

11 $\int \frac{2}{1+4x^2}\,dx,$ $2x = \tan\theta.$

12 $\int \frac{1}{\sqrt{(x+2)}}\,dx,$ $(x+2) = t.$

13 By means of the substitution $x = (\pi/2 - \theta)$, show that

$$\int \frac{1}{1+\sin x}\,dx = \int \frac{-1}{1+\cos\theta}\,d\theta.$$

Deduce that this integral equals $\tan(x/2 - \pi/4) + c$.

14 Show that $\displaystyle\int \frac{1}{x^2 + 4x + 5}\, dx = \arctan(x+2) + c$.

15 Show that $\displaystyle\int \frac{1}{x^2 + 4}\, dx = \tfrac{1}{2}\arctan(x/2) + c$.

16 By means of the substitution $x = \tan\theta$ show that

$$\int \frac{1}{(x^2+1)^{3/2}}\, dx = \frac{x}{(x^2+1)^{1/2}} + c.$$

Standard integrals

$$\int x^n\, dx = \frac{1}{n+1}\, x^{n+1} + c,\ n \neq 1$$
$$\int \sin x\, dx = -\cos x + c$$
$$\int \cos x\, dx = \sin x + c$$
$$\int \sec^2 x\, dx = \tan x + c$$
$$\int \frac{1}{1 + x^2}\, dx = \arctan x + c$$
$$\int \frac{1}{\sqrt{(1 - x^2)}}\, dx = \arcsin x + c,\ -1 < x < 1$$

10.3 Integration by parts

In certain cases the method of integration by parts can be applied to express a given integral in terms of a simpler integral.

The product rule for the derivative of $u(x)v(x)$ is

$$\frac{d}{dx}(u \times v) = u\frac{dv}{dx} + v\frac{du}{dx},$$

or

$$u\frac{dv}{dx} = \frac{d}{dx}(u \times v) - v\frac{du}{dx}.$$

By integrating with respect to x, the rule for integration by parts is obtained.

$$\int u\frac{dv}{dx}dx = u \times v - \int v\frac{du}{dx}dx.$$

The expression to be integrated must first be expressed as the product of two factors, one of which is taken as u and the other as dv/dx.

Example 1
Find $\int x \cos x\, dx$.

Take
$$u = x \quad \text{and} \quad \frac{dv}{dx} = \cos x.$$

Then
$$\frac{du}{dx} = 1 \quad \text{and} \quad v = \sin x.$$

By the rule for integration by parts,
$$\int x \cos x \, dx = x \sin x - \int (1) \ (\sin x) dx$$
$$= x \sin x + \cos x + c.$$

Example 2
Use integration by parts to find $\int 4x(x+1)^3 dx$.

Take
$$u = x \quad \text{and} \quad \frac{dv}{dx} = 4(x+1)^3.$$

Then
$$\frac{du}{dx} = 1 \quad \text{and} \quad v = (x+1)^4.$$

$$\int 4x(x+1)^3 dx = x(x+1)^4 - \int (x+1)^4 dx$$
$$= x(x+1)^4 - (x+1)^5/5 + c$$
$$= (4x-1)(x+1)^4/5 + c.$$

(The substitution $x+1 = t$ would give the result more quickly.)

Example 3
Find $\int \arcsin x \, dx$.

Take
$$u = \arcsin x \quad \text{and} \quad \frac{dv}{dx} = 1.$$

Then
$$\frac{du}{dx} = \frac{1}{\sqrt{(1-x^2)}} \quad \text{and} \quad v = x.$$

$$\int \arcsin x \, dx = x \arcsin x - \int \frac{x}{\sqrt{(1-x^2)}} dx.$$

By the substitution $(1-x^2) = t^2$ the integral on the right-hand side becomes
$$\int \frac{1}{\sqrt{(1-x^2)}} \left(x \frac{dx}{dt} \right) dt = \int \frac{1}{t}(-t) dt = -t + c.$$

This gives
$$\int \arcsin x \, dx = x \arcsin x + \sqrt{(1-x^2)} + c.$$

Exercises 10.3
1 Show that
(a) $\int x \sin x \, dx = \sin x - x \cos x + c$
(b) $\int x^2 \sin x \, dx = (2-x^2)\cos x + 2x \sin x + c.$
2 Find (a) $\int x \cos 2x \, dx$ (b) $\int x \sin 2x \, dx.$
Check your answers by differentiation.

3   Find (a) by integration by parts, (b) by the substitution $(2x + 3) = t$, the
    indefinite integral of $x(2x + 3)^4$.
4   Find $\int x \cos 4x \, dx$.
5   Find $\int x \sin 5x \, dx$.
6   Show that $\int x^2 \cos x \, dx = (x^2 - 2) \sin x + 2x \cos x + c$.
7   Show that $\int 2x \arctan x \, dx = (1 + x^2) \arctan x - x + c$.
8   Show that $\int 4x \sin^2 x \, dx = x^2 - x \sin 2x - \frac{1}{2} \cos 2x + c$.
9   Find $\int 2x/(x + 1)^3 \, dx$
    (a) by the substitution $(x + 1) = t$,
    (b) by integration by parts.
10 Find (a) $\int x \cos (x/2) dx$
         (b) $\int x \sin (x/2) dx$.

10.4   The definite integral

Let the derivative of $F(x)$ be $f(x)$, so that

$$\int f(x) \, dx = F(x) + c,$$

where c is an arbitrary constant. The symbol

$$\int_a^b f(x) \, dx$$

denotes the difference $F(b) - F(a)$, which can also be expressed as $\left[F(x) \right]_a^b$. Thus

$$\int_a^b f(x) \, dx = \left[F(x) \right]_a^b = F(b) - F(a).$$

This is known as the definite integral of $f(x)$ from the lower limit a to the upper
limit b. (a and b are not limits in the sense of section 6.1, but are merely the end-
points of the range of integration.) The effect of interchanging a and b is to change
the sign of the integral, for

$$\int_b^a f(x) \, dx = F(a) - F(b) = - \int_a^b f(x) \, dx.$$

It is easily seen that the definite integral will not contain any arbitrary constant.
Consider for example the definite integral

$$\int_1^2 3x^2 \, dx.$$

We have $\int 3x^2 \, dx = x^3 + c$, and so

$$\int_1^2 3x^2 \, dx = \left[x^3 + c \right]_1^2 = (8 + c) - (1 + c) = 7.$$

There are four steps in the evaluation of $\int_a^b f(x)\,dx$.

Step 1 Find $F(x)$ such that $F'(x) = f(x)$.
Step 2 Evaluate $F(b)$.
Step 3 Evaluate $F(a)$.
Step 4 Simplify $F(b) - F(a)$.

Example 1

Evaluate $\displaystyle\int_{-1}^{3} (6x^2 - 4x - 5)\,dx$.

Let $F'(x) = 6x^2 - 4x - 5$. Then $F(x) = 2x^3 - 2x^2 - 5x$ (there is no need to add a constant).

$$F(3) = 54 - 18 - 15 = 21.$$
$$F(-1) = -2 - 2 + 5 = 1.$$
$$F(3) - F(-1) = 21 - 1 = 20.$$

More briefly,

$$\int_{-1}^{3} (6x^2 - 4x - 5)\,dx = \left[2x^3 - 2x^2 - 5x \right]_{-1}^{3}$$
$$= 21 - 1$$
$$= 20.$$

Example 2

Evaluate $\displaystyle\int_0^{\pi} \sin^2 x\,dx$.

Since $\cos 2x$ equals $(1 - 2\sin^2 x)$, $\sin^2 x$ equals $\frac{1}{2}(1 - \cos 2x)$.
The integral becomes

$$\int_0^{\pi} \tfrac{1}{2}(1 - \cos 2x)\,dx = \left[\tfrac{1}{2}x - \tfrac{1}{4}\sin 2x \right]_0^{\pi} = \pi/2.$$

Exercises 10.4

Evaluate the integrals

1 $\displaystyle\int_{-1}^{3} x\,dx$

2 $\displaystyle\int_{1}^{4} x^2\,dx$

3 $\displaystyle\int_{2}^{4} x^3\,dx$

4 $\displaystyle\int_{-2}^{2} (x^2 + x)\,dx$

5 $\displaystyle\int_{1}^{2} \frac{x^3 + 2x}{x^3}\,dx$

6 $\displaystyle\int_{-1}^{1} x(x+1)(x+2)\,dx$

7 $\displaystyle\int_{-1}^{1} (2x+1)^3\,dx$

8 $\displaystyle\int_{1}^{2} (1 - 8/x^3)\,dx$

9	$\displaystyle\int_0^{\pi/2} \cos 2x \, dx$	11	$\displaystyle\int_0^{\pi/2} \sin 2x \, dx$
10	$\displaystyle\int_0^{\pi/2} \sin (x/2) \, dx$	12	$\displaystyle\int_{-\pi}^{\pi} \cos^2 (x/4) \, dx$

10.5 Change of limits

When a definite integral is evaluated by the method of substitution, there must be a one-to-one correspondence between the values of the old and the new variables. As the old variable moves from its lower limit to its upper limit, the new variable must move in one direction from the new lower limit to the new upper limit. This will be true for the substitution $x = g(t)$ provided that an inverse function g′ exists in a suitable interval. For example, if $x = \tan t$, we restrict t to lie in the interval $-\pi/2 < t < \pi/2$, giving one-to-one correspondence between x and t.

Example

Evaluate $\displaystyle\int_{-1}^{1} (2x-1)(x+1)^3 \, dx$.

Put $(x+1) = t$. Then $dx/dt = 1$. As x moves from -1 to 1, t moves from 0 to 2.

$$\int_0^2 (2t-3)t^3 \frac{dx}{dt} \, dt = \int_0^2 (2t^4 - 3t^3) \, dt$$

$$= \left[2t^5/5 - 3t^4/4 \right]_0^2$$

$$= 64/5 - 48/4 = 4/5.$$

Exercises 10.5

1 By putting $x = 3\tan t$, show that $\displaystyle\int_0^3 \frac{1}{x^2+9} \, dx = \pi/12$.

2 By putting $\sin x = t$, show that $\displaystyle\int_0^{\pi/2} \sin^3 x \cos x \, dx = 1/4$.

3 Evaluate $\displaystyle\int_{-2}^{1} \sqrt{(x+3)} \, dx$ by putting $(x+3) = t$

4 Evaluate $\displaystyle\int_0^1 2x(x^2+1)^3 \, dx$ by putting $(x^2+1) = t$.

5 Evaluate $\displaystyle\int_0^1 \sqrt{(1-x^2)} \, dx$ by putting $x = \sin \theta$.

10.6 The area under a curve

In Fig. 10.1 the points A and B on the x-axis are given by $x = a$ and $x = b$ respectively. The region $ABCD$ is bounded by the x-axis, the lines $x = a$, $x = b$ and the arc DC of the curve $y = f(x)$. The area of this region is called the area

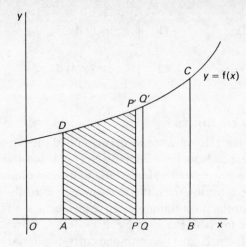

Fig. 10.1

under the curve $y = f(x)$ between $x = a$ and $x = b$. Let P' and Q' be the points (x, y), $(x + \delta x, y + \delta y)$ on the arc DC, and let P, Q be the points $(x, 0)$, $(x + \delta x, 0)$ respectively. The area of the shaded region to the left of PP' depends on the value of x at P. This area will be denoted by $F(x)$. If x is increased to $(x + \delta x)$, the increase in $F(x)$ will be larger than $PP' \times \delta x$ but less than $QQ' \times \delta x$, i.e.

$$y\,\delta x < F(x + \delta x) - F(x) < (y + \delta y)\,\delta x$$

or
$$y < \frac{F(x + \delta x) - F(x)}{\delta x} < y + \delta y.$$

In the limit as δx tends to zero, δy will also tend to zero and it follows that

$$\frac{dF}{dx} = y = f(x).$$

Then $F(x) + c$ is the indefinite integral of $f(x)$.

Now the area of the region $ABCD$ equals the change in the value of $F(x)$ as x increases from a to b. Hence the area beneath the curve from $x = a$ to $x = b$ is given by the definite integral

$$F(b) - F(a) = \int_a^b f(x)\,dx.$$

Notice that the inequalities above are true only when the gradient of the curve is positive. If the gradient is negative, the inequalities are reversed but the same result is obtained.

We have taken y to be positive in the range of integration $a \leqslant x \leqslant b$. If y is negative throughout this range, the definite integral will give a negative value. If y changes sign once in the range, the definite integral will give the difference between two areas.

Example 1
Find the area under the curve $y = 3x^2 - 2x + 1$ between $x = -1$ and $x = 2$.

The curve is shown in Fig. 10.2. The required area equals

$$\int_{-1}^{2} (3x^2 - 2x + 1)\,dx = \left[x^3 - x^2 + x \right]_{-1}^{2}$$
$$= 6 - (-3) = 9.$$

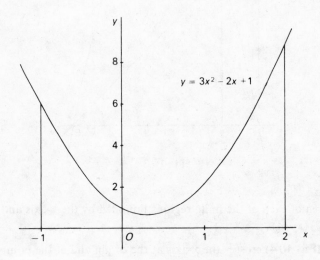

Fig. 10.2

Example 2
Find the area of the region lying above the line $x + y = 5$ and beneath the curve $y = 5x - x^2$.

The curve and the line (Fig. 10.3) meet at the points $A(5, 0)$ and $B(1, 4)$. The area beneath the curve between $x = 1$ and $x = 5$ equals

$$\int_{1}^{5} (5x - x^2)\,dx = \left[5x^2/2 - x^3/3 \right]_{1}^{5}$$
$$= (125/2 - 125/3) - (5/2 - 1/3) = 18\tfrac{2}{3}.$$

The area of the triangle CAB is 8, and so the required area is $10\tfrac{2}{3}$.

Alternatively, the required area is given by

$$\int_{1}^{5} (5x - x^2)\,dx - \int_{1}^{5} (5 - x)\,dx = \int_{1}^{5} (-5 + 6x - x^2)\,dx$$

$$= \left[-5x + 3x^2 - x^3/3 \right]_{1}^{5} = 8\tfrac{1}{3} + 2\tfrac{1}{3} = 10\tfrac{2}{3}.$$

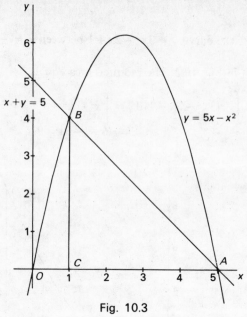

Fig. 10.3

Example 3
Find the area of each of the finite regions bounded by the x-axis and the curve
$y = x(x-2)(x-4)$.

The curve (Fig. 10.4) crosses the x-axis at the origin and at the points given by
$x = 2$ and $x = 4$. For one region we have

$$\int_0^2 y\, dx = \int_0^2 (x^3 - 6x^2 + 8x)\, dx$$

$$= \left[x^4/4 - 2x^3 + 4x^2 \right]_0^2 = 4.$$

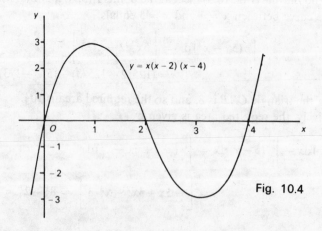

$y = x(x-2)(x-4)$

Fig. 10.4

For the other region we have

$$\int_2^4 y\,dx = \int_2^4 (x^3 - 6x^2 + 8x)\,dx$$

$$= \left[x^4/4 - 2x^3 + 4x^2 \right]_2^4 = -4.$$

The area of each region is 4, but the second integral gives a negative answer because y is negative in the range of integration. If we integrate y from $x = 0$ to $x = 4$, the result will be zero, the difference between the two areas, not their sum.

Example 4

A curve is defined by the parametric equations $x = t^2$, $y = t^3$. Calculate the area under the arc of this curve from $t = 1$ to $t = 2$ (Fig. 10.5).

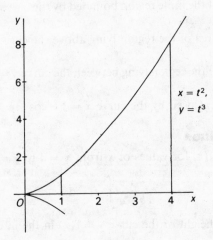

Fig. 10.5

Since $x = 1$ when $t = 1$, and $x = 4$ when $t = 2$, y has to be integrated from $x = 1$ to $x = 4$. The area can be expressed as an integral with respect to t from $t = 1$ to $t = 2$.

$$\text{Area} = \int_1^4 y\,dx = \int_1^2 y\frac{dx}{dt}\,dt$$

$$= \int_1^2 (t^3)(2t)\,dt = \left[2t^5/5 \right]_1^2 = 62/5.$$

Exercises 10.6

1   Find the area beneath the curve $y = 4x^3 - x$ from $x = 1$ to $x = 3$.
2   Find the area beneath the curve $y = 4x + 6x^2$ from $x = \frac{1}{2}$ to $x = 1\frac{1}{2}$.

3 Find the area between the curve $y = (x + 2)(3 - x)$ and the x-axis.

4 Find the area beneath the curve $y = 2x(3x - 1)$ from $x = 2$ to $x = 3$.

5 The points P and Q are given by $t = 1$ and $t = 4$ respectively on the curve $x = at^2$, $y = 2at$. Find the area beneath the arc PQ.

6 Find the area beneath the curve $y = 1/x^2$ from $x = \frac{1}{2}$ to $x = 2$.

7 Find the area beneath the curve $y = 2x^2 + 2x + 3$ from $x = -1$ to $x = 2$.

8 Find the area beneath (a) the curve $y = \sin x$, (b) the curve $y = \cos x$, from $x = 0$ to $x = \pi/2$.

9 Calculate the area of the region in the first quadrant enclosed by the axes and the curve $x = 3 \cos t$, $y = 2 \sin t$.

10 Find the area of the region bounded by the line $y = x$ and the curve $y = 9 + x - x^2$.

11 Find the area of the region lying below the line $y = x$ but above the curve $y = x^2$.

12 Find the area of the finite region bounded by the curves $y = x^2 - 2x + 2$ and $y = 2 - x^2$.

13 Calculate the area of the region lying above the curve $y = x^2 - 4x + 3$ and below the x-axis.

14 Find the area of the region lying between the curve $y = 2 - x^2$ and the curve $y = 2x^2 - 1$.

15 Find the area enclosed by the curve $x = 1 - \cos t$, $y = 4 \sin t$.

10.7 Mean values

The mean value of f(x) for values of x from $x = a$ to $x = b$ is defined to be

$$\frac{1}{b-a} \int_a^b f(x)dx.$$

This is the average height of the curve $y = f(x)$ in the interval $a \leqslant x \leqslant b$.

Example 1

Find the mean value of $\sin^2 x$ from $x = 0$ to $x = \pi$.

Since $2 \sin^2 x = 1 - \cos 2x$, we have

$$\int_0^\pi \sin^2 x \, dx = \int_0^\pi \tfrac{1}{2}(1 - \cos 2x)dx$$

$$= \left[\tfrac{1}{2}(x - \tfrac{1}{2} \sin 2x) \right]_0^\pi = \pi/2.$$

The length of the interval is π, and so the mean value is given by

$$\frac{1}{\pi} \int_0^\pi \sin^2 x \, dx = \tfrac{1}{2}.$$

Example 2

Find the mean value of $(x^3 + 6x^2 + 4x)$ over the interval $-2 \leqslant x \leqslant 2$.

$$\int_{-2}^{2} (x^3 + 6x^2 + 4x)\, dx = \left[x^4/4 + 2x^3 + 2x^2 \right]_{-2}^{2}$$

$$= (28) - (-4) = 32.$$

As the length of the interval is 4, the mean value is 8.

Exercises 10.7

1 Find the mean value of (a) $\sin x$ (b) $\cos x$, over the interval $0 \leqslant x \leqslant \pi/2$.
2 Find the mean value of x^3 from $x = 2$ to $x = 4$.
3 Find the mean value of $1/x^2$ from $x = 1$ to $x = 5$.
4 Find the mean value of $(\cos x - \sin x)$ from $x = 0$ to $x = \pi$.
5 If $u = 2\sin t$ and $v = 1 + \cos t$, find the mean value of uv from $t = 0$ to $t = \pi$.
6 Find the mean value of $1/(1 + x^2)$ for values of x such that $x^2 \leqslant 3$.
7 Find the mean value of $\arcsin x$ from $x = 0$ to $x = \frac{1}{2}$.
8 Find the mean value of $x(x - 2)(x - 4)$ from $x = 2$ to $x = 4$.
9 Find the mean value of $\sin x \cos^2 x$ from $x = 0$ to $x = \pi/2$.
10 Find the mean value of $1/(2x + 1)^3$ from $x = 0$ to $x = 2$.
11 Find the mean value of $\sin 2x \cos 2x$ from $x = 0$ to $x = \pi/4$.
12 Find the mean value of $(1 + \tan^2 x)$ from $x = 0$ to $x = \pi/4$.

10.8 The definite integral as the limit of a sum

So far integration has been treated as the reverse of differentiation. However, of the two processes, integration is very much the older; it originates from the problem of calculating areas and volumes.

In Fig. 10.6, A and B are the points $(a, 0)$ and $(b, 0)$ respectively. The region

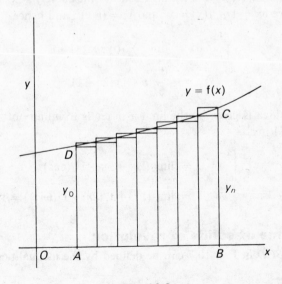

Fig. 10.6

$ABCD$ is bounded by the x-axis, the lines $x = a$, $x = b$ and the arc DC of the curve $y = f(x)$. The gradient of the curve is positive, but the following argument will still apply when the gradient is negative.

The region is divided into n strips each of width δx, where $\delta x = (b-a)/n$. Let the lengths of the vertical edges of the strips be denoted by

$$y_0, y_1, y_2, \ldots, y_r, \ldots, y_n$$

where $y_0 = f(a)$, $y_r = f(a+r\,\delta x)$ and $y_n = f(b)$. The area of the strip with sides of length y_{r-1} and y_r lies between $y_{r-1}\,\delta x$ and $y_r\,\delta x$, and so the area of the region $ABCD$ lies between

and
$$(y_0 + y_1 + \ldots + y_{r-1} + \ldots + y_{n-1})\delta x$$
$$(y_1 + y_2 + \ldots + y_r + \ldots + y_n)\delta x,$$

i.e. between
$$\sum_{r=1}^{n} y_{r-1}\,\delta x \text{ and } \sum_{r=1}^{n} y_r\,\delta x.$$

The difference between these two sums is $(y_n - y_0)\,\delta x$, which tends to zero as n tends to infinity and δx tends to zero. If the first sum tends to a limit as n tends to infinity, the second sum will tend to the same limit. It follows that the area of the region $ABCD$ is given by

$$\int_a^b f(x)\,dx = \lim_{n\to\infty} \sum_{r=1}^{n} y_{r-1}\,\delta x = \lim_{n\to\infty} \sum_{r=1}^{n} y_r\,\delta x,$$

where $y_r = f(a+r\,\delta x)$ and $\delta x = (b-a)/n$.

In this way a definite integral can be defined as the limit of a sum, but apart from a few cases such a limit is difficult to calculate.

As an illustration, let us find the area beneath the curve $y = x^2$ from $x = 0$ to $x = 1$. We have $\delta x = 1/n$, $f(x) = x^2$ and $y_r = (r/n)^2$, and hence

$$\int_0^1 x^2\,dx = \lim_{n\to\infty} \sum_{r=1}^{n} (r^2/n^2)(1/n).$$

Now
$$\sum_{r=1}^{n} r^2 = \frac{1}{6}n(n+1)(2n+1).$$

Therefore the area is given by the limit as n tends to infinity of $n(n+1)(2n+1)/(6n^3)$.

$$\int_0^1 x^2\,dx = \lim_{n\to\infty} (2n^2 + 3n + 1)/(6n^2)$$
$$= \lim_{n\to\infty} (1/3 + 1/(2n) + 1/(6n^2)) = 1/3.$$

10.9 Volume of solids of revolution

The region $ABCD$ in Fig. 10.7 can be defined by the inequalities $a \leqslant x \leqslant b$, $0 \leqslant y \leqslant f(x)$.

If this region is rotated completely about the x-axis, it will sweep out a solid of

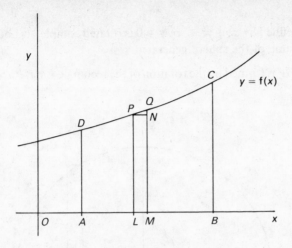

Fig. 10.7

revolution. Let P and Q be the points (x, y) and $(x + \delta x, y + \delta y)$ respectively on the arc DC, so that $LM = \delta x$. Let δV denote the volume swept out by the strip $LMQP$. Provided that the gradient of the curve is positive, δV will be greater than $\pi LP^2 \times LM$ and less than $\pi MQ^2 \times LM$, i.e.

$$\pi y^2 \delta x < \delta V < \pi (y + \delta y)^2 \, \delta x.$$

Then

$$\pi y^2 < \frac{\delta V}{\delta x} < \pi (y + \delta y)^2.$$

When δx tends to zero this gives $dV/dx = \pi y^2$.

The same result follows when the gradient of the curve is negative. It follows that the volume V swept out by the region $ABCD$ is given by

$$V = \int_a^b \pi y^2 \, dx.$$

To express this volume as a limit of a sum, AB is divided into n equal parts each of length δx. A rectangle such as $LMNP$ will sweep out a disc of volume $\pi y^2 \, \delta x$. This disc is called an element of volume. The sum of the n such elements is

$$\sum_{r=0}^{n-1} \pi y_r^2 \, \delta x,$$

where $y_r = f(a + r \, \delta x)$. As n tends to infinity and δx tends to zero, the limit of this sum will give the volume of the solid of revolution.

$$V = \int_a^b \pi y^2 \, dx = \lim_{n \to \infty} \sum_{r=0}^{n-1} \pi y_r^2 \, \delta x.$$

Example 1

The region defined by $x^2 + y^2 \leqslant a^2$, $y \geqslant 0$ is rotated completely about the x-axis. Find the volume of the sphere generated.

The element δV is formed by the rotation of the rectangle $LMNP$ about the x-axis (Fig. 10.8).

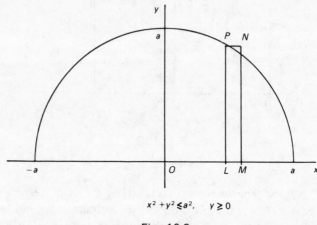

$$x^2 + y^2 \leqslant a^2, \qquad y \geqslant 0$$

Fig. 10.8

$$\delta V = \pi LP^2 \times LM = \pi y^2 \times \delta x.$$

Then
$$V = \int_{-a}^{a} \pi y^2 \, dx = \int_{-a}^{a} \pi(a^2 - x^2) \, dx$$

$$= \pi \left[a^2 x - x^3/3 \right]_{-a}^{a} = 4\pi a^3/3.$$

Example 2

A bowl is formed by rotating completely about the y-axis the arc of the parabola $4ay = x^2$ from the origin to the point $(2a, a)$. Find the volume of the bowl.

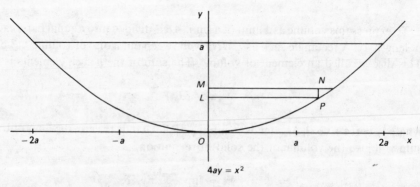

$$4ay = x^2$$

Fig. 10.9

The element of volume δV is formed by rotating the rectangle $LMNP$ about the y-axis (Fig. 10.9).

$$\delta V = \pi LP^2 \times LM = \pi x^2 \times \delta y.$$

$$V = \int_0^a \pi x^2 \, dy = \int_0^a 4\pi a y \, dy = \left[2\pi a y^2 \right]_0^a = 2\pi a^3.$$

Example 3

A curve is given by the equations $x = t^3$, $y = 2t^2$. The region bounded by the arc of this curve from the origin to the point (8, 8), the x-axis and the line $x = 8$ is rotated completely about the x-axis. Find the volume swept out.

The element of volume δV is swept out by the rectangle $LMNP$, where $LM = \delta x$ and $LP = y$ (Fig. 10.10).

$$\delta V = \pi LP^2 \times LM = \pi y^2 \times \delta x.$$

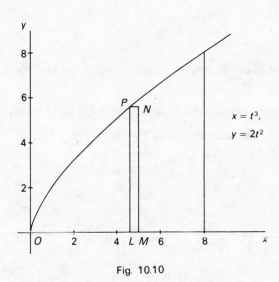

$$x = t^3,$$
$$y = 2t^2$$

Fig. 10.10

Now $y = 2t^2$, $\delta x = 3t^2 \, \delta t$ and t increases from 0 to 2.

Hence $$V = \int_0^2 \pi (4t^4) 3t^2 \, dt = 12\pi \left[t^7/7 \right]_0^2 = 1536\pi/7.$$

Example 4

The segment of the circle $x^2 + y^2 \leq 1$ for which $x \geq \frac{1}{2}$ is rotated completely about the y-axis. Find the volume generated.

In Fig. 10.11, PQ is one edge of a rectangle of width δx at a distance x from the

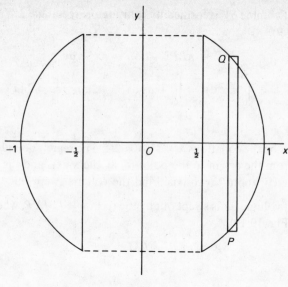

Fig. 10.11

y-axis. The length of the rectangle is $2y$, and it sweeps out an element of volume δV such that

$$(2\pi x)\,(2y)\,\delta x < \delta V < 2\pi(x+\delta x)\,(2y)\,\delta x.$$

By dividing by δx and letting δx tend to zero, we get

$$\frac{dV}{dx} = 4\pi xy.$$

Hence

$$V = 4\pi \int_{\frac{1}{2}}^{1} x\sqrt{(1-x^2)}\,dx = (-4\pi/3)\left[(1-x^2)^{3/2}\right]_{1/2}^{1} = \pi\sqrt{3}/2.$$

Exercises 10.9

1 Find the volume generated when the region beneath the curve $y = x^2 + 3$ from $x = -1$ to $x = 1$ is rotated completely about the x-axis.

2 The sides of a triangle are the x-axis, the line $x = h$ and the line $y = x\tan\alpha$. If this triangle is rotated completely about the x-axis, find the volume swept out.

3 Find the volume generated when the region beneath the curve $y = x^2(1-x)$ between $x = 0$ and $x = 1$ is rotated completely about the x-axis.

4 The region beneath the curve $y = \cos x$ for which $0 \leqslant x \leqslant \pi/2$ is rotated completely about the x-axis. Find the volume of the solid of revolution formed.

5 P and Q are the points on the curve $x = ct$, $y = c/t$ given by $t = \frac{1}{2}$ and $t = 2$.

Find the volume swept out when the region beneath the arc PQ is rotated completely about the x-axis.

6 The arc of the curve $y^2 = x(x-2)^2$ for which $0 \leqslant x \leqslant 2$, $y \geqslant 0$ is rotated completely about the x-axis. Find the volume enclosed by the surface generated.

7 The arc of the curve $y = 3x^2 - x^3$ for which $0 \leqslant x \leqslant 3$ is rotated completely about the x-axis. Show that the volume enclosed is $729\pi/35$.

8 The points A and B on the parabola $x = at^2$, $y = 2at$ are given by $t = 1$ and $t = 3$. Find the volume generated when the area beneath the arc AB is rotated completely about the x-axis.

9 The finite region bounded by the y-axis, the line $y = 8$ and the curve $y = x^3$ is rotated completely about the y-axis. Find the volume swept out.

10 The finite region bounded by the parabola $y^2 = 4ax$ and the line $x = a$ is rotated completely about the y-axis. Find the volume swept out.

Miscellaneous exercises 10

1 Use the identity

$$2 \cos A \cos B = \cos (A + B) + \cos (A - B)$$

to evaluate

$$\int_0^{\pi/4} \cos (3x) \cos (5x)\,dx.$$

Evaluate also

$$\int_0^{\pi/4} \sin (3x) \sin (5x)\,dx.$$

2 Show by a suitable substitution or graph that

(a) $\displaystyle\int_0^{\pi/2} \sin^2 x\,dx = \int_0^{\pi/2} \cos^2 x\,dx$

(b) $\displaystyle\int_0^{\pi/2} \frac{\sin x}{\sin x + \cos x}\,dx = \int_0^{\pi/2} \frac{\cos x}{\sin x + \cos x}\,dx.$

Deduce that the value of each integral is $\pi/4$.

3 The gradient of a curve at the point (x, y) is $4x + 8$. The curve passes through the point $(-1, 1)$. Show that the curve touches the line $4x + 5 = 2y$.

4 Find the area of the loop of the curve given by $x = 1 - t^2$, $y = t - t^3$.

5 Shade in a diagram the region defined by the inequalities $1 \leqslant y \leqslant 2$, $0 \leqslant x \leqslant y^2$. Calculate the area of this region.

6 Show that

(a) $\displaystyle\int_0^{\pi/2} x \sin x \cos x\,dx = \pi/8$

(b) $\displaystyle\int_0^1 \frac{1}{(1+x)\sqrt{x}}\,dx = \pi/2.$

7 Given that $\displaystyle\sum_{r=1}^{n} r^3 = \tfrac{1}{4}n^2(n+1)^2$, show that $\displaystyle\int_0^1 x^3\,dx = \tfrac{1}{4}$ by finding the limit of a sum.

8 Evaluate

(i) $\displaystyle\int_0^{\pi/2} (2\sin^2 x + 5\sin 5x)\,dx$

(ii) $\displaystyle\int_1^2 \left(x + \frac{1}{x}\right)^4\,dx$

(iii) $\displaystyle\int_0^1 \frac{x}{\sqrt{(3x^2+1)}}\,dx$

(iv) $\displaystyle\int_0^5 \left(\frac{1}{\sqrt{(x+4)}} + \sqrt{(x+4)}\right)dx.$

9 Sketch the curve $y = \tfrac{1}{2}(x-2)(8-x)$ and find the coordinates of its turning point.

Find the area of the region enclosed by the curve and the x-axis.

Show that the normal at the point $(3, 2\tfrac{1}{2})$ divides this area in the ratio 91 : 125. [L]

10 Find the x coordinates of the three points of intersection of the curves $y = x^3 - 9x + 11$ and $y = 3(x^2 - x + 1)$.

Prove that the two loops contained between the two curves are of equal area.

11 Find the area of the region bounded by the curve

$$y = \frac{x(x+6)}{(x+3)^2},$$

the x-axis and the ordinates $x = 1$, $x = 3$.

12 Sketch the curve $2y = x(x-4)(2x-5)$.

The line $y = x$ cuts this curve at the origin O and at A and B, where A is between O and B. Find the area of the region bounded by the arc OA of the curve and the line $y = x$. [L]

13 The region enclosed by the curves $y = 2x^2$ and $y = x^4 + 1$ from $x = 0$ to $x = 1$ is revolved completely about the x-axis. Find the volume of the solid of revolution formed.

14 The tangents to the curve $y = \cos x$ at A, where $x = 0$, and at B, where $x = \tfrac{1}{2}\pi$, meet at C. Find the coordinates of the point C and the area of the region enclosed by the lines AC, CB and the arc BA of the curve.

15 Sketch the arc of the curve $y = 2x - x^2$ for which y is positive. Find the area of the region which lies between this arc and the x-axis. If this region is

rotated completely about the x-axis find the volume of the solid of revolution generated.

16 Find the coordinates of the points in which the line $y = 1 - k^2$ cuts the curve $y = 2x - x^2$.

  The region bounded by the portion of this curve for which $y \geqslant 0$, and the x-axis, is to be divided into three regions of equal area by two lines each parallel to the x-axis. Find the equations of the required lines.

17 Integrate with respect to x
  (a) $4 \sin 2x$ (c) $1 + \tan^2 x$ (e) $3x^3 \sqrt{(x^4 + 1)}$
  (b) $x / \sqrt{(1 + x^2)}$ (d) $\tan^2 x$ (f) $8 \sin^2 x \cos^2 x$.

18 Evaluate the integrals

  (a) $\displaystyle\int_0^1 x \sqrt{(1 - x)} dx$ (b) $\displaystyle\int_0^{\pi/2} \sin\left(\frac{\pi}{4} - \frac{x}{2}\right) dx$.

19 Evaluate

  (a) $\displaystyle\int_0^{1/2} (1 - 2x)^7 dx$.   (b) $\displaystyle\int_{1/2}^1 \frac{(x - 1)(x - 2)}{x^4} dx$.

20 Find the area of the region enclosed by the loop of the curve $y^2 = x(4 - x)^2$.

  Also find the volume obtained by revolving the upper half of the loop through four right angles
  (i) about the x-axis
  (ii) about the y-axis.

21 A region is enclosed by the curve $y = x - x^2$ from $x = 0$ to $x = 1$ and the axis of x. Find the area of this region, and the volume swept out when the region is revolved completely round the axis of x.

22 Draw the graph of the curve $9y^2 = x(3 - x)^2$.

  If the loop of the curve is rotated about the x-axis, find the volume of the solid of revolution formed.

23 Obtain the equation of the tangent at the point $(2, 8)$ to the curve $y = x^3$.

  The region in the first quadrant enclosed by the curve, this tangent and the x-axis is rotated through four right angles about the x-axis. Find the volume swept out.

24 The region enclosed by the curve $y = \sin(x + \pi/6)$, from $x = 0$ to $x = 5\pi/6$, and the x- and y-axes is rotated completely about the x-axis. Find the volume swept out.

25 The part of the curve $y = 3(x^2 - 1)$ from $x = 1$ to $x = 3$ is rotated about the y-axis. Find the volume of the solid of revolution so generated.

26 Differentiate $\arcsin(2x - 1)$ and hence evaluate

$$\int_{1/4}^{3/4} \frac{dx}{\sqrt{(x - x^2)}}.$$

27 Show that

$$\int_0^1 \frac{1-2x^2}{(1+x^2)(1+4x^2)}\,dx = \arctan 2 - (\pi/4).$$

28 The region enclosed by the curve $y^4 = x$, the line $x = 1$, and the x-axis is rotated about the x-axis through four right angles. Find the volume of the solid of revolution thus produced.

29 Sketch the curve with equation $y = x - 1/x$.

   The region bounded by the curve, the x-axis and the lines $x = 2$, $x = 3$ is rotated through 2π radians about the x-axis. Calculate the volume of the solid of revolution so formed.

30 The curved surface of an open bowl with a flat circular base may be traced out by the complete revolution of a portion of the curve $ay = x^2$ about the vertical axis Oy. The radius of the top rim of the bowl is twice that of the base, and the capacity of the bowl is $\frac{5}{6}\pi a^3$ units^3. Find the vertical height of the bowl.   [JMB]

31 Find the mean value of $x \sin x$, (a) from $x = 0$ to $x = \pi$, (b) from $x = \pi$ to $x = 2\pi$, (c) from $x = 0$ to $x = 2\pi$.

32 Find the area A of the region in the x–y plane bounded by the x-axis, the arc of the curve $x = u^2$, $y = u^3$ from the origin O to the point $P(p^2, p^3)$, where $p > 0$, and the ordinate through P. Find also the volume V swept out when this region is rotated through an angle of 2π about the x-axis.

   If P moves along the curve in such a way that the rate of increase of the area A with respect to time is constant and equal to 3 units, show that at the instant when $p = 1$, the rate of increase of p is $3/2$. At this instant, find the rates of increase of:

   (i) the volume V
   (ii) the distance between O and P
   (iii) the angle between OP and the x-axis.   [JMB]

11 Natural logarithms and the exponential function

11.1 Indices and logarithms

For any real number a and any positive integer n the product of n factors each equal to a is denoted by a^n. Then if m is a positive integer

$$a^m \times a^n = a^{m+n}$$

and

$$(a^m)^n = a^{mn}.$$

We define a^{-n} to be the reciprocal of a^n, and it follows that

$$a^m / a^n = a^{m-n}$$

for any integers m and n, positive or negative. When m and n are equal this gives $a^m / a^n = a^0$, so that we need to define the value of a^0 to be 1.

It will be seen in section 11.7 that these laws of indices are true for all real values of m and n provided that a is positive. For the rest of this section it will be assumed that a^x is defined for $a > 0$, $x \in \mathbb{R}$, and that a^x satisfies the index laws above.

Consider now the function f defined for $a > 0$ and x real by

$$f : x \mapsto a^x.$$

The range of this function will be $\mathbb{R}^+$, and we have $f(1) = a$, $f(0) = 1$, $f(-1) = 1/a$, $f(2) = a^2$.

The inverse f^{-1} of this function is denoted by $\log_a$

i.e.

$$f^{-1} : x \mapsto \log_a x.$$

The domain of this function is $\mathbb{R}^+$ and its range is $\mathbb{R}$. We have $f^{-1}(1) = 0$, $f^{-1}(a) = 1$, $f^{-1}(a^2) = 2$, $f^{-1}(1/a) = -1$.

We call $\log_a x$ the logarithm of x to the base a. Notice especially the identity

$$x \equiv a^{\log_a x}.$$

The properties of $\log_a x$ can be deduced from the index laws above.

Let	$\log_a x = p$ and $\log_a y = q,$
i.e.	$a^p = x$ and $a^q = y.$
Then	$xy = a^p \times a^q = a^{p+q}.$
Hence	$\log_a (xy) = p + q,$
i.e.	$\log_a (xy) = \log_a x + \log_a y.$
Also	$x/y = a^p / a^q = a^{p-q}.$
Hence	$\log_a (x/y) = p - q,$

i.e.
$$\log_a(x/y) = \log_a x - \log_a y.$$

For any real number m we have $x^m = (a^p)^m = a^{mp}$

Hence
$$\log_a(x^m) = mp = m \log_a x.$$

It is sometimes necessary to change from one base to another. Let a, b and x be positive and let $x = b^m$, $(a \neq 1, b \neq 1)$

Then
$$\log_a x = m \log_a b.$$
But
$$\log_b x = m,$$
and hence
$$\log_a x = (\log_a b)(\log_b x)$$
i.e.
$$\log_b x = (\log_a x)/(\log_a b).$$

Example 1

Solve the equation $4(3^{2x+1}) + 17(3^x) - 7 = 0$.

The substitution $y = 3^x$ gives the equation

$$4(3y^2) + 17y - 7 = 0,$$
$$12y^2 + 17y - 7 = 0,$$
$$(3y - 1)(4y + 7) = 0,$$
$$y = 1/3 \quad \text{or} \quad y = -7/4.$$

If $y = 1/3$, we have $3^x = 1/3$, giving $x = -1$. If $y = -7/4$ the value of 3^x will be negative, which is not possible if x is real.

Example 2

Solve the simultaneous equations

$$\log_y x - 4 \log_x y = 0,$$
$$\log_{10} x + \log_{10} y = 1.$$

Let $x = y^m$, so that $y = x^{1/m}$.

Then
$$\log_y x = m \quad \text{and} \quad \log_x y = 1/m.$$

From the first equation we find

$$m - 4/m = 0,$$
i.e.
$$m = 2 \text{ or } -2.$$

From the second equation

$$xy = 10.$$

When $m = 2$, $x = y^2$ and hence $y^3 = 10$. This gives $y = 10^{1/3}$ and $x = 10^{2/3}$. When $m = -2$, $x = y^{-2}$ and hence $1/y = 10$. This gives $y = 1/10$ and $x = 100$.

Example 3

The variables x and y are related by a law of the form $y = ab^x$. The table gives approximate values of y corresponding to integral values of x. By drawing a linear graph estimate to two significant figures the values of the constants a and b.

x	2	3	4	5	6
y	5·37	8·11	12·16	18·20	27·35

Since $y = ab^x$, we have by taking logarithms

$$\log_{10} y = \log_{10} a + x \log_{10} b.$$

If we put $u = \log_{10} y$, $m = \log_{10} b$, $c = \log_{10} a$, this equation becomes

$$u = mx + c.$$

We plot u against x and draw the straight line that appears to fit the points best. The slope of this line will be m, which equals $\log_{10} b$. The intercept made by the line on the u-axis will be c, which equals $\log_{10} a$. Once we have found m and c from the graph we can calculate b and a.

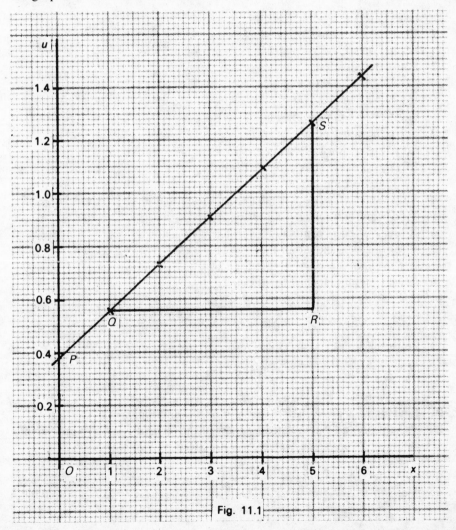

Fig. 11.1

We find the following values for u.

x	2	3	4	5	6
u	0·73	0·91	1·08	1·26	1·43

To find the slope we draw the right-angled triangle QRS and find that $QR = 4$, $RS = 0·7$. Then

$$\log_{10} b = m = RS/QR = 0·175.$$

Hence $b = 1·5$, to two significant figures.

The intercept on the u-axis is OP, which equals 0·38. This gives

$$\log_{10} a = c = 0·38.$$

Hence $a = 2·4$, to two significant figures.

Exercises 11.1

1 Find the values of
   (a) $\log_4 (32)$ (b) $\log_5 (0.04)$ (c) $\log_2 25 \times \log_5 8$.
2 Solve the equations
   (a) $3^{2x+1} - 28(3^x) + 9 = 0$
   (b) $(2)^{2x+1} = 3(2^x) - 1$
   (c) $(5)^{2x-1} - 6(5^x) + 25 = 0$.
3 Solve the equations
   (a) $2 \log x - \log 2 = \log (5x - 12)$
   (b) $\log_3 (2 - 3x) = \log_9 (6x^2 - 19x + 2)$
   (c) $\log_3 x - 2 \log_x 3 = 1$.
4 Find values of x and y greater than 1 such that $x^y = y^{2x}$ and $y^2 = x^3$.
5 Find x and y if $\log_x y = 3$ and $\log_x 4y = 5$.
6 If $\log_x y = \log_y a$, express $\log_a x$ in terms of x and y.
7 If $\log_{10} 2 = a$, show that $\log_8 5 = (1 - a)/3a$.
   If also $\log_{10} 3 = b$, show that $\log_5 24 = (3a + b)/(1 - a)$.
8 Solve the simultaneous equations

$$\log (x - 2) + \log 2 = 2 \log y,$$
$$\log (x - 3y + 3) = 0.$$

9 Solve the simultaneous equations

$$\log_2 x - \log_4 y = 4,$$
$$\log_2 (x - 2y) = 5.$$

10 Find x and y given that $18 \log_x y = 6y - x$ and $\log_y x = 2$.
11 Variables x and y are related by a law of the form of $y = kx^n$. Approximate values of y for various values of x are given by the table:

x	3	$4\frac{1}{4}$	$5\frac{1}{2}$	8	10	11	12
y	22	26	30	36	40	42	44

From the graph of log y against log x, estimate the values of the constants k and n.

12 Two variables x and y are connected by a relation of the form $xy^n = C$, where n and C are constants. Approximate corresponding values are given by the table:

x	1·5	2·0	2·5	3·0	3·5
y	424	284	206	160	128

By drawing a graph of log y against log x, estimate the values of n and C.

11.2 Natural logarithms

Figure 11.2 shows the curve $y = 1/t$ for positive values of t. The area of the shaded region bounded by the curve, the t-axis and the lines $t = 1$, $t = x$ is equal to the integral

$$\int_1^x \frac{1}{t}\,\mathrm{d}t.$$

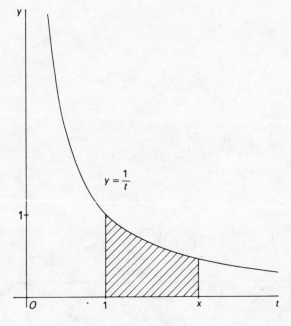

Fig. 11.2

The function ln for positive values of x is defined by

$$\ln : x \mapsto \int_1^x \frac{1}{t}\,\mathrm{d}t.$$

The area of the shaded region in Fig. 11.2 is then equal to ln x. It is clear that ln $1 = 0$ and that ln x is positive when x is greater than 1. When x lies between 0 and 1, the area beneath the curve $y = 1/t$ for values of t between x and 1 equals

$$\int_x^1 \frac{1}{t}\,dt = -\int_1^x \frac{1}{t}\,dt = -\ln\,x.$$

This shows that ln x is negative when x lies between 0 and 1.

In order to differentiate a function f, the limit as h tends to zero of

$$\frac{f(x+h) - f(x)}{h}$$

has to be found. The derivative of ln x will be given by

$$\lim_{h \to 0} \frac{\ln(x+h) - \ln x}{h}.$$

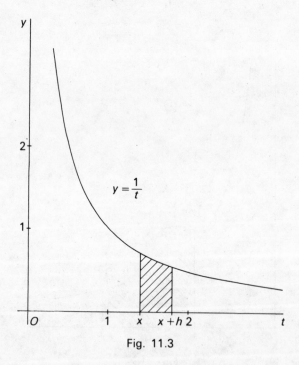

Fig. 11.3

In Fig. 11.3 the shaded region is bounded by the curve $y = 1/t$, the t-axis and the lines $t = x$, $t = x + h$. The area of this region equals

$$\int_x^{x+h} \frac{1}{t}\,dt = \int_1^{x+h} \frac{1}{t}\,dt - \int_1^x \frac{1}{t}\,dt$$

$$= \ln(x+h) - \ln\,x.$$

Now this area is less than the area of a rectangle of base h and height $1/x$, and is greater than the area of a rectangle of base h and height $1/(x+h)$,

i.e.
$$\frac{h}{x+h} < \ln (x+h) - \ln x < \frac{h}{x}.$$

We divide by h and let h tend to zero. This gives

$$\lim_{h \to 0} \frac{\ln (x+h) - \ln x}{h} = \frac{1}{x}$$

i.e.
$$\frac{d}{dx} \ln x = \frac{1}{x}.$$

(It has been assumed that h is positive. If h is negative, the same result is obtained.)

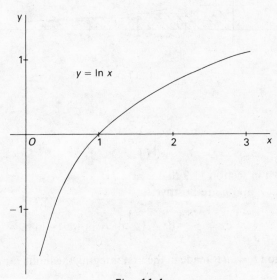

Fig. 11.4

The gradient at any point on the curve $y = \ln x$ is therefore $1/x$. This curve (Fig. 11.4) will pass through the point $(1, 0)$, its gradient there being 1. For values of x between 0 and 1, the gradient is larger than 1 and it increases as x grows smaller. For x greater than 1, the gradient is less than 1 and decreases as x increases.

By considering rectangles of unit width beneath the curve $y = 1/t$, as in Fig. 11.5, it can be seen that

$$\int_1^4 \frac{1}{t} dt > \frac{1}{2} + \frac{1}{3} + \frac{1}{4} > 1,$$

i.e.
$$\ln 4 > 1.$$

It follows that $\ln x$ takes the value 1 for some value of x less than 4. The constant e is defined to be such that $\ln e = 1$,

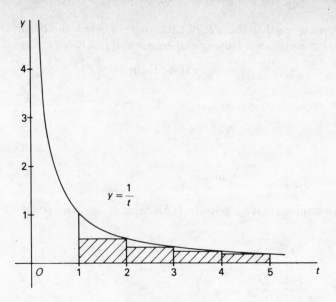

Fig. 11.5

i.e.
$$\int_1^e \frac{1}{t}\,dt = 1.$$

(It will be shown in section 11.9 that e ≈ 2·718.)

Provided that a and b are positive,

$$\int_a^{ab} \frac{1}{t}\,dt = \int_1^{ab} \frac{1}{t}\,dt - \int_1^a \frac{1}{t}\,dt = \ln(ab) - \ln a.$$

If the substitution $t = au$ is made in the first integral, the limits for u will be 1 and b, and hence

$$\int_a^{ab} \frac{1}{t}\,dt = \int_1^b \left(\frac{1}{au}\right)\left(\frac{dt}{du}\right)du = \int_1^b \frac{1}{u}\,du = \ln b,$$

since
$$\int_1^b \frac{1}{u}\,du = \int_1^b \frac{1}{t}\,dt.$$

This shows that

$$\ln(ab) - \ln a = \ln b,$$

i.e.
$$\ln(ab) = \ln a + \ln b.$$

By putting $b = 1/a$ we deduce that

$$\ln 1 = \ln a + \ln(1/a),$$

i.e.
$$\ln(1/a) = -\ln a.$$

By putting $b = 1/c$ we deduce that

$$\ln (a/c) = \ln a + \ln (1/c),$$

i.e.
$$\ln (a/c) = \ln a - \ln c.$$

These results are the justification for calling $\ln x$ the natural logarithm of x. It will be shown in section 11.7 that for any positive number x the natural logarithm $\ln x$ equals the logarithm of x to the base e, i.e. $\ln x = \log_e x$.

Example 1

If $f(x)$ is positive, show that $\dfrac{d}{dx}\ln f(x) = \dfrac{f'(x)}{f(x)}$, and hence

differentiate $\ln (x^2 + 1)$ and $\ln (\cos^2 x)$ with respect to x.
Let $y = f(x)$, so that $dy/dx = f'(x)$. Then

$$\frac{d}{dx}\ln f(x) = \frac{d}{dx}\ln y = \frac{1}{y}\frac{dy}{dx} = \frac{f'(x)}{f(x)}.$$

If $f(x) = x^2 + 1$, $f'(x) = 2x$ and so

$$\frac{d}{dx}\ln (x^2 + 1) = \frac{2x}{x^2 + 1}.$$

If $f(x) = \cos^2 x$, $f'(x) = -2 \sin x \cos x$ and so

$$\frac{d}{dx}\ln (\cos^2 x) = -\frac{2 \sin x \cos x}{\cos^2 x} = -2 \tan x.$$

Example 2
Show that, for any positive integer n, $\ln (2^n) = n \ln 2$.

We make the substitution $t = u^n$ in the integral giving $\ln (2)$. The limits for u will be 1 and 2.

$$\ln (2^n) = \int_1^{2^n} \frac{1}{t}\,dt = \int_1^2 \frac{1}{u^n}(nu^{n-1})\,du$$

$$= n\int_1^2 \frac{1}{u}\,du$$

$$= n \ln 2.$$

Example 3
Find the derivative of $\log_{10} x$.

We have

$$\log_{10} x = (\log_e x)(\log_{10} e)$$
$$= (\ln x)(\log_{10} e),$$

from which it follows that

$$\frac{d}{dx}(\log_{10} x) = (\log_{10} e)/x \approx (0 \cdot 4343)/x.$$

While logarithms to the base 10 are convenient for arithmetical calculations, the awkward factor in the derivative is a great disadvantage when they are used for other work.

Exercises 11.2

1 Draw the curve $y = 1/t$ on squared paper for $0 \leqslant t \leqslant 5$. Use it to evaluate approximately $\log_e 2, \log_e 3, \log_e 4$ and $\log_e 5$. Check your values by means of a calculator.

2 Find correct to three decimal places $[\log_e 256 - \log_e 192]/[\log_e 4]$.

3 Differentiate with respect to x
   (a) $\ln (4x)$ (b) $\ln (x^2)$ (c) $\ln (1/x)$ (d) $\ln (x + 2)$.

4 State the domain of the function f when
   (a) $f(x) = \ln x$
   (b) $f(x) = \ln (1 + x)$
   (c) $f(x) = \ln [(1 + x)/(1 - x)]$.

5 Differentiate with respect to x
   (a) $\log_e (3x + 4)$ (c) $\ln (3x) - \ln (2x)$
   (b) $\log_e (2 - x)$ (d) $x \ln x$.

6 Sketch the curves
   (a) $y = \ln (1 + x)$ (b) $y = \ln (2x)$.

7 Find the maximum value of
   (a) $(1/x) \ln x$ (b) $x^2 \ln (1/x)$.

8 Differentiate with respect to x
   (a) $\ln (\sin^2 x)$ (c) $\log_2 x$
   (b) $\ln (x^4 + x)$ (d) $\ln (\ln x)$.

9 Use integration by parts to show that $\int_1^e \ln x \, dx = 1$.

10 Integrate by parts (a) $x \ln x$ (b) $(\ln x)/x$.

11.3 Integrals of the form $\int \dfrac{1}{ax + b} dx$

In section 10.1 it was shown that

$$\int x^n \, dx = \frac{1}{n + 1} x^{n+1} + c,$$

where n is any rational number except -1.

The case when $n = -1$ can now be dealt with. Since the derivative of $\ln x$ is $1/x$, we have

$$\int \frac{1}{x}\, dx = \ln x + c,$$

provided that x is positive. For $x > a$,

$$\int \frac{1}{x-a}\, dx = \ln (x-a) + c,$$

but for $x < a$,

$$\int \frac{1}{x-a}\, dx = -\int \frac{1}{a-x}\, dx = \ln (a-x) + c.$$

When $(ax+b)$ is positive,

$$\frac{d}{dx} \ln (ax+b) = \frac{a}{ax+b}.$$

It follows that

$$\int \frac{1}{ax+b}\, dx = \frac{1}{a} \ln (ax+b) + c.$$

This enables us to find integrals of the form

$$\int \frac{f(x)}{ax+b}\, dx,$$

where $f(x)$ is a polynomial, for the integrand can be reduced to the sum of a polynomial and a fraction with a constant numerator, as in example 3 overleaf.

Example 1

(a) $\displaystyle \int_3^4 \frac{1}{x-2}\, dx = \left[\ln (x-2) \right]_3^4 = \ln 2 - \ln 1 = \ln 2.$

(b) $\displaystyle \int_0^1 \frac{1}{x-2}\, dx = -\int_0^1 \frac{1}{2-x}\, dx = \left[\ln (2-x) \right]_0^1 = \ln (\tfrac{1}{2}).$

Example 2

(a) $\displaystyle \int \frac{1}{4x+1}\, dx = \frac{1}{4} \ln (4x+1) + c$, provided that $4x+1 > 0$.

(b) $\displaystyle \int \frac{6}{3x+2}\, dx = 2 \ln (3x+2) + c$, provided that $3x+2 > 0$.

Example 3

(a) $\int \dfrac{x}{x+1} dx = \int \left(1 - \dfrac{1}{x+1}\right) dx = x - \ln(x+1) + c,$ if $x+1 > 0.$

(b) $\int \dfrac{x^2+1}{x+1} dx = \int \left(x - 1 + \dfrac{2}{x+1}\right) dx$

$\qquad = \frac{1}{2}x^2 - x + 2\ln(x+1) + c,$ if $x+1 > 0.$

(c) $\int \dfrac{4x^2+3}{2x-1} dx = \int \left(2x + 1 + \dfrac{4}{2x-1}\right) dx$

$\qquad = x^2 + x + 2\ln(2x-1) + c,$ if $2x > 1.$

Exercises 11.3

1 Show that

(a) $\displaystyle\int_{-1}^{1} \dfrac{1}{x+2} dx = \ln 3$ (b) $\displaystyle\int_{-1}^{0} \dfrac{1}{x-1} dx = -\ln 2.$

2 Evaluate

(a) $\displaystyle\int_{1}^{3} \dfrac{1}{x+1} dx$ (b) $\displaystyle\int_{0}^{1} \dfrac{1}{2x+1} dx.$

3 Given that $x > 1$, find the indefinite integral of
(a) $x/(x-1)$ (c) $2x/(x+1)$
(b) $2x/(2x-1)$ (d) $4x/(2x+1)$.

4 Show that

$$\int_{0}^{2} \dfrac{2x}{x+2} dx = 4 - 4\ln 2.$$

5 Evaluate

(a) $\displaystyle\int_{0}^{1} \dfrac{2x+1}{x+1} dx$ (b) $\displaystyle\int_{0}^{1} \dfrac{1-x}{x+2} dx.$

6 Find

(a) $\displaystyle\int \dfrac{x^2}{x+1} dx$ for $x > -1$ (b) $\displaystyle\int \dfrac{x^3}{x-1} dx$ for $x > 1.$

7 Show that

$$\int_{1}^{2} \dfrac{4-4x}{2x-1} dx = \ln 3 - 2.$$

8 Find the indefinite integral of
(a) $(x^2+1)/(x-1)$ for $x > 1$ (b) $(x^2+x)/(x+2)$ for $x+2 > 0.$
9 Show that, if $2x+1 > 0,$

$$\int \dfrac{4x^2-5}{2x+1} dx = x^2 - x - 2\ln(2x+1) + c.$$

10 Show that

$$\int_{-1}^{1} \frac{x^4 - 4x^2 + 1}{x+2} dx = \log_e 3 - 4/3.$$

11.4 Integrals of the form $\int \frac{px+q}{(x-a)(x-b)} dx$

First the integrand is expressed in partial fractions:

$$\frac{px+q}{(x-a)(x-b)} \equiv \frac{A}{x-a} + \frac{B}{x-b}.$$

Then, provided that $x > a$ and $x > b$,

$$\int \frac{px+q}{(x-a)(x-b)} dx = A \ln (x-a) + B \ln (x-b).$$

Example 1
Since

$$\frac{x-4}{(x-1)(x-2)} \equiv \frac{3}{x-1} - \frac{2}{x-2},$$

$$\int \frac{x-4}{(x-1)(x-2)} dx = 3 \ln (x-1) - 2 \ln (x-2) + c \quad \text{for } x > 2.$$

Example 2
Since

$$\frac{2x-1}{(x-2)^2} \equiv \frac{2}{x-2} + \frac{3}{(x-2)^2},$$

$$\int \frac{2x-1}{(x-2)^2} dx = 2 \ln (x-2) - \frac{3}{x-2} + c, \quad \text{for } x > 2.$$

Example 3
Evaluate $\int_{1}^{2} \frac{5}{x^2 - 3x - 4} dx.$

The integrand must first be expressed in partial fractions.

$$\frac{5}{x^2 - 3x - 4} \equiv \frac{1}{x-4} - \frac{1}{x+1}$$

$$\equiv \frac{-1}{4-x} - \frac{1}{x+1}.$$

The first fraction has been rearranged so that the denominator is positive in the range of integration. This now gives

$$\left[\ln (4-x) - \ln (x+1) \right]_{1}^{2} = 2 \log_e (2/3).$$

Exercises 11.4

1 Find (a) $\displaystyle\int \frac{4}{x^2-4}dx \quad (x>2)$ (b) $\displaystyle\int \frac{x+4}{x^2+3x+2}dx \quad (x>-1)$.

2 Find (a) $\displaystyle\int \frac{2x+2}{2x^2+x}dx \quad (x>0)$ (b) $\displaystyle\int \frac{x+3}{x^2+x}dx \quad (x>0)$.

3 Find (a) $\displaystyle\int \frac{2x+3}{(x+2)^2}dx \quad (x>-2)$ (b) $\displaystyle\int \frac{x}{(x-1)^2}dx \quad (x>1)$.

4 Find (a) $\displaystyle\int \frac{1}{2x^2+3x+1}dx \ (2x>-1)$ (b) $\displaystyle\int \frac{1}{6x^2+5x+1}dx \ (3x>-1)$.

5 Show that $\displaystyle\int_0^1 \frac{1}{x^2+3x+2}dx = \log_e(4/3)$.

6 Show that $\displaystyle\int_0^1 \frac{1}{x^2-5x+6}dx = \log_e(4/3)$.

7 Show that $\displaystyle\int_0^1 \frac{3x-1}{x^2+2x-15}dx = \log_e(0\cdot96)$.

8 Sketch the curve $y = 1/(4-x^2)$. Find the area of the region defined by $-1 \leqslant x \leqslant 1, 0 \leqslant y \leqslant 1/(4-x^2)$.

9 Sketch the curve $y = x/(x^2+5x+6)$. Show that the area of the finite region bounded by the curve, the x-axis and the line $x = 3$ is $\log_e(1\cdot28)$.

10 Sketch the curve $y = (x-4)/[(x-3)(x+2)]$. Show that the area of the finite region bounded by the x-axis, the curve and the lines $x = -1$, $x = 2$ is approximately $1\cdot94$.

11.5 Integrals of the form $\displaystyle\int \frac{f'(x)}{f(x)}dx$

It has already been shown that when $f(x)$ is positive

$$\frac{d}{dx}\ln f(x) = \frac{f'(x)}{f(x)}.$$

It follows that

$$\int \frac{f'(x)}{f(x)}dx = \ln f(x) + c.$$

Example 1

Find $\displaystyle\int \frac{2x}{1+x^2}dx$.

If $f(x) = 1+x^2$, $f'(x) = 2x$, and so the integral equals $\ln(1+x^2)+c$.

Example 2

Find $\int \tan x \, dx$.

Put $f(x) = \cos x$, to give $f'(x) = -\sin x$. Then, if $\cos x > 0$,

$$\int \tan x \, dx = -\int \frac{-\sin x}{\cos x} \, dx = -\log (\cos x) + c.$$

This can be expressed as $\log (\sec x) + c$.

Example 3

Evaluate $\displaystyle\int_0^1 \frac{1+x}{1+x^2} \, dx$.

This equals the sum of the two integrals

$$\int_0^1 \frac{1}{1+x^2} \, dx + \int_0^1 \frac{x}{1+x^2} \, dx = \left[\arctan x + \tfrac{1}{2} \ln (1+x^2) \right]_0^1 = \pi/4 + \tfrac{1}{2} \ln 2.$$

Example 4

Find $\displaystyle\int \frac{x+1}{x^2 - 2x + 5} \, dx$.

As $x^2 - 2x + 5$ has no real factors, the integrand cannot be expressed in real partial fractions. In this case the first step is to 'complete the square':

$$x^2 - 2x + 5 = (x-1)^2 + 4.$$

Next we make the substitution $x - 1 = u$.

$$\int \frac{x+1}{x^2 - 2x + 5} \, dx = \int \frac{(x-1) + 2}{(x-1)^2 + 4} \, dx$$

$$= \int \frac{u}{u^2 + 4} \, du + \int \frac{2}{u^2 + 4} \, du$$

$$= \tfrac{1}{2} \ln (u^2 + 4) + \arctan (u/2) + c$$

$$= \tfrac{1}{2} \ln (x^2 - 2x + 5) + \arctan \tfrac{1}{2} (x-1) + c.$$

Example 5

Evaluate $\displaystyle\int_0^2 \frac{x^3 + 2x}{x^2 + x + 1} \, dx$.

In the integrand, the degree of the numerator is higher than the degree of the denominator. By dividing the numerator by the denominator we obtain the identity

$$\frac{x^3 + 2x}{x^2 + x + 1} \equiv x - 1 + \frac{2x + 1}{x^2 + x + 1}.$$

Now if $f(x) = x^2 + x + 1$, $f'(x) = 2x + 1$. Therefore,

$$\int_0^2 \frac{x^3 + 2x}{x^2 + x + 1} \, dx = \left[\tfrac{1}{2} x^2 - x \right]_0^2 + \left[\ln (x^2 + x + 1) \right]_0^2 = \ln 7.$$

Exercises 11.5

1 Integrate with respect to x
   (a) $3x^2/(x^3+1)$ $(x > -1)$ (c) $x/(2x^2+1)$
   (b) $(2x+2)/(x^2+2x)$ $(x > 0)$ (d) $4x/(x^2+4)$.

2 Evaluate

   (a) $\displaystyle\int_{-1}^{1} \frac{x+1}{x^2+2x+5}\,dx$ (b) $\displaystyle\int_{-1}^{1} \frac{4x+3}{2x^2+3x+4}\,dx$.

3 Evaluate

   (a) $\displaystyle\int_{0}^{\pi/4} \tan x\,dx$ (b) $\displaystyle\int_{\pi/6}^{\pi/4} \cot x\,dx$.

4 Evaluate $\displaystyle\int_{0}^{\pi/2} \frac{\sin x}{1+\cos x}\,dx$.

5 Evaluate

   (a) $\displaystyle\int_{0}^{1} \frac{2x+4}{x^2+4x+5}\,dx$ (b) $\displaystyle\int_{1}^{3} \frac{2x+1}{x^2+x}\,dx$.

6 Evaluate $\displaystyle\int_{0}^{\pi/4} \frac{\sec^2\theta}{2+\tan\theta}\,d\theta$.

7 Show that
$$\int \frac{x}{x^2+2x+2}\,dx = \tfrac{1}{2}\ln\,(x^2+2x+2) - \arctan\,(x+1)+c.$$

8 Show that
$$\int \frac{x^2}{x^2+2x+2}\,dx = x - \ln\,(x^2+2x+2)+c.$$

9 Evaluate $\displaystyle\int_{0}^{1} \frac{2x-1}{1+x^2}\,dx$.

10 Evaluate $\displaystyle\int_{0}^{2} \frac{4x+2}{x^2+4}\,dx$.

11.6 Logarithmic differentiation

The work entailed in finding a derivative can sometimes be reduced by considering the logarithm of the expression to be differentiated. This method, known as logarithmic differentiation, is useful when the expression is a product or when the variable appears as an index.

Example 1

If $y = (x+1)^2(1+\sin x)\sec x$, find the value of dy/dx when $x = 0$.

We have

$$\ln y = 2\ln\,(x+1) + \ln\,(1+\sin x) + \ln\sec x.$$

Now
$$\frac{d}{dx}\ln y = \left(\frac{d}{dy}\ln y\right)\frac{dy}{dx} = \frac{1}{y}\frac{dy}{dx}.$$

Hence
$$\frac{1}{y}\frac{dy}{dx} = \frac{2}{x+1} + \frac{\cos x}{1+\sin x} + \tan x.$$

When $x = 0$, $y = 1$ and we find that $dy/dx = 3$.

Example 2
Find dy/dx when $y = 10^x$.

From $\ln y = x \ln 10$, we have $\dfrac{1}{y}\dfrac{dy}{dx} = \ln 10$.

Hence
$$\frac{dy}{dx} = y \ln 10 = 10^x \times \ln 10.$$

Example 3
Find dy/dx when $y = \sqrt{[(1+x)/(1-x)]}$.

We have
$$\ln y = \tfrac{1}{2}\ln(1+x) - \tfrac{1}{2}\ln(1-x).$$
$$\frac{1}{y}\frac{dy}{dx} = \frac{\tfrac{1}{2}}{1+x} + \frac{\tfrac{1}{2}}{1-x} = \frac{1}{1-x^2}.$$
$$\frac{dy}{dx} = \left(\frac{1}{1-x^2}\right)\bigg/\sqrt{\left(\frac{1+x}{1-x}\right)}.$$

Exercises 11.6
1 Given that $y = 2^x$, find dy/dx and d^2y/dx^2.
2 If $y = (2x+1)^3(2-\tan x)^2$, find the value of dy/dx when $x = 0$.
3 Show that $x^{1/x}$ has a maximum value when $x = e$.
4 Find dy/dx when $y = \sqrt{[(1-x^2)/(1+x^2)]}$.
5 Find the gradient of the curve $y = x^x$ at the point $(1, 1)$.

11.7 The exponential function
The exponential function exp is defined as the inverse of the natural logarithm. If $x = \ln y$ where $y > 0$, then $y = \exp x$ and *vice versa*. This can be expressed as
$$(x = \ln y) \Leftrightarrow (y = \exp x).$$

The natural logarithm maps the set of positive real numbers on to the set of real numbers. The exponential function maps the set of real numbers on to the set of positive real numbers. For all real values of x, $\exp x$ is positive.

By differentiating $x = \ln y$ with respect to x we obtain
$$1 = \frac{1}{y}\frac{dy}{dx}, \quad \text{i.e. } \frac{dy}{dx} = y.$$

Alternatively, we can differentiate $x = \ln y$ with respect to y. This will give

$$\frac{dx}{dy} = \frac{1}{y}, \quad \text{i.e.} \quad \frac{dy}{dx} = y.$$

Either method gives

$$\frac{d}{dx}(\exp x) = \exp x.$$

When $y = \exp x$, the rate of change of y with respect to x is proportional to y. If x represents time, y is said to grow exponentially. This law of growth occurs frequently in nature.

The gradient at any point of the curve $y = \exp x$ is equal to y. Now since $\ln 1 = 0$, $\exp 0 = 1$. Hence the curve $y = \exp x$ passes through the point $(0, 1)$, and its gradient at this point is 1 (Fig. 11.6). Since $\ln e = 1$, we have $\exp 1 = e$, and so the curve $y = \exp x$ will pass through the point $(1, e)$. This curve is the reflexion of the curve $y = \ln x$ in the line $y = x$.

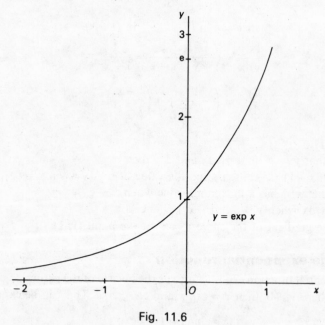

Fig. 11.6

From the relation

$$\ln (ab) = \ln a + \ln b$$

we can deduce that

$$(\exp x)(\exp y) = \exp (x + y).$$

For let $\ln a = x$ and $\ln b = y$, where a and b are positive. Then $a = \exp x$ and $b = \exp y$. Also

$$x + y = \ln a + \ln b = \ln (ab).$$

It follows that

$$\exp (x + y) = ab = (\exp x)(\exp y).$$

The exponential function can be used to define the value of a^x, where a is positive and x is any real number. We already know what is meant by 3^2 or $4^{2/3}$, but the value of an expression such as $5^{\sqrt{2}}$ has not yet been made clear. We define $5^{\sqrt{2}}$ to be equal to $\exp(\sqrt{2}\ln 5)$. For any positive number a and any real number x we define a^x by

$$a^x = \exp (x \ln a).$$

Then
$$\begin{aligned} a^x \times a^y &= \exp (x \ln a) \times \exp (y \ln a) \\ &= \exp\left[(x+y)\ln a\right], \end{aligned}$$

i.e.
$$a^x \times a^y = a^{x+y}.$$

This index law is true for all real values of x and y.

By putting $a = e$ in the definition of a^x we obtain

$$e^x = \exp (x \ln e) = \exp x.$$

This shows that the exponential function can be denoted by e^x, with the properties

$$e^x \times e^y = e^{x+y},$$

and
$$\frac{d}{dx} e^x = e^x.$$

In example 3, section 6.8 it was shown that for any rational number k

$$\frac{d}{dx} x^k = k x^{k-1}.$$

It will now be shown that this is true for any real number k. Since $x^k = \exp (k \ln x)$ we have

$$\frac{d}{dx} x^k = \frac{k}{x} \exp (k \ln x)$$

$$= \frac{k}{x} x^k = k x^{k-1}.$$

This result enables us to justify equating $\ln x$ and $\log_e x$. For by substituting $t = u^k$ we have

$$\int_1^{e^k} \frac{1}{t}\,dt = \int_1^e \frac{1}{u^k}(k u^{k-1})\,du = \int_1^e \frac{k}{u}\,du.$$

i.e. $\ln(e^k) = k \ln e = k = \log_e(e^k)$.

If $x = e^k$, this gives $\ln x = \log_e x$.

Example 1

Evaluate $\displaystyle\int_0^1 e^{-4x}\,dx$.

Since the derivative of e^{-4x} is $-4e^{-4x}$, we have

$$\int_0^1 e^{-4x}\,dx = \left[-\frac{1}{4}e^{-4x} \right]_0^1 = \frac{1}{4}(1 - e^{-4}).$$

Example 2

At time t s, $y = Ae^t$. Show that the value of y is doubled in any time interval of length $\ln 2$ s.

Let $y = y_1$ when $t = t_1$, and let $y = y_2$ when $t = t_1 + \ln 2$.
Then $y_2 = Ae^{t_1 + \ln 2} = (Ae^{t_1})(e^{\ln 2}) = y_1 \times 2$.

Example 3

Find $\int xe^{-x/2}\,dx$.

We take

$$u = x \quad \text{and} \quad \frac{dv}{dx} = e^{-x/2}.$$

Then

$$\frac{du}{dx} = 1 \quad \text{and} \quad v = -2e^{-x/2}.$$

By integration by parts

$$\int xe^{-x/2}\,dx = -2xe^{-x/2} - \int(-2e^{-x/2})\,dx$$

$$= -2xe^{-x/2} - 4e^{-x/2} + c.$$

Exercises 11.7

1 Simplify (a) $\exp(2\ln x)$ (b) $\exp(x\ln 2)$ (c) $\ln(2\exp x)$.
2 Sketch the curves (a) $y = e^{-x}$ (b) $y = e^{x/2}$.
3 Differentiate with respect to x
   (a) $e^x \sin x$ (c) $(e^x + e^{-x})^2$ (e) $(x+1)e^{-x}$
   (b) $e^{2x}(\sin x + 2\cos x)$ (d) $\exp(x + \log_e x)$ (f) $e^{3x}\cos 2x$.
4 Show that the maximum value of xe^{-x} is $1/e$.
5 Integrate with respect to x
   (a) e^{-x} (b) e^{4x} (c) e^{2-x} (d) $e^{x/2}$.
6 Evaluate

(a) $\displaystyle\int_0^1 e^{-2x}\,dx$ (c) $\displaystyle\int_1^e \exp(-\log_e u)\,du$

(b) $\displaystyle\int_{-1}^1 \exp(t+1)\,dt$ (d) $\displaystyle\int_0^1 \ln\left[(\exp x)^2 \right]\,dx$.

7   Evaluate

(a) $\displaystyle\int_0^1 \frac{\exp x}{1+\exp x}\,dx$ (b) $\displaystyle\int_0^1 (2^x + 2^{-x})\,dx.$

8   Show that, if $a = \ln 2$, $\displaystyle\int_0^a xe^{2x}\,dx = 2\ln 2 - 3/4.$

9   Use integration by parts to find

(a) $\displaystyle\int xe^x\,dx$ (b) $\displaystyle\int xe^{-x}\,dx.$

10   Find the two possible values of n given that the equation

$$\frac{d^2y}{dx^2} - 5\frac{dy}{dx} + 6y = 0$$

is satisfied by $y = e^{nx}$.

11.8   Maclaurin series

Let f be a function such that $f(x)$ can be differentiated as many times as we please at $x = 0$. Under certain circumstances a polynomial $P(x)$ can be constructed of any required degree n such that the difference $P(x) - f(x)$ will be small near $x = 0$. Let $P(x)$ be the polynomial

$$a_0 + a_1x + a_2x^2 + \ldots + a_nx^n.$$

By taking $a_0 = f(0)$ we ensure that $P(x)$ and $f(x)$ are equal at $x = 0$. Now the first derivative of $P(x)$ is given by

$$P'(x) = a_1 + 2a_2x + 3a_3x^2 + \ldots + na_nx^{n-1}.$$

At $x = 0$, $P'(x)$ equals a_1. We take $a_1 = f'(0)$, so that $P'(x)$ and $f'(x)$ are equal at $x = 0$. Next we have

$$P''(x) = 2a_2 + 6a_3x + 12a_4x^2 + \ldots + n(n-1)a_nx^{n-2}.$$

At $x = 0$, $P''(x)$ equals $2a_2$. We take $a_2 = \frac{1}{2}f''(0)$ so that $P''(x)$ and $f''(x)$ are equal at $x = 0$. By continuing in this way we make each of the first n derivatives of $P(x) - f(x)$ equal to zero at $x = 0$. We find $a_3 = f'''(0)/3!$ and $a_n = f^{(n)}(0)/n!$.
  The required polynomial is given by

$$P(x) = f(0) + xf'(0) + \frac{x^2}{2!}f''(0) + \frac{x^3}{3!}f'''(0) + \ldots + \frac{x^n}{n!}f^{(n)}(0).$$

If now n tends to infinity, an infinite series is obtained, the sum of which under certain conditions will equal $f(x)$. In this case we have the expansion of $f(x)$ in a series of ascending powers of x

$$f(x) = f(0) + xf'(0) + \frac{x^2}{2!}f''(0) + \ldots + \frac{x^r}{r!}f^{(r)}(0) + \ldots.$$

This is known as the Maclaurin series for $f(x)$.

Example

Obtain the expansion of $\sin x$ in powers of x as far as the term in x^5.

When $f(x) = \sin x$, $f(0) = 0$, and $a_0 = 0$.

$\quad f'(x) = \cos x$, $f'(0) = 1$, and $a_1 = 1$.

$\quad f''(x) = -\sin x$, $f''(0) = 0$, and $a_2 = 0$.

$\quad f'''(x) = -\cos x$, $f'''(0) = -1$, and $a_3 = -1/3!$.

$\quad f^{(4)}(x) = \sin x$, $f^{(4)}(0) = 0$, and $a_4 = 0$.

$\quad f^{(5)}(x) = \cos x$, $f^{(5)}(0) = 1$, and $a_5 = 1/5!$.

This gives the expansion

$$\sin x \approx x - \frac{x^3}{3!} + \frac{x^5}{5!},$$

a very good approximation for $\sin x$ when x is small.

Exercises 11.8

1   Show that when x is small

$$\cos x \approx 1 - x^2/2! + x^4/4! - x^6/6!.$$

2   Explain why, when $f(x) = (1 + x)^6$, the Maclaurin series for $f(x)$ is a finite series. Show that the series gives the binomial expansion of $f(x)$.

3   Show that, when x is small, $\tan x \approx x + x^3/3$.

4   Obtain the approximation $\arcsin x \approx x + x^3/6$.

11.9   The exponential series

When f is the exponential function, we have $f(x) = e^x$, $f'(x) = e^x$ and every derivative of $f(x)$ equals $f(x)$. Consequently $f(0) = 1$, $f'(0) = 1$, $f''(0) = 1$, and so on. In this case, the polynomial

$$P(x) = 1 + x + x^2/2! + x^3/3! + \ldots + x^n/n!$$

has the same value 1 for each of its first n derivatives at $x = 0$, as the exponential function. Thus the Maclaurin series for e^x is given by

$$e^x = 1 + x + x^2/2! + x^3/3! + \ldots + x^r/r! + \ldots.$$

This series can be shown to converge to e^x for all values of x, and is known as the exponential series.

Example 1

Find the sum of the first seven terms of the exponential series when $x = 1$. Show that this sum gives the value of e correct to three decimal places.

When $x = 1$, the first seven terms are

$$1 + 1 + 1/2! + 1/3! + 1/4! + 1/5! + 1/6!.$$

The sum of these terms is $2\frac{517}{720}$ or $2\cdot718\,0\dot{5}$. The terms remaining form the series

$$1/7! + 1/8! + 1/9! + \ldots .$$

Comparing this series with the geometric series with first term $1/7!$ and common ratio $1/8$, the sum of the terms neglected cannot be larger than the sum of this geometric series, which is $8/(7 \times 7!)$. This shows that the sum of the neglected terms is less than $0\cdot000\,23$, and so the value of e lies between $2\cdot718$ and $2\cdot7183$. This gives $e = 2\cdot718$ correct to three decimal places.

Example 2

In the expansion of $(1 + ax + bx^2)(e^x + e^{-x})$ in a series of ascending powers of x, the coefficients of x and x^2 are zero. Find the constants a and b. Find also the coefficients of x^3 and x^4.

Since

$$e^x = 1 + x + x^2/2! + x^3/3! + x^4/4! + \ldots ,$$
$$e^{-x} = 1 - x + x^2/2! - x^3/3! + x^4/4! - \ldots$$

and

$$e^x + e^{-x} = 2(1 + x^2/2! + x^4/4! + \ldots).$$

Therefore
$$(1 + ax + bx^2)(e^x + e^{-x})$$

$$= (1 + ax + bx^2)(2 + x^2 + x^4/12 + \ldots)$$
$$= 2 + 2ax + (1 + 2b)x^2 + ax^3 + (1/12 + b)x^4 + \ldots .$$

Since the coefficients of x and x^2 are zero, $a = 0$ and $b = -1/2$. The coefficient of x^3 will be zero and the coefficient of x^4 will be $-5/12$.

Exercises 11.9

1 Expand $(2e^{2x} + 2e^{-x} + e^{-2x})$ in a series of ascending powers of x as far as the term in x^3.

2 Find the sum of the series in which the nth term is
   (a) $2^n/n!$ (b) $(-1)^n 2^n/n!$ (c) $2^{2n+1}/n!$.

3 Find the coefficient of x^3 in the expansion of
   (a) $(1 + x)e^{-x}$ (b) $(1 - 2x)e^x$ (c) $(3 - x)^2 e^{2x}$.

4 Show that the coefficient of x^n in the expansion of $(x^2 + 1)e^x$ is $(n^2 - n + 1)/n!$.

5 Find the general term in the expansion of 2^x in a series of ascending powers of x.

6 Show that when x is small
   (a) $e^x - e^{-x} - 2\sin x \approx 2x^3/3$
   (b) $e^x + e^{-x} - 2\cos x \approx 2x^2$.

7 Find constants a and b such that when x is small
   $(1 + 2x + ax^2)e^{bx} \approx 1 + x^2 - 10x^3/3$.

8 Use the exponential series to show that correct to four decimal places
   $\exp(0\cdot1) = 1\cdot1052$.

9 Calculate the square root of e correct to four decimal places by putting $x = \frac{1}{2}$ in the exponential series.

10 Show that the sum of the first ten terms of the exponential series when $x = 1$ is greater than $2 \cdot 718\,281\,5$. By considering the sum of the remaining terms, show that e lies between this number and $2 \cdot 718\,281\,9$.

11.10 The logarithmic series

Since $\ln x$ tends to infinity as x tends to zero, it is not possible to expand $\ln x$ in a series of ascending powers of x. However, we can obtain such an expansion for $\ln (1 + x)$. If $P(x)$ is the polynomial

$$a_0 + a_1 x + a_2 x^2 + \ldots + a_n x^n,$$

we have $P(0) = a_0$, $P'(0) = a_1$ and $P''(0) = 2a_2$. For any positive integer r not greater than n, the value of the rth derivative of $P(x)$ at $x = 0$ will be $(r!)a_r$.
If $f(x) = \ln (1 + x)$, $f(0) = 0$ and so we put $a_0 = 0$.
  $f'(x) = 1/(1 + x)$, $f'(0) = 1$ and we take $a_1 = 1$.
  $f''(x) = -1/(1 + x)^2$, $f''(0) = -1$ and we take $a_2 = -1/2$.
  $f'''(x) = 2/(1 + x)^3$, $f'''(0) = 2$ and we take $a_3 = 1/3$.
The rth derivative of $f(x)$ is $(-1)^{r-1}(r-1)!/(1+x)^r$, which equals $(-1)^{r-1}(r-1)!$ at $x = 0$. This will equal the rth derivative of $P(x)$ at $x = 0$ if we put $a_r = (-1)^{r-1}/r$.

Then

$$P(x) = x - x^2/2 + x^3/3 - x^4/4 + \ldots + (-1)^{n-1} x^n/n.$$

If n tends to infinity an infinite series is obtained which can be shown to converge to $\ln (1 + x)$ provided that x lies in the interval $-1 < x \leqslant 1$.
  This series is known as the logarithmic series and is given by

$$\ln (1 + x) = x - x^2/2 + x^3/3 - x^4/4 + \ldots + (-1)^{r-1} x^r/r + \ldots$$

for $-1 < x \leqslant 1$. By substituting $-x$ for x we obtain the series for $\ln (1 - x)$, which is valid provided that x lies in the interval $-1 \leqslant x < 1$.

$$\ln (1 - x) = -x - x^2/2 - x^3/3 - x^4/4 - \ldots - x^r/r - \ldots .$$

By subtracting this series from the series for $\ln (1 + x)$ we find

$$\ln \left(\frac{1+x}{1-x} \right) = 2(x + x^3/3 + x^5/5 + \ldots + x^{2r-1}/(2r-1) + \ldots).$$

This expansion is valid for $-1 < x < 1$.
  Now if $x = 1/(2n + 1)$ we have $(1 + x)/(1 - x) = (n + 1)/n$ so that

$$\ln \left(\frac{n+1}{n} \right) = 2 \left[\frac{1}{2n + 1} + \frac{1}{3(2n + 1)^3} + \frac{1}{5(2n + 1)^5} + \ldots \right].$$

This expansion is valid when n is positive, for then x will lie between 0 and 1. As this series converges rapidly it is useful for the calculation of logarithms. By putting $n = 1$ we find

$$\ln 2 = 2\left[\frac{1}{3} + \frac{1}{3}\left(\frac{1}{3}\right)^3 + \frac{1}{5}\left(\frac{1}{3}\right)^5 + \cdots\right].$$

By putting $n = 2$ we find

$$\ln (3/2) = 2\left[\frac{1}{5} + \frac{1}{3}\left(\frac{1}{5}\right)^3 + \frac{1}{5}\left(\frac{1}{5}\right)^5 + \cdots\right],$$

from which we can find $\ln 3$ if $\ln 2$ is known.

Example 1
Obtain the first four terms in the expansion of $\ln (1 - 3x + 2x^2)$ in a series of ascending powers of x. Find the coefficient of x^n and state the set of values of x for which the expansion is valid.

$$\ln (1 - 3x + 2x^2) = \ln (1 - x) + \ln (1 - 2x).$$
$$\ln (1 - x) = -x - x^2/2 - x^3/3 - x^4/4 - \cdots .$$
$$\ln (1 - 2x) = -2x - 2x^2 - 8x^3/3 - 4x^4 - \cdots .$$

Hence

$$\ln (1 - 3x + 2x^2) = -3x - 5x^2/2 - 3x^3 - 17x^4/4 - \cdots .$$

The coefficient of x^n in the expansion of $\ln (1 - x)$ is $-1/n$, while in the expansion of $\ln (1 - 2x)$ it is $-2^n/n$. In the required series it will be $-(2^n + 1)/n$.

The expansion of $\ln (1 - x)$ is valid for $-1 \leqslant x < 1$, and the expansion of $\ln (1 - 2x)$ is valid for $-1 \leqslant 2x < 1$. Therefore both expansions are valid when x lies in the interval $-\frac{1}{2} \leqslant x < \frac{1}{2}$, and this is the set of values of x for which the expansion of $\ln (1 - 3x + 2x^2)$ is valid.

Example 2
Given that $\log_{10} e = 0.434\,29 \ldots$, find $\log_{10} 11$ correct to five decimal places.

By substituting $n = 10$ in the expansion

$$\ln \left(\frac{n+1}{n}\right) = 2\left[\frac{1}{2n+1} + \frac{1}{3(2n+1)^3} + \cdots\right]$$

we have

$$\ln 11 - \ln 10 = 2(1/21 + 1/(3 \times 21^3) + \cdots)$$
$$= 0.095\,310.$$

The sum of the terms neglected is less than the sum of a geometrical series with first term $2/(5 \times 21^5)$ and common ratio $1/21^2$. This sum is less than 10^{-7}.

Now $\log_{10} 11 = (\ln 11)/(\ln 10)$
$$= 1 + (0.095\,310)/(\ln 10)$$

$$\log_{10} 11 = 1 + (0\cdot095\,310) \times \log_{10} e$$
$$= 1 + 0\cdot041\,392.$$

Hence, to five decimal places, $\log_{10} 11 = 1\cdot041\,39$.

Exercises 11.10

1 Write down the first four terms in the Maclaurin series for
(a) $\ln(1+2x)$ (c) $\ln(1+x^2)$
(b) $\ln(2+x)$ (d) $\ln(1+x)^2$.
State the set of values of x for which each expansion is valid.

2 Find to four places of decimals the sum of the first three terms in the series for $\ln 2$ given in section 11.10. Show that the sum of the remaining terms is less than 2×10^{-4}.

3 Show that

$$\ln(5/4) = 2[1/9 + 1/(3 \times 9^3) + 1/(5 \times 9^5) + \ldots 1/(2r-1)9^{2r-1} + \ldots].$$

Given that to five decimal places $\ln 4 = 1\cdot386\,29$, show that to four decimal places $\ln 5 = 1\cdot6094$.

4 Show that, if x is small,

$$e^x \ln(1+x) + \ln(1-x) \approx -x^4/4.$$

5 If $Y = (y-1)/(y+1)$ and $y > 0$, show that

$$\ln y = 2[Y + Y^3/3 + Y^5/5 + \ldots + Y^{2r-1}/(2r-1) + \ldots].$$

6 Show that

$$\ln 8 = 2(1 + 1/(3 \times 9) + 1/(5 \times 9^2) + 1/(7 \times 9^3) + \ldots).$$

7 Find the coefficient of x^n in the Maclaurin series for
(a) $\ln(2+3x)$ (b) $\ln(2-x-x^2)$.

8 Find the value of the constant k given that when x is small

$$(2+x)\ln(1+x) + (2-x)\ln(1-x) \approx kx^4.$$

9 Sketch the curve $y = f(x)$ when $f(x)$ equals
(a) $\ln(1+x)$ (b) $\ln(1-x)$ (c) $\ln[(1+x)/(1-x)]$.

10 Assuming that the series

$$x - x^2/2 + x^3/3 - x^4/4 + \ldots + (-1)^{n-1}x^n/n + \ldots$$

converges to $\ln(1+x)$ for all values of x in the interval $-1 < x \leqslant 1$, find the sum of the series

(a) $\dfrac{1}{2} - \dfrac{1}{2}\left(\dfrac{1}{2}\right)^2 + \dfrac{1}{3}\left(\dfrac{1}{2}\right)^3 - \ldots + (-1)^{n-1}\dfrac{1}{n}\left(\dfrac{1}{2}\right)^n + \ldots$

(b) $\dfrac{1}{3} + \dfrac{1}{2}\left(\dfrac{1}{3}\right)^2 + \dfrac{1}{3}\left(\dfrac{1}{3}\right)^3 + \ldots + \dfrac{1}{n}\left(\dfrac{1}{3}\right)^n + \ldots$

(c) $1/2 - 1/6 + 1/12 - \ldots + (-1)^{n-1}/[n(n+1)] + \ldots$.

11.11 Separable differential equations

Consider an equation involving x, y and an arbitrary constant c, such that for each value of c the equation represents a curve in the $(x–y)$ plane. The set of curves obtained by assigning different values to the constant c is called a family of curves.

For example, the equation

$$x^2 + y^2 = c$$

corresponds to the family of circles with centre at the origin. Each positive value of c gives a circle belonging to the family. If this equation is differentiated with respect to x, the constant c is eliminated and the differential equation

$$2x + 2y\frac{dy}{dx} = 0 \quad \text{or} \quad y\frac{dy}{dx} = -x.$$

is obtained. This equation is satisfied at every point on each circle.

When a differential equation involving x, y and dy/dx is given, the problem arises of finding the equation of the family of curves from which it is derived. If the differential equation can be put in the form

$$f(y)\frac{dy}{dx} = g(x),$$

as in the example above, the variables x and y are said to be separable. Now if two functions F and G can be found such that $F' = f$ and $G' = g$, the *general solution* of the differential equation will be

$$F(y) = G(x) + c,$$

where c is an arbitrary constant. The general solution always contains one arbitrary constant, giving a family of curves. If the solution is required to satisfy some specific condition, the value of the constant c can be found. A *particular solution* is then obtained, giving one curve of the family.

Example 1
Find the general solution of the differential equation

$$2x^2 y\frac{dy}{dx} - x^2 = 1.$$

Divide by x^2 and rearrange to obtain

$$2y\frac{dy}{dx} = 1 + \frac{1}{x^2}.$$

The variables are now separated, and integration with respect to x gives the general solution

$$y^2 = x - \frac{1}{x} + c.$$

Example 2

Find the equation of a curve given that it passes through the point (1, 2) and that its gradient at any point (x, y) is equal to y.

It is given that $\dfrac{dy}{dx} = y$, which can be put in the form

$$\frac{1}{y}\frac{dy}{dx} = 1.$$

The general solution will be

$$\ln y = x + c.$$

By substituting $x = 1$ and $y = 2$, we find the value of c to be $(\ln 2 - 1)$. This gives the particular solution

$$\ln y = x - 1 + \ln 2 \quad \text{or} \quad y = 2e^{x-1}.$$

Exercises 11.11

1   Form a differential equation (not involving c) satisfied by each member of the family of curves given by
   (a) $x^2 - y^2 = c$ (c) $y^2 = 4x + c$
   (b) $xy = c$ (d) $y = cx + 1$.

2   Find the general solution of the differential equations
   (a) $dy/dx = x/y$ (b) $(xy)dy/dx = 1$.

3   Find the particular solution of the equation

$$dy/dx - 1 = (2x - 2y)/(2y + 1)$$

such that $y = 2$ when $x = 0$.

4   Find the particular solution of the equation

$$dy/dx = x^2 + x^2 y^2$$

such that $y = 1$ when $x = 0$.

5   Solve the differential equation $dx/dt = 2x$ given that $x = 4$ when $t = 0$.

6   Solve the differential equation $dy/dx = -y$ given that $y = e$ when $x = 3$.

7   Solve the differential equation $du/dt + 2ut = 0$ given that $u = 2e$ when $t = 0$.

8   Each member of a family of circles touches the y-axis at the origin. Find the differential equation satisfied by this family.

9   Show that the differential equation of the family of straight lines which pass through the point $(0, 1)$ is $x\, dy/dx + 1 = y$.

10   Each member of a family of parabolas passes through the origin and has the x-axis for its axis. Find the differential equation of this family.

Miscellaneous examples 11

1   Find $\displaystyle\int_0^{\pi/2} \frac{1}{1 + \cos\theta + \sin\theta}\, d\theta.$

Make the substitution $t = \tan(\theta/2)$. Then

$$\sin\theta = \frac{2t}{1+t^2}, \quad \cos\theta = \frac{1-t^2}{1+t^2}$$

and so

$$1 + \cos\theta + \sin\theta = \frac{2+2t}{1+t^2}.$$

Since $\theta = 2\arctan t$, we have $\dfrac{d\theta}{dt} = \dfrac{2}{1+t^2}$.

The limits for t are $t = 0$ and $t = 1$, and the integral becomes

$$\int_0^1 \left(\frac{1+t^2}{2+2t}\right)\left(\frac{2}{1+t^2}\right) dt = \int_0^1 \frac{1}{1+t}\, dt$$

$$= \left[\ln(1+t)\right]_0^1$$

$$= \ln 2.$$

2 Find the sum of the infinite series

$$\frac{3}{2!} + \frac{5}{3!} + \frac{7}{4!} + \ldots + \frac{2n+1}{(n+1)!} + \ldots.$$

First the general term must be expressed as the sum of fractions which do not have n in the numerator. By replacing $2n+1$ by $2(n+1)-1$ we have

$$\frac{2n+1}{(n+1)!} \equiv \frac{2(n+1)}{(n+1)!} - \frac{1}{(n+1)!} \equiv \frac{2}{n!} - \frac{1}{(n+1)!}.$$

The sum of the first n terms in the series becomes

$$\left(\frac{2}{1} - \frac{1}{2!}\right) + \left(\frac{2}{2!} - \frac{1}{3!}\right) + \left(\frac{2}{3!} - \frac{1}{4!}\right) + \ldots + \left(\frac{2}{n!} - \frac{1}{(n+1)!}\right).$$

This can be simplified to

$$\left[2 + \frac{1}{2!} + \frac{1}{3!} + \frac{1}{4!} + \ldots + \frac{1}{n!}\right] - \frac{1}{(n+1)!}.$$

As n tends to infinity, the sum of the series in the bracket tends to e while the final term tends to zero. Hence the sum of the infinite series is e.

Miscellaneous exercises 11

1 Use the substitution $t = \tan(\theta/2)$ to find
   (a) $\int \sec\theta\, d\theta$ (b) $\int \operatorname{cosec}\theta\, d\theta$.

   Show that $\displaystyle\int_0^{\pi/2} \frac{5}{4\cos\theta + 3\sin\theta}\, d\theta = \ln 6.$

2 Find the sum of the infinite series in which the nth term is

(a) $(n+1)/n!$ (b) $n^2/(n+1)!$ (c) $2^n/n!$.

3 Sketch the curve $y = (x+2)e^{-x}$. Show that the point $(0, 2)$ is a point of inflexion of the curve and find the equation of the tangent at this point.

4 Differentiate with respect to x

(a) $\ln(\tan x)$ (b) $\ln(\sec x + \tan x)$ (c) x^x.

5 Show that

(a) $\int \ln x \, dx = x \ln x - x + c$

(b) $\int (\ln x)^2 \, dx = x(\ln x)^2 - 2x \ln x + 2x + c$.

6 Find the gradient of the curve $y = e^{2x}(x+1)^2 \cos x$ at the point $(0, 1)$.

7 Show that

(a) $\displaystyle\int_0^{\pi/2} \frac{1}{1+\sin\theta} \, d\theta = 1$

(b) $\displaystyle\int_0^{\pi} \frac{1}{2+\cos\theta} \, d\theta = \pi/\sqrt{3}$.

8 Find

(a) $\displaystyle\int \frac{4}{x^4-1} \, dx$ (b) $\displaystyle\int \frac{4}{x(4-x^2)} \, dx$.

9 Show that

(a) $\int x^2 e^x \, dx = (x^2 - 2x + 2)e^x + c$

(b) $\int x^2 e^{-x} \, dx = -(x^2 + 2x + 2)e^{-x} + c$.

10 Show that

$$\int e^{3x} \sin(2x) \, dx = e^{3x}[3\sin(2x) - 2\cos(2x)]/13.$$

11 Determine the values of the constants a and b so that the coefficients of x^3 and x^4 in the expansion of

$$(1 + ax + bx^2)e^{-x}$$

in ascending powers of x may vanish. Find the first four non-zero terms of the expansion and hence show that, neglecting powers of x higher than the fifth, the error in taking $\dfrac{12 - 6x + x^2}{12 + 6x + x^2}$ for e^{-x} is $-x^5/720$.   [AEB]

12 If $e^{2y}(1+x) = (1-2x)$, express y in terms of x. Expand y as a series in ascending powers of x up to and including the term in x^3 and state the set of values of x for which the expansion is valid.   [AEB]

13 Show that the curve $xy = \log_e(1/x)$ passes through the point $(1, 0)$ and that y has a minimum value at $x = e$. Show further that the tangent to the curve at the point where $x = \sqrt{e}$ passes through the origin of coordinates and find the area included between this tangent, the curve and the x-axis.   [AEB]

14 If x is so small that terms in x^n $(n \geqslant 4)$ may be neglected, show that $e^x \sin x - e^{-x} \cos x$ can be expressed in the form $a + bx + x^2$ where a and b are constants whose values are to be found. Show that under the same conditions, $e^x \cos x - e^{-x} \sin x = 1 - ax^2 + bx^3/(a-b)$.   [AEB]

15 Solve the differential equation $\dfrac{dy}{dx} = \dfrac{y^3}{x^2}$, given that $y = 1$ when $x = 2$.

Sketch the curve represented by the solution, and obtain the equation of the normal at the point where $y = \frac{1}{2}$.

16 Solve the equation $y\dfrac{dy}{dx} = 2 - x$, given that $y = 3$ when $x = -2$.

Find the equations of the tangents to the curve represented by the solution at the points where $x = -1$.

17 The gradient at any point (x, y) on a curve which passes through the origin is given by

$$(x+1)^2\frac{dy}{dx} - y = 1/y.$$

Find the equation of the curve.

18 (i) Sketch on the same diagram the graphs of $y = e^{2x}$ and $y = 1 + 2x$, showing clearly any points of contact or points of intersection.

(ii) Assuming the series expansion for $\ln(1 + x)$, show that, for $m > n > 0$,

$$\ln\left(\frac{m}{n}\right) = 2\left[\frac{m-n}{m+n} + \frac{1}{3}\left(\frac{m-n}{m+n}\right)^3 + \frac{1}{5}\left(\frac{m-n}{m+n}\right)^5 + \ldots\right].$$

Hence calculate $\ln(13/12)$ to 6 decimal places.   [L]

19 (i) If the first two terms in the expansion in ascending powers of x of $(a + bx)e^{-2x}$ are $1 - 2x^2$, find a, b and show that the coefficient of x^r is

$$\frac{(-1)^{r-1}2^r(r-1)}{r!}.$$

(ii) For each of the series

$$\sum_{r=1}^{\infty}\frac{(-1)^{r-1}}{r}\frac{x^r}{2^r} \quad \text{and} \quad \sum_{r=1}^{\infty}\frac{1}{r}\frac{x^r}{3^r},$$

state the set of real values of x for which the series converges and find the sum of the series when it does converge. If the series have the same sum, find the two possible values of x.   [L]

20 (i) The functions f and g are defined by

$$f(x) = 1 + x^2/2! + x^4/4! + \ldots + x^{2n}/(2n)! + \ldots,$$
$$g(x) = x + x^3/3! + x^5/5! + \ldots + x^{2n+1}/(2n+1)! + \ldots.$$

Express $f(x)$ and $g(x)$ in terms of e^x, and hence obtain the expansion of $[f(x)]^2 + [g(x)]^2$ in a series of powers of x, giving the coefficient of x^{2n}.

(ii) Find the first four terms in the expansion of

$$\ln[(1+2x)/(1-x)]$$

in a series of ascending powers of x, and state the set of values of x for which this expansion is valid. If a_n and a_{n+1} are the coefficients of x^n and x^{n+1} respectively in the expansion, show that the value of $(n+1)a_{n+1} + 2na_n$ is independent of n.   [L]

21 Integrate the equation

$$y - x\frac{dy}{dx} = \frac{dy}{dx} - y^2,$$

given that $y = 1$ when $x = 1$, and find the value of y when $x = 2$.

22 Prove that

(i) $\displaystyle\int_1^2 x \ln x \, dx = 2 \ln 2 - 3/4$

(ii) $\displaystyle\int_0^1 \frac{x^2 dx}{(x+1)^2} = 3/2 - 2 \ln 2.$

23 Evaluate

$$\int_2^4 \frac{(x+1)^2}{x^4 - 1} \, dx.$$

24 If $y = e^{-x} \cos x$, find $\dfrac{dy}{dx}$ and $\dfrac{d^2y}{dx^2}$ and show that

$$\frac{d^2y}{dx^2} + 2\frac{dy}{dx} + 2y = 0.$$

Show that y has a minimum value when $x = 3\pi/4$.

25 If $y = e^{2x} \sin 3x$, prove that

$$\frac{d^2y}{dx^2} = 13e^{2x} \cos(3x + \alpha),$$

where $\tan \alpha = 5/12$.

26 Find the value of x to one decimal place, given that

$$2 + \log_{10} x = \log_e x.$$

27 A particle is moving in a straight line. The displacement x m, from an origin O on the line, is given at time t s by the equation

$$x = e^{-\frac{3}{4}t}(a \sin t + b \cos t).$$

Initially $t = 0$, $x = 4$ and $dx/dt = 0$. Find the constants a and b. Determine also

(i) the time elapsing from the start before the particle first reaches O

(ii) the time taken from O to attain the greatest displacement on the negative side of the origin.   [JMB]

28 The rate of decay at any instant of a radioactive substance is proportional to the amount of the substance remaining at that instant. If the initial amount of the substance is A and the amount remaining after time t is x, prove that

$$x = Ae^{-kt},$$

where k is a constant.

If the amount remaining is reduced from $\frac{1}{2}A$ to $\frac{1}{3}A$ in 8 hours, prove that the initial amount of the substance was halved in about 13·7 hours.

[JMB]

29 Solve the equation $e^x + 2e^{-x} = 3$.

30 Show that

$$e^x(\cos x - \sin x) = 1 - x^2 - 2x^3/3 - x^4/6 + \ldots.$$

31 Find the sum of the infinite series in which the nth term is $2^{2n}/(2n-1)!$.

32 Sketch the family of curves given by the differential equation
$$dy/dx = y/(1+x).$$

33 Evaluate

(a) $\displaystyle\int_0^\pi 5e^x \sin x \cos x \, dx$

(b) $\displaystyle\int_1^2 x \log_e(1+x)\, dx.$

34 If x is so small that powers above the fourth may be neglected, show that

$$\log_e\left(\frac{1}{1-x}\right) - e^x \log_e(1+x)$$

is approximately equal to $\frac{1}{4}x^4$.

35 Prove that

$$\log_e\left(1 + \frac{3}{x} + \frac{2}{x^2}\right) = \frac{2+1}{x} - \frac{2^2+1}{2x^2} + \frac{2^3+1}{3x^3} - \ldots,$$

stating the condition to be satisfied by x for the expansion to be valid.

Calculate $\log_e 1\cdot32$ to three places of decimals.

36 Prove that, if $-1 \leqslant x < 1$,

$$\frac{1-x}{x}\log_e(1-x) + 1 = \sum_{n=1}^{\infty} \frac{x^n}{n(n+1)}.$$

Deduce, by putting $x = -1$, that

$$\log_e 2 - \frac{1}{2} = \frac{1}{(1)(2)(3)} + \frac{1}{(3)(4)(5)} + \frac{1}{(5)(6)(7)} + \ldots \quad \text{[L]}$$

37 (i) Find the coefficients of x, x^2 and x^3 in the expansion of $e^{nx}/(1+x)^n$ in a series of ascending powers of x.

(ii) Assuming the expansion of $\ln(1+x)$ in ascending powers of x, obtain the expansion of $f(x)$, where

$$f(x) = \frac{1}{2}\ln\left(\frac{1+x}{1-x}\right),$$

giving the general term. State the set of values of x for which this expansion is valid, and sketch the graph of $f(x)$. [L]

38 Solve the differential equation $3y^2 \dfrac{dy}{dx} = 4x^3$ and find the particular solution

for which $y = 1$ when $x = 1$.

Find the equations of the tangent and normal to the curve represented by the solution, at the point where $x = -1$. [L]

39 The rate of cooling of a kettle of hot water is proportional to the difference between its temperature and that of the room. When the room temperature is $15°\,C$ the kettle cools from $95°\,C$ to $55°\,C$ in ten minutes. Find the time taken for the kettle to cool from $55°\,C$ to $25°\,C$.

40 (i) Find the coordinates of the point of intersection of the curves $y = 2e^x$ and $y = 3 + 2e^{-x}$.

(ii) The curve $y = e^x(ax^2 + bx + c)$ passes through the point $(0, 9)$. The tangents to the curve at the points given by $x = 1$ and $x = 3$ are parallel to the x-axis. Find the values of a, b and c.

41 Find the stationary values of $(3e^{3x} - 9e^{2x} + 8e^x)$ and determine their nature.

42 The amount x of radioactive substance present at time t in a reaction is given by the differential equation

$$\frac{dx}{dt} = a - bx,$$

where a and b are positive constants. Solve the differential equation for x in terms of t, given that when $t = 0$ the amount of the substance present is $a/(4b)$, and find the time taken for x to attain the value $a/(2b)$.

Find the limiting value of x as t tends to infinity. [JMB]

43 The volume, x litres, of water present in a solution during a chemical process varies with time, t seconds, and satisfies the relation

$$\frac{dx}{dt} = \frac{-3x}{(1+t)^2}.$$

Initially, at $t = 0$, $x = 1000$. Show that at time t the volume is given by

$$x = 1000\,\exp\left[-3t/(1+t)\right].$$

Prove that the volume of water tends to a limit as t tends to infinity and find that limit to the nearest litre. [JMB]

12 Vectors

12.1 Free vectors

Quantities such as distance and volume, which can be specified by their magnitude, are scalar quantities or *scalars*. Displacement, velocity and force, which possess a direction and a magnitude in that direction, are vector quantities or *vectors*.

A directed line segment $\overrightarrow{AB}$, drawn from A and B, is a vector, with magnitude and direction. The vector $\overrightarrow{AB}$ represents the displacement of the point A to the point B. If every point in the x–y plane is moved the same distance in the same direction, the transformation is known as a translation. Any other directed line segment $\overrightarrow{PQ}$, of same length as $\overrightarrow{AB}$ and in the same direction, will represent the same translation. We regard $\overrightarrow{AB}$ and $\overrightarrow{PQ}$ as equal vectors and write $\overrightarrow{AB} = \overrightarrow{PQ}$. A single letter in bold-face type, such as $\mathbf{a}$, is used to represent any vector equal to $\overrightarrow{AB}$. Then $\mathbf{a}$ is a free vector, since it can be placed anywhere in the plane. (A force is not a free vector since it has a specified line of action.)

The vector $\overrightarrow{BA}$ represents the reverse translation to that given by $\overrightarrow{AB}$. If $\overrightarrow{AB} = \mathbf{a}$, we put $\overrightarrow{BA} = -\mathbf{a}$. The magnitude of a vector $\mathbf{a}$ is known as its modulus, and is denoted by $|\mathbf{a}|$ or by a. The modulus is a scalar quantity and is always positive; except in the case of the zero vector.

12.2 Vector addition

To find the sum of the vectors $\mathbf{p}$ and $\mathbf{q}$, we represent them by the directed line segments $\overrightarrow{OP}$ and $\overrightarrow{OQ}$ respectively and then complete the parallelogram $OPRQ$. The diagonal $\overrightarrow{OR}$ represents the sum of $\mathbf{p}$ and $\mathbf{q}$ in magnitude and direction. This is the parallelogram rule (Fig. 12.1).

Alternatively, we can regard $\mathbf{p}$ as representing the displacement of the point O to the point P, and $\mathbf{q}$ as representing the displacement of the point P to the point R. Then $\mathbf{p} = \overrightarrow{OP}$, $\mathbf{q} = \overrightarrow{PR}$, and the third side $\overrightarrow{OR}$ of the triangle OPR represents the sum of $\mathbf{p}$ and $\mathbf{q}$. This is the triangle rule.

We use the plus sign to denote the addition of vectors. Since $(\overrightarrow{OP} + \overrightarrow{PR})$ and $(\overrightarrow{OQ} + \overrightarrow{QR})$ are both equal to $\overrightarrow{OR}$, it follows that

$$\mathbf{p} + \mathbf{q} = \mathbf{q} + \mathbf{p}.$$

The order in which two vectors are added together makes no difference, i.e. vector addition is commutative. The difference $\mathbf{p} - \mathbf{q}$ is defined as the sum of $\mathbf{p}$ and

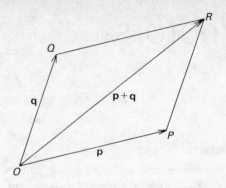

Fig. 12.1

$(-\mathbf{q})$. Since $-\mathbf{q}$ is represented in Fig. 12.1 by $\overrightarrow{QO}$, $\mathbf{p}-\mathbf{q}$ is represented by $\overrightarrow{QO}$ $+\overrightarrow{OP}$, i.e. $\overrightarrow{QP}$. When $\mathbf{p}$ and $\mathbf{q}$ are equal, their difference will be the zero vector.

Example 1
Show that $\mathbf{a}+(\mathbf{b}+\mathbf{c}) = (\mathbf{a}+\mathbf{b})+\mathbf{c}.$

Let $\overrightarrow{AB} = \mathbf{a}$, $\overrightarrow{BC} = \mathbf{b}$, $\overrightarrow{CD} = \mathbf{c}$ (Fig. 12.2).

Then
$$\mathbf{a}+(\mathbf{b}+\mathbf{c}) = \overrightarrow{AB}+(\overrightarrow{BC}+\overrightarrow{CD})$$
$$= \overrightarrow{AB}+\overrightarrow{BD} = \overrightarrow{AD}.$$

Also
$$(\mathbf{a}+\mathbf{b})+\mathbf{c} = (\overrightarrow{AB}+\overrightarrow{BC})+\overrightarrow{CD}$$
$$= \overrightarrow{AC}+\overrightarrow{CD} = \overrightarrow{AD}.$$

Hence
$$\mathbf{a}+(\mathbf{b}+\mathbf{c}) = (\mathbf{a}+\mathbf{b})+\mathbf{c}.$$

Thus vector addition is associative.

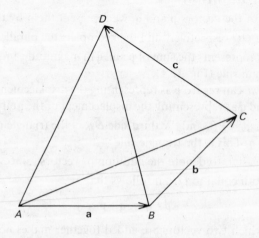

Fig. 12.2

The product of a real number (or scalar) k and a vector $\mathbf{p}$ is written $k\mathbf{p}$. When k is positive, $k\mathbf{p}$ is a vector in the same direction as $\mathbf{p}$ with length k times the length of $\mathbf{p}$. When k is negative, $k\mathbf{p}$ is a vector in the opposite direction to $\mathbf{p}$ with length $(-k)$ times the length of $\mathbf{p}$.

If the vectors $\mathbf{p}$ and $\mathbf{q}$ are in the same direction or are in opposite directions, $\mathbf{p}$ can be expressed as a multiple of $\mathbf{q}$. When $\mathbf{p}$ is not a multiple of $\mathbf{q}$, the two vectors are said to be independent, and the vector $m\mathbf{p} + n\mathbf{q}$ will be zero only when both m and n are zero.

Example 2
In the triangle ABC, M and N are the mid-points of the sides AB and AC respectively. Show that MN is parallel to BC and that $MN = \frac{1}{2}BC$.

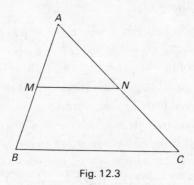

Fig. 12.3

Let $\overrightarrow{AB} = \mathbf{b}$ and $\overrightarrow{AC} = \mathbf{c}$. Then $\overrightarrow{AM} = \frac{1}{2}\mathbf{b}$ and $\overrightarrow{AN} = \frac{1}{2}\mathbf{c}$.
Since $\overrightarrow{AB} + \overrightarrow{BC} = \overrightarrow{AC}$, $\overrightarrow{BC} = \overrightarrow{AC} - \overrightarrow{AB} = \mathbf{c} - \mathbf{b}$.
Since $\overrightarrow{AM} + \overrightarrow{MN} = \overrightarrow{AN}$, $\overrightarrow{MN} = \overrightarrow{AN} - \overrightarrow{AM}$
$$= \tfrac{1}{2}(\mathbf{c} - \mathbf{b})$$
$$= \tfrac{1}{2}\overrightarrow{BC},$$
  i.e. MN is parallel to BC, and $MN = \frac{1}{2}BC$.

Example 3
If $\mathbf{p}$ and $\mathbf{q}$ are independent and are coplanar, show that any vector $\mathbf{r}$ in the same plane as $\mathbf{p}$ and $\mathbf{q}$ can be expressed in the form $m\mathbf{p} + n\mathbf{q}$.

Draw a parallelogram $OARB$ with the diagonal $\overrightarrow{OR}$ equal to $\mathbf{r}$, and with $\overrightarrow{OA}$ parallel to $\mathbf{p}$ and $\overrightarrow{OB}$ parallel to $\mathbf{q}$. Then $\overrightarrow{OA}$ is a multiple of $\mathbf{p}$ and $\overrightarrow{OB}$ is a multiple of $\mathbf{q}$, while $\mathbf{r} = \overrightarrow{OR} = \overrightarrow{OA} + \overrightarrow{OB}$.

Example 4

Show that for any vectors **p** and **q**,

$$|\mathbf{p}+\mathbf{q}| \leqslant |\mathbf{p}|+|\mathbf{q}|.$$

In the triangle OPR in Fig. 12.1, the length of the side OR cannot be greater than the sum of the lengths of the sides OP and PR, i.e. $|\mathbf{p}+\mathbf{q}|$ cannot be greater than the sum of $|\mathbf{p}|$ and $|\mathbf{q}|$ giving the result.

Clearly $|\mathbf{p}+\mathbf{q}|$ is equal to $|\mathbf{p}|+|\mathbf{q}|$ only when **p** and **q** are in the same direction (or when one of them is zero). Compare with Section 8.10.

Exercises 12.2

1 In the rectangle $ABCD$, $|\overrightarrow{AB}| = 4$ and $|\overrightarrow{AD}| = 3$. Find the values of $|\overrightarrow{AB}+\overrightarrow{AD}|$ and $|\overrightarrow{AB}-\overrightarrow{AD}|$.

2 Find the value of $|\mathbf{a}+\mathbf{b}|$, given that $|\mathbf{a}| = 7$, $|\mathbf{b}| = 8$ and that the angle between **a** and **b** is 60°.

3 $OPQRST$ is a regular hexagon in which $\overrightarrow{OP} = \mathbf{p}$ and $\overrightarrow{PQ} = \mathbf{q}$. Show that $\overrightarrow{OQ} = \mathbf{p}+\mathbf{q}$, $\overrightarrow{OR} = 2\mathbf{q}$, $\overrightarrow{OS} = 2\mathbf{q}-\mathbf{p}$, $\overrightarrow{OT} = \mathbf{q}-\mathbf{p}$.

4 Find the modulus of $(\mathbf{a}-\mathbf{b})$, given that $|\mathbf{a}| = |\mathbf{b}| = 1$ and $|\mathbf{a}+\mathbf{b}| = 1$.

5 Show how to construct a parallelogram $ABCD$ in which $\overrightarrow{AC} = \mathbf{a}$ and $\overrightarrow{BD} = \mathbf{b}$, where **a** and **b** are known vectors.

6 OPQ is an equilateral triangle in which $\overrightarrow{OP} = \mathbf{p}$ and $\overrightarrow{OQ} = \mathbf{q}$. Express $\overrightarrow{PQ}$, $\overrightarrow{PM}$, $\overrightarrow{OM}$, where M is the mid-point of PQ, in terms of **p** and **q**.

7 A, B, C and D are the mid-points of the sides OP, PQ, QR and RO respectively of the square $OPQR$. If $\overrightarrow{OP} = \mathbf{p}$ and $\overrightarrow{PQ} = \mathbf{q}$, express $\overrightarrow{PR}$, $\overrightarrow{AB}$, $\overrightarrow{PC}$ and $\overrightarrow{CD}$ in terms of **p** and **q**.

8 Show geometrically that for any vectors **a** and **b**,

$$|\mathbf{a}-\mathbf{b}| \geqslant |\mathbf{a}|-|\mathbf{b}|.$$

9 In the quadrilateral $ABCD$, the points K, L, M and N are the mid-points of AB, BC, CD and DA respectively. Prove that $KLMN$ is a parallelogram.

10 The direction of the vector **p** is due east and $|\mathbf{p}| = 4$. The direction of the vector **q** is N 30° E and $|\mathbf{q}| = 3$. Find graphically the modulus and the direction of
(a) $3\mathbf{p}+4\mathbf{q}$ (b) $3\mathbf{p}-4\mathbf{q}$.
Check your results by calculation.

12.3 Position vectors

The position vector of a point P with respect to a fixed origin O is the directed line segment $\overrightarrow{OP}$ drawn from O to P. This is not a free vector. The sum of two position vectors $\overrightarrow{OP}$ and $\overrightarrow{OQ}$ is given by the vector $\overrightarrow{OR}$, where $OPRQ$ is a parallelogram. If M is the mid-point of PQ (Fig. 12.4), the position vector of M is $\frac{1}{2}(\overrightarrow{OP}+\overrightarrow{OQ})$.

For
$$\overrightarrow{OM} = \overrightarrow{OP} + \tfrac{1}{2}\overrightarrow{PQ}$$
$$= \overrightarrow{OP} + \tfrac{1}{2}(\overrightarrow{OQ} - \overrightarrow{OP})$$
$$= \tfrac{1}{2}(\overrightarrow{OP} + \overrightarrow{OQ}).$$

This shows that M is also the mid-point of OR, so that PQ and OR bisect one another.

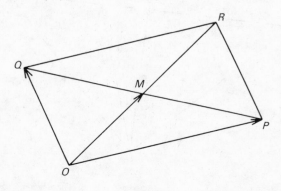

Fig. 12.4

Example

In the quadrilateral $ABCD$, the points P, Q, R and S are the mid-points of the sides AB, BC, CD and DA respectively. Show that PR and QS bisect each other.

Let $\mathbf{a}$, $\mathbf{b}$, $\mathbf{c}$ and $\mathbf{d}$ be the position vectors of the points A, B, C and D. Then the position vector of P is $\tfrac{1}{2}(\mathbf{a}+\mathbf{b})$, and that of R is $\tfrac{1}{2}(\mathbf{c}+\mathbf{d})$. The position vector of the mid-point of PR will be $\tfrac{1}{4}(\mathbf{a}+\mathbf{b}+\mathbf{c}+\mathbf{d})$. Also the position vectors of Q and S are $\tfrac{1}{2}(\mathbf{b}+\mathbf{c})$ and $\tfrac{1}{2}(\mathbf{d}+\mathbf{a})$, so that the position vector of the mid-point of QS is $\tfrac{1}{4}(\mathbf{a}+\mathbf{b}+\mathbf{c}+\mathbf{d})$. Hence the mid-points of PR and QS coincide.

Exercises 12.3

1 In the equilateral triangle OPQ the position vectors of P and Q with respect to O are $\mathbf{p}$ and $\mathbf{q}$. Show in a diagram the points with position vectors $\mathbf{p}+\mathbf{q}$, $\mathbf{p}-\mathbf{q}$, $\mathbf{q}-\mathbf{p}$, $2\mathbf{p}$, $2\mathbf{p}-\mathbf{q}$, $2\mathbf{q}$, $2\mathbf{q}-\mathbf{p}$.
2 In the square $OPQR$, the position vectors of P and Q with respect to O are $\mathbf{p}$ and $\mathbf{q}$ respectively. The diagonal OQ is produced to S so that $OS = 2OQ$. Find the position vectors of the mid-points of PS and RS.
3 The point O is the centre of the regular hexagon $ABCDEF$, and the position vectors with respect to O of A and B are $\mathbf{a}$ and $\mathbf{b}$ respectively. Find the position vectors of C, D, E and F.
4 The points A, B, C, D, E and F lie at equal intervals on a straight line. The position vectors of C and D are $\mathbf{p}$ and $\mathbf{q}$ respectively. Find the position vectors of A, B, E and F.

12.4 The ratio theorem

Let **a** and **b** be the position vectors of the points A and B with respect to a fixed origin O. Let **r** be the position vector of the point R which divides AB internally in the ratio λ to μ (Fig. 12.5).

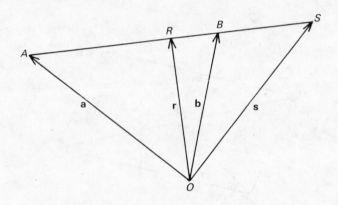

Fig. 12.5

Then $\overrightarrow{AR} = \mathbf{r} - \mathbf{a}$ and $\overrightarrow{RB} = \mathbf{b} - \mathbf{r}$. Since these two vectors are in the same direction with lengths in the ratio λ to μ, we have

$$\mu(\mathbf{r} - \mathbf{a}) = \lambda(\mathbf{b} - \mathbf{r})$$
$$(\lambda + \mu)\mathbf{r} = \lambda\mathbf{b} + \mu\mathbf{a}$$
$$\mathbf{r} = \frac{\lambda\mathbf{b} + \mu\mathbf{a}}{\lambda + \mu}.$$

Alternatively we can substitute

$$\overrightarrow{AR} = \frac{\lambda(\mathbf{b} - \mathbf{a})}{\lambda + \mu}.$$

in the equation $\overrightarrow{OR} = \overrightarrow{OA} + \overrightarrow{AR}$, giving

$$\mathbf{r} = \mathbf{a} + \frac{\lambda(\mathbf{b} - \mathbf{a})}{\lambda + \mu} = \frac{\lambda\mathbf{b} + \mu\mathbf{a}}{\lambda + \mu}.$$

Now let S be the point which divides AB externally in the ratio λ to μ (the case $\lambda > \mu$ is shown in Fig. 12.5). If **s** is the position vector of S, then $\overrightarrow{AS} = \mathbf{s} - \mathbf{a}$, and $\overrightarrow{BS} = \mathbf{s} - \mathbf{b}$. Hence

$$\mu(\mathbf{s} - \mathbf{a}) = \lambda(\mathbf{s} - \mathbf{b})$$
$$(\lambda - \mu)\mathbf{s} = \lambda\mathbf{b} - \mu\mathbf{a}$$
$$\mathbf{s} = \frac{\lambda\mathbf{b} - \mu\mathbf{a}}{\lambda - \mu}.$$

Example 1

The position vectors of the points P and Q are $\mathbf{p}$ and $\mathbf{q}$ respectively. Find the position vector of

(a) the point R dividing PQ internally in the ratio $3:1$

(b) the point S dividing PQ externally in the ratio $3:1$.

Show that $4\overrightarrow{RS} = 3\overrightarrow{PQ}$.

The position vector of R is $(3\mathbf{q}+\mathbf{p})/4$, and the position vector of S is $(3\mathbf{q}-\mathbf{p})/2$. It follows that

$$\overrightarrow{RS} = \frac{3\mathbf{q}-\mathbf{p}}{2} - \frac{3\mathbf{q}+\mathbf{p}}{4} = \frac{3\mathbf{q}-3\mathbf{p}}{4}.$$

Then

$$4\overrightarrow{RS} = 3(\mathbf{q}-\mathbf{p}) = 3\overrightarrow{PQ}.$$

Example 2

Show that the medians of a triangle trisect one another.

Let the position vectors of the points A, B and C be $\mathbf{a}$, $\mathbf{b}$ and $\mathbf{c}$ respectively. The position vector of M, the mid-point of BC, is $\frac{1}{2}(\mathbf{b}+\mathbf{c})$. Let G be the point which divides AM internally in the ratio $2:1$. The position vector of G will be

$$\{\mathbf{a}+2(\tfrac{1}{2}\mathbf{b}+\tfrac{1}{2}\mathbf{c})\}/3 = (\mathbf{a}+\mathbf{b}+\mathbf{c})/3.$$

This is symmetrical in $\mathbf{a}$, $\mathbf{b}$ and $\mathbf{c}$, so that the result would be the same if either the median through B or the median through C were used. Hence the point G lies on each median and trisects each median.

Exercises 12.4

1 The position vectors of the points A and B are $\mathbf{a}$ and $\mathbf{b}$ respectively. The point C divides AB internally in the ratio $2:1$, the point D divides AB externally in the ratio $1:4$ and the point E divides CD internally in the ratio $2:1$. Find the position vectors of C, D and E.

2 In the parallelogram $OPQR$, L and M are the mid-points of PQ and QR. Show that OL and OM trisect PR.

3 Show that the points with position vectors $\mathbf{a}$, $\mathbf{b}$, $3\mathbf{b}-2\mathbf{a}$ and $3\mathbf{a}-2\mathbf{b}$ lie in a straight line.

4 The position vectors of P and Q with respect to O are $\mathbf{p}$ and $\mathbf{q}$ respectively. The point A divides OP internally in the ratio $1:2$ and the point B divides OQ internally in the ratio $1:2$. Show that the position vector of the point of intersection of AQ and BP is $(\mathbf{p}+\mathbf{q})/4$.

5 The position vectors of the points A, B and C are $\mathbf{q}$, $2\mathbf{p}-2\mathbf{q}$, $6\mathbf{p}-8\mathbf{q}$ respectively. Show that the points are collinear, and find the position vector of the point D which divides AC externally in the ratio $AB:BC$.

12.5 The vector equation of a straight line

Consider the straight line drawn through the two points with position vectors $\mathbf{a}$

and **b**. In Fig. 12.5 the vector $\overrightarrow{BR}$ is a multiple of $\overrightarrow{BA}$, i.e. $\overrightarrow{BR} = t(\mathbf{a}-\mathbf{b})$ where t is a scalar. Since $\overrightarrow{OR} = \overrightarrow{OB} + \overrightarrow{BR}$, the position vector of R is

$$\mathbf{r} = \mathbf{b} + t(\mathbf{a}-\mathbf{b}),$$
or
$$\mathbf{r} = t\mathbf{a} + (1-t)\mathbf{b}.$$

This is the vector equation of the line AB. Each value of the parameter t corresponds to a point on the line. For $0 < t < 1$, the point lies between A and B. For $t > 1$, the point lies on BA produced, while for $t < 0$, the point lies on AB produced.

Next consider the straight line drawn through the point with position vector **p** in a direction parallel to the vector **q**. The position vector **r** of any point on this line will be such that $(\mathbf{r}-\mathbf{p})$ will be a multiple of **q**. We can write

$$\mathbf{r} - \mathbf{p} = t\mathbf{q},$$
or
$$\mathbf{r} = \mathbf{p} + t\mathbf{q}.$$

This is the vector equation of a line drawn through a given point in a given direction. The equation of a straight line through the origin can be put in the simple form $\mathbf{r} = t\mathbf{q}$.

Example

A straight line is drawn through the point with position vector $2\mathbf{p} - \mathbf{q}$ parallel to the vector $\mathbf{p} + \mathbf{q}$. Find the position vector of the point in which this line meets the line with equation $\mathbf{r} = \mathbf{p} + t\mathbf{q}$ (**p** and **q** are independent vectors).

The equation of the first line is

$$\mathbf{r} = 2\mathbf{p} - \mathbf{q} + s(\mathbf{p} + \mathbf{q}),$$

where s is a parameter. (Since t is used as a parameter in the equation of the second line, we have to use another letter such as s.) We now equate the two expressions for **r**.

$$2\mathbf{p} - \mathbf{q} + s(\mathbf{p} + \mathbf{q}) = \mathbf{p} + t\mathbf{q}$$
or
$$(1 + s)\mathbf{p} = (1 - s + t)\mathbf{q}.$$

Now a multiple of the vector **p** cannot equal a multiple of the vector **q**, and so each side of this equation must be zero. This gives $s = -1$ and $t = -2$, showing that the lines meet at the point with position vector $\mathbf{p} - 2\mathbf{q}$.

Exercises 12.5

1 The line drawn through the point with position vector $2\mathbf{a}$ parallel to the vector **b** meets the line drawn through the point with position vector $3\mathbf{b}$ parallel to the vector **a** at a point P. Find the equations of the lines and the position vector of P.

2 If **p** and **q** are the position vectors with respect to O of the vectices P and Q of the square $OPRQ$, obtain the vector equation of

(a) the side PR (b) the diagonal PQ (c) the diagonal OR

(d) the straight line drawn through R parallel to PQ.

3   The position vectors of the points A, B and C are $\mathbf{a}$, $\mathbf{b}$ and $\mathbf{c}$ respectively. Find the equations of the line through B parallel to CA and the line through C parallel to AB. Find also the position vector of the point in which the lines meet.

4   The position vectors of the points P and Q with respect to the origin O are $\mathbf{p}$ and $\mathbf{q}$ respectively. G is the centroid of the triangle OPQ. Find the vector equation

(a) of the line through G parallel to PQ

(b) of the line through P parallel to OG.

Find also the position vector of the point of intersection of the lines.

5   Find the position vector of the point of intersection of the lines with equations

$$\mathbf{r} = 2\mathbf{a} + t(\mathbf{a} + \mathbf{b}),$$
$$\mathbf{r} = \mathbf{b} + s(\mathbf{a} - 2\mathbf{b}).$$

12.6 Unit vectors in two dimensions

A unit vector is a vector of unit magnitude. To obtain the unit vector in the direction of a given vector $\mathbf{p}$, divide $\mathbf{p}$ by its modulus $|\mathbf{p}|$. Two unit vectors in the x–y plane can be defined; $\mathbf{i}$ in the direction of the positive x-axis and $\mathbf{j}$ in the direction of the positive y-axis. These are free vectors, and any vector $\mathbf{p}$ in the plane can be expressed in the form $a\mathbf{i} + b\mathbf{j}$, where a and b are scalars. $a\mathbf{i}$ and $b\mathbf{j}$ are the component vectors of $\mathbf{p}$, and a and b are the components of $\mathbf{p}$. If P is the point (a, b), the position vector of P with respect to the origin is $a\mathbf{i} + b\mathbf{j}$. Corresponding to each point (a, b) there is a vector $a\mathbf{i} + b\mathbf{j}$.

A column vector in two dimensions is defined to be a number pair, written in the form $\begin{pmatrix} a \\ b \end{pmatrix}$, satisfying the two relations

$$\begin{pmatrix} a \\ b \end{pmatrix} + \begin{pmatrix} c \\ d \end{pmatrix} = \begin{pmatrix} a+c \\ b+d \end{pmatrix}$$

and

$$k\begin{pmatrix} a \\ b \end{pmatrix} = \begin{pmatrix} ka \\ kb \end{pmatrix}$$

where k is a scalar. The first of these relations is equivalent to the parallelogram law.

Example 1

Express the vector $\begin{pmatrix} 1 \\ -1 \end{pmatrix}$ as the sum of multiples of the vectors $\begin{pmatrix} 2 \\ 1 \end{pmatrix}$ and $\begin{pmatrix} 1 \\ 3 \end{pmatrix}$.

Scalars m and n have to be found such that

$$m(2\mathbf{i} + \mathbf{j}) + n(\mathbf{i} + 3\mathbf{j}) = \mathbf{i} - \mathbf{j},$$

i.e.

$$(2m + n - 1)\mathbf{i} + (m + 3n + 1)\mathbf{j} = \mathbf{0}.$$

This can be true only if

$$2m + n - 1 = 0 \quad \text{and} \quad m + 3n + 1 = 0.$$

These two equations give $m = 4/5$, $n = -3/5$, from which we have

$$\begin{pmatrix} 1 \\ -1 \end{pmatrix} = \tfrac{4}{5}\begin{pmatrix} 2 \\ 1 \end{pmatrix} - \tfrac{3}{5}\begin{pmatrix} 1 \\ 3 \end{pmatrix}.$$

Example 2

Find the unit vector which bisects the angle between the vectors $\begin{pmatrix} 4 \\ 3 \end{pmatrix}$ and $\begin{pmatrix} 6 \\ 8 \end{pmatrix}$.

The modulus of the first vector is 5, and that of the second vector is 10. If each vector is divided by its modulus, two unit vectors, $\begin{pmatrix} 4/5 \\ 3/5 \end{pmatrix}$ and $\begin{pmatrix} 3/5 \\ 4/5 \end{pmatrix}$, are obtained.

Their sum gives a vector which bisects the angle, and so the required unit vector is $\begin{pmatrix} 1/\sqrt{2} \\ 1/\sqrt{2} \end{pmatrix}$.

Exercises 12.6

1 Find the unit vector in the direction of the vectors

(a) $\begin{pmatrix} 5 \\ 12 \end{pmatrix}$ (b) $\begin{pmatrix} 3 \\ -4 \end{pmatrix}$ (c) $\begin{pmatrix} 20 \\ 21 \end{pmatrix}$ (d) $\begin{pmatrix} -5 \\ 5 \end{pmatrix}$.

2 Express each of the following vectors in the form $m\mathbf{a} + n\mathbf{b}$, where $\mathbf{a} = \begin{pmatrix} 4 \\ -3 \end{pmatrix}$ and $\mathbf{b} = \begin{pmatrix} -2 \\ 5 \end{pmatrix}$.

(a) $\begin{pmatrix} 8 \\ 1 \end{pmatrix}$ (b) $\begin{pmatrix} 10 \\ 3 \end{pmatrix}$ (c) $\begin{pmatrix} -1 \\ 6 \end{pmatrix}$ (d) $\begin{pmatrix} 0 \\ 1 \end{pmatrix}$.

3 Find the moduli of the vectors $\mathbf{p}$, $\mathbf{q}$ and $\mathbf{r}$ where

$$\mathbf{p} = \begin{pmatrix} 4 \\ -3 \end{pmatrix} - \begin{pmatrix} 1 \\ 1 \end{pmatrix},$$

$$\mathbf{q} = 2\begin{pmatrix} 2 \\ 5 \end{pmatrix} + 3\begin{pmatrix} -1 \\ -3 \end{pmatrix},$$

$$\mathbf{r} = 4\begin{pmatrix} 3 \\ 2 \end{pmatrix} + 3\begin{pmatrix} -2 \\ 1 \end{pmatrix} - 5\begin{pmatrix} 1 \\ 2 \end{pmatrix}.$$

4 The position vectors of the points P and Q are $\begin{pmatrix} 7 \\ 9 \end{pmatrix}$ and $\begin{pmatrix} -1 \\ 3 \end{pmatrix}$ respectively. Find the position vector of

   (a) the point R which divides PQ internally in the ratio $3:2$

   (b) the point S which divides QP externally in the ratio $1:4$.

5 Show that the vector equation $\mathbf{r} = \mathbf{i} + t(\mathbf{i}+\mathbf{j})$ represents the straight line through the points with position vectors $\mathbf{i}$ and $-\mathbf{j}$. Find the vector equations of the straight lines drawn through

   (a) the point with position vector $\mathbf{i}$ parallel to the vector $\mathbf{j}$

   (b) the points with position vectors $\mathbf{i}$ and $\mathbf{j}$.

6 Show that the points with position vectors $\mathbf{i}+\mathbf{j}$, $5\mathbf{i}+4\mathbf{j}$, $3\mathbf{i}+2\mathbf{j}$, $3\mathbf{i}+3\mathbf{j}$ are the vertices of a parallelogram.

7 Each point in the x–y plane is moved through a distance of 15 units in the direction of the vector $3\mathbf{i}-4\mathbf{j}$. Find the position vectors of the points on to which the points with position vectors $-3\mathbf{i}+4\mathbf{j}$ and $-4\mathbf{i}+8\mathbf{j}$ are mapped.

8 The distance of the point P with position vector $(x\mathbf{i}+y\mathbf{j})$ from the point with position vector $4\mathbf{i}-4\mathbf{j}$ is twice its distance from the point with position vector $\mathbf{i}-\mathbf{j}$. Show that the locus of P is a circle with radius $2\sqrt{2}$.

12.7 Scalar products

The scalar product $\mathbf{a}\,.\,\mathbf{b}$ of two vectors $\mathbf{a}$ and $\mathbf{b}$ is a real number, defined by

$$\mathbf{a}\,.\,\mathbf{b} = ab\cos\theta,$$

where θ is the angle between the vectors. In words, the scalar product of $\mathbf{a}$ and $\mathbf{b}$ is the product of their moduli multiplied by the cosine of the angle between the vectors. This implies that $\mathbf{a}\,.\,\mathbf{b} = \mathbf{b}\,.\,\mathbf{a}$. Since $\mathbf{a}\,.\,\mathbf{b}$ is read as 'vector a dot vector b', the scalar product is often called the dot product. When $\mathbf{a}$ and $\mathbf{b}$ are in the same direction, their scalar product is ab; in particular

$$\mathbf{a}\,.\,\mathbf{a} = a^2 = |\mathbf{a}|^2.$$

In Fig. 12.6, $\mathbf{a}$ and $\mathbf{b}$ are represented by $\overrightarrow{OA}$ and $\overrightarrow{OB}$, and BL is perpendicular to OA. The projection of OB on OA is OL, which equals $OB\cos\theta$. This is positive when θ is acute and negative when θ is obtuse. Thus the scalar product $\mathbf{a}\,.\,\mathbf{b}$ equals OA multiplied by the projection of OB and OA.

When two vectors are at right angles, the cosine of the angle between them is

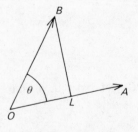

(a) θ acute, $\mathbf{a}\,.\,\mathbf{b} > 0$

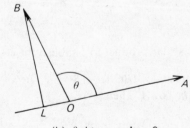

(b) θ obtuse, $\mathbf{a}\,.\,\mathbf{b} < 0$

Fig. 12.6

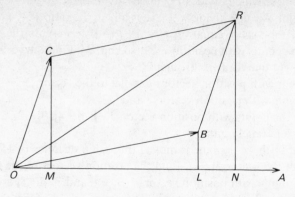

Fig. 12.7

zero and their scalar product is zero. Conversely, if $\mathbf{a} \cdot \mathbf{b} = 0$ and neither vector is zero, $\mathbf{a}$ and $\mathbf{b}$ are at right angles.

In Fig. 12.7, the vectors $\mathbf{a}$, $\mathbf{b}$ and $\mathbf{c}$ are represented by $\overrightarrow{OA}$, $\overrightarrow{OB}$ and $\overrightarrow{OC}$, and in the parallelogram $OBRC$, $\overrightarrow{OR}$ equals $(\mathbf{b}+\mathbf{c})$. OL, OM and ON are the projections on OA of OB, OC, OR respectively. Then

$$\mathbf{a} \cdot (\mathbf{b}+\mathbf{c}) = OA \times OR \cos AOR = OA \times ON,$$
$$\mathbf{a} \cdot \mathbf{b} = OA \times OB \cos AOB = OA \times OL,$$
$$\mathbf{a} \cdot \mathbf{c} = OA \times OC \cos AOC = OA \times OM.$$

Since $OL+OM = OL+LN = ON$, it follows that

$$\mathbf{a} \cdot (\mathbf{b}+\mathbf{c}) = \mathbf{a} \cdot \mathbf{b}+\mathbf{a} \cdot \mathbf{c}.$$

This is the distributive law for scalar products.

This can now be applied to find the scalar product of $\mathbf{a}$ and $\mathbf{b}$ when they are expressed in terms of $\mathbf{i}$ and $\mathbf{j}$.

Let $\qquad \mathbf{a} = a_1\mathbf{i} +a_2\mathbf{j}$ and $\mathbf{b} = b_1\mathbf{i}+b_2\mathbf{j}$.

Then $\qquad \mathbf{a} \cdot \mathbf{b} = (a_1\mathbf{i}+a_2\mathbf{j}) \cdot (b_1\mathbf{i}+b_2\mathbf{j})$
$$= a_1b_1\mathbf{i} \cdot \mathbf{i}+a_1b_2\mathbf{i} \cdot \mathbf{j}+a_2b_1\mathbf{j} \cdot \mathbf{i}+a_2b_2\mathbf{j} \cdot \mathbf{j}.$$

Now $\mathbf{i} \cdot \mathbf{i} = \mathbf{j} \cdot \mathbf{j} = 1$, and since $\mathbf{i}$ and $\mathbf{j}$ are at right angles $\mathbf{i} \cdot \mathbf{j} = \mathbf{j} \cdot \mathbf{i} = 0$, so that

$$\mathbf{a} \cdot \mathbf{b} = a_1b_1+a_2b_2.$$

Example 1

Use the scalar product to establish the cosine rule.

In the triangle ABC let $\overrightarrow{AB} = \mathbf{c}$, $\overrightarrow{BC} = \mathbf{a}$, $\overrightarrow{AC} = \mathbf{b}$, so that $\mathbf{a} = \mathbf{b}-\mathbf{c}$ and $\mathbf{b} \cdot \mathbf{c} = bc \cos A$. Then

$$\mathbf{a} \cdot \mathbf{a} = (\mathbf{b}-\mathbf{c}) \cdot (\mathbf{b}-\mathbf{c})$$
$$= \mathbf{b} \cdot \mathbf{b}+\mathbf{c} \cdot \mathbf{c}-\mathbf{b} \cdot \mathbf{c}-\mathbf{c} \cdot \mathbf{b},$$
i.e. $\qquad a^2 = b^2 +c^2 -2bc \cos A.$

Example 2

Find the equation of the line drawn through the origin at right angles to the line

$$\mathbf{r} = 3\mathbf{i} + 2\mathbf{j} + t(2\mathbf{i} - \mathbf{j}).$$

The given line is parallel to the vector $2\mathbf{i} - \mathbf{j}$. The vector $a\mathbf{i} + b\mathbf{j}$ will be at right angles to this vector if

$$(a\mathbf{i} + b\mathbf{j}) \cdot (2\mathbf{i} - \mathbf{j}) = 0,$$

i.e. if $\qquad\qquad 2a - b = 0, \quad \text{or} \quad b = 2a.$

Hence any vector $a(\mathbf{i} + 2\mathbf{j})$ is perpendicular to the given line, and the equation required is

$$\mathbf{r} = t(\mathbf{i} + 2\mathbf{j}).$$

Exercises 12.7

1   If $\mathbf{a} = 4\mathbf{i} - \mathbf{j}$, $\mathbf{b} = \mathbf{i} + \mathbf{j}$, $\mathbf{c} = 2\mathbf{i} - 4\mathbf{j}$ show that

$$(\mathbf{a} - \mathbf{b}) \cdot (\mathbf{a} - \mathbf{c}) = 0.$$

2   Find the scalar product of the vectors
   (a) $3\mathbf{i} + 2\mathbf{j}$ and $5\mathbf{i} - 4\mathbf{j}$
   (b) $-2\mathbf{i} + 5\mathbf{j}$ and $3\mathbf{i} + \mathbf{j}$
   (c) $3\mathbf{i} - 5\mathbf{j}$ and $3\mathbf{i} + 5\mathbf{j}$
   (d) $6\mathbf{i} + 5\mathbf{j}$ and $5\mathbf{i} - 6\mathbf{j}$.

3   Find the angle between the vectors
   (a) $2\mathbf{i} + \mathbf{j}$ and $3\mathbf{i} - \mathbf{j}$
   (b) $3\mathbf{i} + 6\mathbf{j}$ and $4\mathbf{i} - 2\mathbf{j}$
   (c) $\sqrt{3}\mathbf{i} + \mathbf{j}$ and $\mathbf{i} + \sqrt{3}\mathbf{j}$
   (d) $\sqrt{3}\mathbf{i} - \mathbf{j}$ and $\sqrt{3}\mathbf{i} + \mathbf{j}$
   (e) $\mathbf{i} - 3\mathbf{j}$ and $-2\mathbf{i} + \mathbf{j}$.

4   If $|\mathbf{a}| = |\mathbf{b}|$, show that $(\mathbf{a} - \mathbf{b}) \cdot (\mathbf{a} + \mathbf{b}) = 0$.
   Interpret this result geometrically.

5   Find the vector equation of the line drawn through the points with position vectors $3\mathbf{i} - \mathbf{j}$ and $5\mathbf{i} + \mathbf{j}$. Find the position vector of the point on this line which is nearest to the origin.

12.8   Differentiation of vectors

When a point P is moving in a straight line, its position vector $\mathbf{r}$ satisfies an equation of the form $\mathbf{r} = \mathbf{a} + t\mathbf{b}$, so that $\mathbf{r}$ is a function of the parameter t. This can be indicated by putting $\mathbf{r} = \mathbf{r}(t)$.

When a point Q moves along a parabola, its coordinates can be taken as $x = at^2$, $y = 2at$. Its position vector can be expressed as

$$\mathbf{r} = (at^2)\mathbf{i} + (2at)\mathbf{j}.$$

Again **r** is a function of t and this is shown by putting $\mathbf{r} = \mathbf{r}(t)$

In such cases as these **r** can be differentiated with respect to t. The derivative is defined by

$$\frac{d\mathbf{r}}{dt} = \lim_{\delta t \to 0} \frac{\mathbf{r}(t + \delta t) - \mathbf{r}(t)}{\delta t}.$$

It is very important to remember that $d\mathbf{r}/dt$ is itself a vector. When t denotes time, the vector $d\mathbf{r}/dt$ gives the velocity of the point. The second derivative of **r**, also a vector, gives the acceleration. For the point P above, $\mathbf{r} = \mathbf{a} + \mathbf{b}t$ and $d\mathbf{r}/dt = \mathbf{b}$, i.e. the velocity is constant in magnitude and direction. For the point Q the velocity is given by

$$\frac{d\mathbf{r}}{dt} = (2at)\mathbf{i} + (2a)\mathbf{j},$$

and the acceleration by

$$\frac{d^2\mathbf{r}}{dt^2} = (2a)\mathbf{i}.$$

Example

At time t s, the position vector **r** of a point P with respect to the origin O is

$$\mathbf{r} = [(3 \cos 2t)\mathbf{i} + (3 \sin 2t)\mathbf{j}] \text{ m}.$$

Show that the velocity of P is always at right angles to OP and that its acceleration is always in the direction PO.

The velocity **v** of P is given by

$$\mathbf{v} = \frac{d\mathbf{r}}{dt} = [(-6 \sin 2t)\mathbf{i} + (6 \cos 2t)\mathbf{j}] \text{ m s}^{-1}.$$

The scalar product $\mathbf{v} \cdot \mathbf{r}$ is zero, showing that **v** is at right angles to OP. The acceleration **a** of P is given by

$$\mathbf{a} = \frac{d^2\mathbf{r}}{dt^2} = [(-12 \cos 2ty)\mathbf{i} + (-12 \sin 2t)\mathbf{j}] \text{ m s}^{-2}.$$

This shows that the acceleration is in the opposite direction to OP.

Exercises 12.8

1 Differentiate with respect to t
  (a) $(1 + t)\mathbf{i} + (1 - t)\mathbf{j}$ (c) $(\cos t)\mathbf{i} - (\sin t)\mathbf{j}$
  (b) $2\mathbf{i} + (3t^2)\mathbf{j}$ (d) $(2t^2)\mathbf{i} + t^3\mathbf{j}$.

2 Find the speed when $t = 2$ of the point P given that the position vector of P at time t s is
  (a) $[\sin(\pi t)\mathbf{i} + \cos(\pi t)\mathbf{j}]$ m   (c) $[t^3\mathbf{i} - (5t)\mathbf{j}]$ m
  (b) $[(3t)\mathbf{i} + t^2\mathbf{j}]$ m   (d) $[\cos^2(\pi t)\mathbf{i} + \sin^2(\pi t)\mathbf{j}]$ m

3 Given that $\mathbf{r} = t\mathbf{i} + t^2\mathbf{j}$, find expressions in terms of t for

$$\frac{d}{dt}|\mathbf{r}| \quad \text{and} \quad \left|\frac{d\mathbf{r}}{dt}\right|.$$

Show that they are equal only when $t = 0$.

4 The position vector of a point P at time t s is given by

$$\mathbf{r} = [(1 - \sin t)\mathbf{i} - (\cos t)\mathbf{j}] \text{ m}.$$

Show that the acceleration of P is always at right angles to its velocity.

5 The acceleration of a point P is constant and is given by the vector $(\mathbf{i} + \mathbf{j})$ m s^{-2}. At time $t = 0$ the point passes through the origin with velocity $(\mathbf{i} - \mathbf{j})$ m s^{-1}. Show that at time $t = 2$, the point P passes through the point with position vector $4\mathbf{i}$ m at a velocity of $(3\mathbf{i} + \mathbf{j})$ m s^{-1}.

6 At time t s, the position vectors of the points A and B are $[(\sin 2t)\mathbf{i} + (\cos 2t)\mathbf{j}]$ m and $[2t\mathbf{i} + t^2\mathbf{j}]$ m respectively. Sketch the path of each point, and show that when A passes through the point with position vector $\mathbf{j}$ its acceleration is in the opposite direction to the acceleration of B.

12.9 Unit vectors in three dimensions

Three unit vectors $\mathbf{i}, \mathbf{j}$ and $\mathbf{k}$ can be defined in the directions of the positive x, y and z axes respectively (Fig. 12.8). Any vector in three dimensions can then be expressed in the form $\mathbf{r} = a\mathbf{i} + b\mathbf{j} + c\mathbf{k}$, or as a column vector

$$\mathbf{r} = \begin{pmatrix} a \\ b \\ c \end{pmatrix},$$

where $a\mathbf{i}, b\mathbf{j}, c\mathbf{k}$ are the component vectors of $\mathbf{r}$, and the modulus of $\mathbf{r}$ equals $\sqrt{(a^2 + b^2 + c^2)}$.

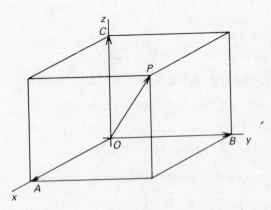

Fig. 12.8

Let $\mathbf{r}$ be the position vector of the point P with respect to the origin O. Then if OA, OB, OC are the projections of OP on the axes

$$\cos AOP = OA/OP, \cos BOP = OB/OP, \cos COP = OC/OP.$$

These ratios are called the direction cosines of $\overrightarrow{OP}$ and are denoted by l, m and n respectively. This gives

$$l = a/r; m = b/r, n = c/r,$$

where r is the modulus of the vector $\mathbf{r}$. Also

$$l^2 + m^2 + n^2 = (a^2 + b^2 + c^2)/r^2 = 1.$$

The ratios $a : b : c$ are called the direction ratios of $\overrightarrow{OP}$.

Since the scalar product $\mathbf{a} \cdot \mathbf{b}$ of two vectors is defined to be $ab \cos \theta$ where θ is the angle between $\mathbf{a}$ and $\mathbf{b}$,

$$\mathbf{i} \cdot \mathbf{i} = \mathbf{j} \cdot \mathbf{j} = \mathbf{k} \cdot \mathbf{k} = 1$$

and

$$\mathbf{i} \cdot \mathbf{j} = \mathbf{j} \cdot \mathbf{k} = \mathbf{k} \cdot \mathbf{i} = 0.$$

If

$$\mathbf{a} = a_1\mathbf{i} + a_2\mathbf{j} + a_3\mathbf{k} \quad \text{and} \quad \mathbf{b} = b_1\mathbf{i} + b_2\mathbf{j} + b_3\mathbf{k},$$

the scalar product of $\mathbf{a}$ and $\mathbf{b}$ is given by

$$\mathbf{a} \cdot \mathbf{b} = a_1b_1 + a_2b_2 + a_3b_3.$$

Example 1

The position vector of the point P with respect to the origin O is $7\mathbf{i} + 6\mathbf{j} - 6\mathbf{k}$. Find the direction cosines of the vector $\overrightarrow{OP}$.

The direction ratios of $\overrightarrow{OP}$ are $7 : 6 : -6$, and

$$OP^2 = 7^2 + 6^2 + (-6)^2 = 121.$$

Hence the length of OP is 11 and the direction cosines of $\overrightarrow{OP}$ are

$$l = 7/11, m = 6/11, n = -6/11.$$

Example 2

The position vectors of the points P, Q and R are

$$\mathbf{p} = 2\mathbf{i} - 4\mathbf{j} + \mathbf{k}, \quad \mathbf{q} = 4\mathbf{i} + 3\mathbf{j} - 2\mathbf{k}, \quad \mathbf{r} = 6\mathbf{i} + 2\mathbf{j} - 3\mathbf{k}$$

respectively. Show that the angle PQR is a right angle.

$$\overrightarrow{PQ} = \mathbf{q} - \mathbf{p} = 2\mathbf{i} + 7\mathbf{j} - 3\mathbf{k},$$
$$\overrightarrow{QR} = \mathbf{r} - \mathbf{q} = 2\mathbf{i} - \mathbf{j} - \mathbf{k},$$

The scalar product of these two vectors is

$$(\mathbf{q} - \mathbf{p}) \cdot (\mathbf{r} - \mathbf{q}) = 4 - 7 + 3 = 0.$$

Hence $\overrightarrow{PQ}$ and $\overrightarrow{QR}$ are at right angles.

Example 3

At time t s, the position vector of the point P is

$$\mathbf{r} = [(\cos 2t)\mathbf{i} + (\sin 2t)\mathbf{j} + kt]\text{ m}.$$

Show that the speed of P is constant, and that the acceleration of P is of constant magnitude.

The velocity and the acceleration of P are given by

$$\frac{d\mathbf{r}}{dt} = [-(2\sin 2t)\mathbf{i} + (2\cos 2t)\mathbf{j} + \mathbf{k}]\text{ m s}^{-1},$$

and

$$\frac{d^2\mathbf{r}}{dt^2} = [-(4\cos 2t)\mathbf{i} - (4\sin 2t)\mathbf{j}]\text{ m s}^{-2}.$$

Then speed $\left|\dfrac{d\mathbf{r}}{dt}\right|$ is equal to

$$\sqrt{(4\sin^2 2t + 4\cos^2 2t + 1)}\text{ m s}^{-1} = \sqrt{5}\text{ m s}^{-1}.$$

The magnitude of the acceleration equals

$$\sqrt{(16\cos^2 2t + 16\sin^2 2t)}\text{ m s}^{-2} = 4\text{ m s}^{-2}.$$

Exercises 12.9

1 Find the unit vector in the direction of the vector
(a) $3\mathbf{i} + 4\mathbf{j} + 12\mathbf{k}$ (b) $2\mathbf{i} - 5\mathbf{j} - 14\mathbf{k}$ (c) $4\mathbf{i} - 12\mathbf{j} + 6\mathbf{k}$.
Find also the direction cosines of each vector.

2 The vectors $\mathbf{i}$, $\mathbf{j}$ and $\mathbf{k}$ drawn through the origin form three edges of a cube. Find the unit vector in the direction of the diagonal of the cube drawn through the origin. Find the angle between this vector and each edge.

3 The position vectors of the points P and Q are $8\mathbf{i} + \mathbf{j} + 3\mathbf{k}$ and $2\mathbf{i} + 8\mathbf{j} - 3\mathbf{k}$. Find the length of PQ and the vector equation of the line PQ.

4 The position vectors of the points A and B are $2\mathbf{i} + 5\mathbf{j} + 3\mathbf{k}$ and $6\mathbf{i} - 3\mathbf{j} + 7\mathbf{k}$ respectively. Find the position vectors of the points P, Q and R such that $\overrightarrow{AP} = \overrightarrow{PQ} = \overrightarrow{QR} = \overrightarrow{RB}$.

5 Find the unit vector which bisects the angle between the vectors $4\mathbf{i} - 4\mathbf{j} + 7\mathbf{k}$ and $-\mathbf{i} + 2\mathbf{j} - 2\mathbf{k}$.

6 Find the value of the scalar c if the vectors $c\mathbf{i} + 4\mathbf{j} + 3\mathbf{k}$ and $3\mathbf{i} + 3\mathbf{j} + c\mathbf{k}$ are at right angles.

7 The position vectors of the points A, B, C and H are $\mathbf{i} + 4\mathbf{j} - 8\mathbf{k}$, $6\mathbf{i} - 3\mathbf{j} + 6\mathbf{k}$, $7\mathbf{i} + 4\mathbf{j} + 4\mathbf{k}$, $14\mathbf{i} + 5\mathbf{j} + 2\mathbf{k}$ respectively. Show that AH, BH and CH are perpendicular to BC, CA and AB respectively.

8 Find the direction ratios of a line which is at right angles to each of the vectors $4\mathbf{i} + 3\mathbf{j} + 3\mathbf{k}$ and $\mathbf{i} - 3\mathbf{j} + 2\mathbf{k}$.

Miscellaneous examples 12

1 Show that the three vectors $\mathbf{p} = 2\mathbf{i} + \mathbf{j} + 3\mathbf{k}$, $\mathbf{q} = 6\mathbf{i} + 5\mathbf{j} + 7\mathbf{k}$, $\mathbf{r} = 3\mathbf{i} + 2\mathbf{j} + 4\mathbf{k}$ are not linearly independent.

It must be shown that there are numbers, a, b and c, not all zero, such that

$$a\mathbf{p} + b\mathbf{q} + c\mathbf{r} = 0,$$

i.e. that

$$(2a + 6b + 3c)\mathbf{i} + (a + 5b + 2c)\mathbf{j} + (3a + 7b + 4c)\mathbf{k} = 0.$$

This can be true only if

$$2a + 6b + 3c = 0,$$
$$a + 5b + 2c = 0,$$
$$3a + 7b + 4c = 0.$$

By subtracting the first equation from twice the second equation, we find

$$4b + c = 0.$$

By substituting $-4b$ for c in the second equation we find that $a = 3b$. It follows that a, b and c are in the ratio $3:1:-4$. If we take $a = 3$, $b = 1$ and $c = -4$, the coefficients of $\mathbf{i}$, $\mathbf{j}$ and $\mathbf{k}$ in the vector $a\mathbf{p} + b\mathbf{q} + c\mathbf{r}$ are zero. This shows that the vectors $\mathbf{p}$, $\mathbf{q}$ and $\mathbf{r}$ are not linearly independent, and that each of them can be expressed in terms of the other two. For example,

$$\mathbf{p} = -(\mathbf{q} - 4\mathbf{r})/3.$$

2 A perpendicular is drawn from each vertex of a triangle to the opposite side. Show that the three perpendiculars meet at a point.

Let the position vectors of the points A, B and C be $\mathbf{a}$, $\mathbf{b}$ and $\mathbf{c}$ respectively. Let H, with position vector $\mathbf{h}$, be the point in which the perpendicular AL meets the perpendicular BM (Fig. 12.9).

Since $\overrightarrow{AH}$ and $\overrightarrow{BC}$ are at right angles, their scalar product is zero, i.e.

$$(\mathbf{h} - \mathbf{a}) \cdot (\mathbf{c} - \mathbf{b}) = 0,$$

or

$$\mathbf{h} \cdot \mathbf{c} - \mathbf{h} \cdot \mathbf{b} - \mathbf{a} \cdot \mathbf{c} + \mathbf{a} \cdot \mathbf{b} = 0$$

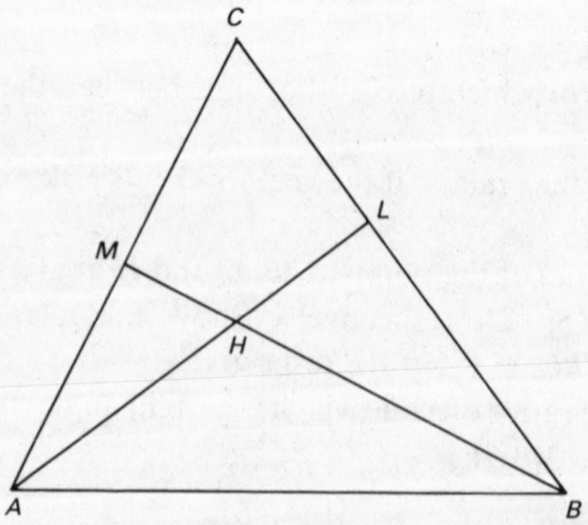

Fig. 12.9

Also $\overrightarrow{BH}$ and $\overrightarrow{CA}$ are at right angles and hence

$$(\mathbf{h}-\mathbf{b}) \cdot (\mathbf{a}-\mathbf{c}) = 0$$

or

$$\mathbf{h} \cdot \mathbf{a} - \mathbf{h} \cdot \mathbf{c} - \mathbf{b} \cdot \mathbf{a} + \mathbf{b} \cdot \mathbf{c} = 0$$

Adding the two equations together gives

$$\mathbf{h} \cdot \mathbf{a} - \mathbf{h} \cdot \mathbf{b} - \mathbf{a} \cdot \mathbf{c} + \mathbf{b} \cdot \mathbf{c} = 0$$

or

$$(\mathbf{h}-\mathbf{c}) \cdot (\mathbf{a}-\mathbf{b}) = 0.$$

This means that the scalar product of $\overrightarrow{CH}$ and $\overrightarrow{BA}$ is zero, so that $\overrightarrow{CH}$ is perpendicular to $\overrightarrow{BA}$. It follows that the three perpendiculars are concurrent at H.

3   The position vectors of the points A and B are

$$\mathbf{a} = -4\mathbf{i}+10\mathbf{j}+13\mathbf{k}, \quad \mathbf{b} = 0\mathbf{i}+7\mathbf{j}+10\mathbf{k}$$

respectively. The line AB cuts the sphere $|\mathbf{r}| = 9$ in the points P and Q. Find the ratios in which the points P and Q divide AB.

The vector equation of the line AB is
$$\begin{aligned}\mathbf{r} &= \mathbf{a}+t(\mathbf{b}-\mathbf{a})\\ &= -4\mathbf{i}+10\mathbf{j}+13\mathbf{k}+t(4\mathbf{i}-3\mathbf{j}-3\mathbf{k}).\end{aligned}$$

The modulus of $\mathbf{r}$ is given by
$$\begin{aligned}|\mathbf{r}|^2 &= (-4+4t)^2+(10-3t)^2+(13-3t)^2\\ &= 34t^2-170t+285.\end{aligned}$$

When $|\mathbf{r}| = 9$,

$$\begin{aligned}34t^2-170t+285 &= 81,\\ 34t^2-170t+204 &= 0,\\ 34(t^2-5t+6) &= 0.\end{aligned}$$

The roots of this equation are $t = 2$ and $t = 3$.
When $t = 2$, $\mathbf{r} = 4\mathbf{i}+4\mathbf{j}+7\mathbf{k}$.
When $t = 3$, $\mathbf{r} = 8\mathbf{i}+\mathbf{j}+4\mathbf{k}$.
Let these be the position vectors of P and Q respectively.

Then

$$\begin{aligned}\overrightarrow{AP} &= 8\mathbf{i}+6\mathbf{j}+6\mathbf{k},\\ \overrightarrow{BP} &= 4\mathbf{i}+3\mathbf{j}+3\mathbf{k}.\\ \overrightarrow{AP} &= 2\overrightarrow{BP}.\end{aligned}$$

Thus P divides AB externally in the ratio $2:1$.

Also

$$\begin{aligned}\overrightarrow{AQ} &= 12\mathbf{i}+9\mathbf{j}+9\mathbf{k},\\ \overrightarrow{BQ} &= 8\mathbf{i}+6\mathbf{j}+6\mathbf{k}.\\ 2\overrightarrow{AQ} &= 3\overrightarrow{BQ}.\end{aligned}$$

Thus Q divides AB externally in the ratio $3:2$.

Miscellaneous exercises 12

1 Find the unit vector in the direction of the vector $(\mathbf{p}+\mathbf{q}+\mathbf{r})$ given that $\mathbf{p} = 2\mathbf{i}+\mathbf{j}+3\mathbf{k}$, $\mathbf{q} = \mathbf{i}-2\mathbf{j}+2\mathbf{k}$, $\mathbf{r} = \mathbf{i}+\mathbf{j}-2\mathbf{k}$. Find also a vector which is perpendicular to $\mathbf{q}$ and to $\mathbf{r}$.

2 Show that the vectors $\mathbf{p} = 6\mathbf{i}+4\mathbf{j}-4\mathbf{k}$, $\mathbf{q} = 6\mathbf{i}+5\mathbf{j}+\mathbf{k}$, $\mathbf{r} = 4\mathbf{i}+3\mathbf{j}-\mathbf{k}$ lie in the same plane by finding scalars λ and μ such that $\mathbf{p} = \lambda\mathbf{q}+\mu\mathbf{r}$.

3 Find the unit vector which bisects the angle between the vectors $6\mathbf{i}-6\mathbf{j}+7\mathbf{k}$ and $-2\mathbf{i}+2\mathbf{j}-\mathbf{k}$.

4 The acceleration $\mathbf{a}$ m s^{-2} of a point P at time t s is given by

$$\mathbf{a} = \cos t\, \mathbf{i}+\sin t\, \mathbf{j}.$$

When $t = 0$, the velocity of P is $\mathbf{i}$ m s^{-1} and its position vector is $\mathbf{j}$ m. Find its position vector when $t = 2\pi$.

5 Find the scalar product of the vectors $\mathbf{a}$ and $\mathbf{b}$ where $\mathbf{a} = 4\mathbf{i}-5\mathbf{k}$ and $\mathbf{b} = 2\mathbf{i}+\mathbf{j}-2\mathbf{k}$. Deduce that the component vector of $\mathbf{a}$ in the direction of $\mathbf{b}$ is $4\mathbf{i}+2\mathbf{j}-4\mathbf{k}$.

6 At time t s, the position vector of a point P is given by

$$\mathbf{r} = [e^t\mathbf{i} + \cos t\, \mathbf{j}+\sin t\, \mathbf{k}] \text{ m}.$$

Find the angle between the velocity and the acceleration of P when $t = 0$.

7 The position vectors of the points P and Q are $(\mathbf{i}+\mathbf{j})$ and $(\mathbf{j}+\mathbf{k})$ respectively. Calculate the angle POQ and find a unit vector perpendicular to both OP and OQ. Find a vector equation for the line PQ.

8 A flight controller at an airport is about to 'talk down' an aircraft A whose position vector relative to the control tower is $(10\mathbf{i}+20\mathbf{j}+5\mathbf{k})$ km and whose constant velocity is $(-210\mathbf{i}-50\mathbf{j})$ km h^{-1}. At this instant a second aircraft B appears on his radar screen with position vector $(-20\mathbf{i}-10\mathbf{j}+3\mathbf{k})$ km and constant velocity $(150\mathbf{i}+250\mathbf{j}+60\mathbf{k})$ km h^{-1}. Find

(i) the velocity of A relative to B

(ii) the position vector of A relative to B, t minutes after B first appeared on the radar screen

(iii) the time that elapses, to the nearest 10 seconds, until the two aircraft are nearest to one another. [JMB]

9 The position vectors of the points A and B are given by $\mathbf{a} = 5\mathbf{i}+\mathbf{j}+2\mathbf{k}$ and $\mathbf{b} = -\mathbf{i}+7\mathbf{j}+8\mathbf{k}$ respectively. Show that the two lines $\mathbf{r} = \mathbf{a}+\lambda\mathbf{l}$ and $\mathbf{r} = \mathbf{b}+\mu\mathbf{m}$, where $\mathbf{l} = -4\mathbf{i}+\mathbf{j}-\mathbf{k}$ and $\mathbf{m} = 2\mathbf{i}-5\mathbf{j}-7\mathbf{k}$, intersect, and find the position vector of the point of intersection C.

If D has position vector $3\mathbf{i}+7\mathbf{j}-2\mathbf{k}$ show that the line CD is perpendicular to the line AB. [JMB]

10 Find unit vectors perpendicular to both of the vectors $\mathbf{j}+4\mathbf{k}$ and $3\mathbf{i}+2\mathbf{j}+4\mathbf{k}$. Find also the angles between these unit vectors and the vector $\frac{1}{2}(\mathbf{i}-\mathbf{j}-\mathbf{k})$. [JMB]

11 Given two vectors $\overrightarrow{OP}$ and $\overrightarrow{OQ}$, show how to construct geometrically the sum $(\overrightarrow{OP}+\overrightarrow{OQ})$ and the difference $(\overrightarrow{OP}-\overrightarrow{OQ})$.

If X, Y, Z are the mid-points of the lines BC, CA, AB respectively and O is any point in the plane of the triangle ABC, show that

$$\overrightarrow{OA} + \overrightarrow{OB} + \overrightarrow{OC} = \overrightarrow{OX} + \overrightarrow{OY} + \overrightarrow{OZ},$$

and find the position of the point D such that

$$\overrightarrow{OA} + \overrightarrow{OB} - \overrightarrow{OC} = \overrightarrow{OD}. \hspace{2cm} \text{[L]}$$

12 In a triangle ABC the mid-points of BC, CA, AB are O, P, Q respectively. Express $OA + OC$ in terms of $\overrightarrow{OP}$ and show that $\overrightarrow{OA} - \overrightarrow{OC} = 2\overrightarrow{OQ}$.

In a plane quadrilateral $EFGH$ the mid-points of EG, FH are X, Y respectively. Prove that the resultant of forces represented completely by $\overrightarrow{EF}$, $\overrightarrow{GF}$, $\overrightarrow{EH}$ and $\overrightarrow{GH}$ is represented completely by $4\overrightarrow{XY}$. [L]

13 (i) Sketch the locus of the point with position vector

$$\mathbf{r} = (t \cos p)\mathbf{i} + (\mathbf{t} \sin p)\mathbf{j}$$

(a) when t is constant and p is a parameter
(b) when p is constant and t is a parameter.

(ii) The position vectors of the points A and B are $2\mathbf{i} - \mathbf{j}$ and $-\mathbf{i} + 3\mathbf{j}$ respectively with respect to the origin O. Find the position vector of the point H in the plane OAB such that OH is perpendicular to AB and AH is perpendicular to OB. [L]

14 State a relation which exists between the vectors $\mathbf{p}$ and $\mathbf{q}$ when these vectors are (a) parallel, (b) perpendicular.

The position vectors of the vertices of a tetrahedron $ABCD$ are

A: $-5\mathbf{i} + 22\mathbf{j} + 5\mathbf{k},$ $\hspace{1cm}$ B: $\mathbf{i} + 2\mathbf{j} + 3\mathbf{k},$
C: $4\mathbf{i} + 3\mathbf{j} + 2\mathbf{k},$ $\hspace{1.2cm}$ D: $-\mathbf{i} + 2\mathbf{j} - 3\mathbf{k}.$

Find the angle CBD and show that AB is perpendicular to both BC and BD. Calculate the volume of the tetrahedron. If $ABDE$ is a parallelogram, find the position vector of E. [L]

15 Show that the straight line with vector equation

$$\mathbf{r} = \mathbf{i} + t\mathbf{j} + t\mathbf{k}$$

passes through the points P and Q which have position vectors $(\mathbf{i} - \mathbf{j} - \mathbf{k})$ and $(\mathbf{i} + \mathbf{j} + \mathbf{k})$ respectively. Find the vector equation of the straight line through the points R and S which have position vectors $(\mathbf{i} + \mathbf{j})$ and $(2\mathbf{i} + \mathbf{j} + \mathbf{k})$ respectively. Show that PQ and RS are inclined at $60°$ to one another.

Show that the vector

$$\mathbf{a} = \mathbf{i} + \mathbf{j} - \mathbf{k}$$

is perpendicular to both PQ and RS. Find the scalar product of $\mathbf{a}$ with $\overrightarrow{PR}$, and hence find the shortest distance between PQ and RS. [L]

16 The vector $\mathbf{p}$ is a multiple of the vector $2\mathbf{i} + \mathbf{j}$, and $\mathbf{q}$ is a multiple of $\mathbf{i} + \mathbf{j}$. Their sum $\mathbf{p} + \mathbf{q}$ is a multiple of the vector $4\mathbf{i} + 3\mathbf{j}$. Given that $|\mathbf{p} + \mathbf{q}| = 10$, find $|\mathbf{p}|$ and $|\mathbf{q}|$.

17 The coordinates of the points P and Q are $(5, 0)$ and $(3, 4)$ respectively. Find the vector equation of the perpendicular bisector of PQ.

18 At time t s, the position vector of the point P is $[\mathbf{i}t + \mathbf{j}(2t^2) + \mathbf{k}t^3]$ m. Find the acceleration of P at the time when it is moving parallel to the vector $3\mathbf{i} + 4\mathbf{j} + \mathbf{k}$.

19 At time $t = 0$, the position vectors of the points P and Q are $[\mathbf{i} + \mathbf{j} + 3\mathbf{k}]$ m and $[4\mathbf{i} + 5\mathbf{j} + \mathbf{k}]$ m respectively. The points have constant velocity vectors $[\mathbf{j} + 2\mathbf{k}]$ ms^{-1} and $[-4\mathbf{i} + 3\mathbf{k}]$ ms^{-1} respectively. Show that the distance between the two points is least at time $t = 1$ s.

20 The position vector of a point P at time t s is $[e^{2t}(\mathbf{i} \cos 2t + \mathbf{j} \sin 2t + \mathbf{k})]$ m. Show that the velocity of P makes the constant angle $\arctan \sqrt{2}$ with the vector $\mathbf{k}$.

21 Obtain the vector equation of the line which is parallel to the vector $\mathbf{i} + 2\mathbf{j} + 2\mathbf{k}$ and which passes through the point P with position vector $2\mathbf{i} + \mathbf{j} + \mathbf{k}$. Find the distance from the origin to this line.

22 The points P, Q, R and S have position vectors $6\mathbf{i} + 8\mathbf{j}$, $12\mathbf{i} - \mathbf{j}$, $6\mathbf{i} + 5\mathbf{j}$, $10\mathbf{i}$ respectively. Find the position vector of the point of intersection of PS and QR.

23 Find the vector equation of the line which passes through the points with position vectors $\mathbf{i}$ and $\mathbf{i} + \mathbf{j} + \mathbf{k}$. Show that this line meets the line $\mathbf{r} = 3\mathbf{j} + t(\mathbf{i} + 3\mathbf{k})$.

24 Show that the lines with vector equations

$$\mathbf{r} = 2\mathbf{i} + \lambda(\mathbf{i} + 3\mathbf{j} + 4\mathbf{k}) \quad \text{and} \quad \mathbf{r} = \mathbf{k} + \mu(\mathbf{i} + \mathbf{j} + \mathbf{k})$$

are in the same plane. Find the position vector of their common point.

25 Show that the equation

$$(\mathbf{r} + \mathbf{i} - 2\mathbf{k}) \cdot (\mathbf{r} - 3\mathbf{i} + 2\mathbf{j}) = 0$$

represents a sphere and find the position vector of its centre.

26 In tetrahedron $PQRS$, PQ is perpendicular to RS and PR is perpendicular to QS. Use position vectors to show that PS is perpendicular to QR.

27 The acceleration $\mathbf{a}$ of a particle A moving in the x–y plane is given by $\mathbf{a} = 2\mathbf{i}$, where distances are measured in metres and time in seconds. Given that the particle starts with a velocity $\mathbf{i} + \mathbf{j}$ from the origin at time $t = 0$, find the velocity vector and position vector of the particle at time t. Show that the particle moves along the curve $x = y^2 + y$.

A particle B moves with constant velocity in the direction $\mathbf{j}$ and is at the point $6\mathbf{i} - 4\mathbf{j}$ at $t = 1$. If the particles A and B collide, find the speed of particle B and the time of collision. [AEB]

28 Show that the vectors $\mathbf{u}$, $\mathbf{v}$ and $\mathbf{w}$, where

$$\mathbf{u} = \mathbf{i} + \mathbf{j} + \mathbf{k},$$
$$\mathbf{v} = \mathbf{i} - \tfrac{1}{2}\mathbf{j} - \tfrac{1}{2}\mathbf{k},$$
$$\mathbf{w} = \mathbf{j} - \mathbf{k},$$

are mutually perpendicular and find the unit vector in the direction of the vector **u**.

Find constants α, β and γ such that

$$\mathbf{i} = \alpha\mathbf{u} + \beta\mathbf{v} + \gamma\mathbf{w}.$$

If P, Q and R have position vectors $5\mathbf{i} - \mathbf{j} - \mathbf{k}$, **u** and **v** respectively with respect to the origin O, show that O, P, Q and R are coplanar. Find the cosine of the angle POQ.

29 Show that the equation

$$\mathbf{r} = (3 + 2\cos\theta)\mathbf{i} - (2\sin\theta)\mathbf{j}$$

represents a circle. Find the equations of the tangents to the circle which are parallel to the vector **i**.

30 Find a unit vector which is perpendicular to the vector $8\mathbf{i} - 3\mathbf{j} - \mathbf{k}$ and also to the vector $2\mathbf{i} + 4\mathbf{j} - 5\mathbf{k}$.

31 Define the scalar (or dot) product of two vectors and interpret geometrically the vanishing of this product.

If **a** and **b** are the position vectors of two points A and B in the plane and if the position vector **p** of a variable point P is required to satisfy the condition $(\mathbf{p} - \mathbf{a}) \cdot (\mathbf{p} - \mathbf{b}) = 0$, describe the locus of P.

If $\mathbf{a} = \mathbf{i} + \mathbf{j}$ and $\mathbf{b} = 5\mathbf{i} + 7\mathbf{j}$, where **i** and **j** are two perpendicular unit vectors, find the position vectors of the other two vertices of the square of which A and B are one pair of opposite vertices. [L]

32 Write down the expression for the scalar product $\mathbf{p} \cdot \mathbf{q}$ of two vectors (in two dimensions) in terms of their components p_1, p_2 and q_1, q_2 in two perpendicular directions. Prove that, for three coplanar vectors **p**, **q**, **r**,

$$(\mathbf{p} + \mathbf{q}) \cdot \mathbf{r} = \mathbf{p} \cdot \mathbf{r} + \mathbf{q} \cdot \mathbf{r}.$$

Show also that

$$|\mathbf{p} + \mathbf{q}|^2 + |\mathbf{p} - \mathbf{q}|^2 = 2|\mathbf{p}|^2 + 2|\mathbf{q}|^2.$$

If $ABCD$ is a rectangle and O is any other point in its plane, show that $OA^2 + OC^2 = OB^2 + OD^2$. [L]

33 The points A, B have position vectors **a**, **b** respectively when referred to an origin O. Show that the vector equation of the line AB can be written in the form

$$\mathbf{r} = \mathbf{a} + t(\mathbf{b} - \mathbf{a}),$$

where **r** is the position vector of a point on the line and t is a parameter.

If C, D are the points with position vectors $\mathbf{a}/2$, $3\mathbf{b}/2$, find the position vector of the point of intersection of the lines AB and CD. [L]

34 Two particles A and B move with constant velocity vectors $(4\mathbf{i} + \mathbf{j} - 2\mathbf{k})$ and $(6\mathbf{j} + 3\mathbf{k})$ respectively, the unit of speed being the metre per second. At time

$t = 0$, A is at the point with position vector $(-\mathbf{i}+20\mathbf{j}+21\mathbf{k})$ and B is at the point with position vector $(\mathbf{i}+3\mathbf{k})$, the unit of distance being the metre.

Find the value of t for which the distance between A and B is least and find also this least distance.　　　　　　　　　　　　　　　　　　　　　　　　[L]

35 The square lamina $ABCD$ of side $2a$ is rotating in a horizontal plane with constant clockwise angular speed ω about the vertex A, while the position vector of the point A at time t is $\mathbf{r} = (a \sin \omega t)\mathbf{i}$. Initially the position vector of the point B is $2a\mathbf{i}$ and that of D is $-2a\mathbf{j}$. Find the position vector of B at time t. Find also the velocity vector $\mathbf{v}$ and the acceleration vector $\mathbf{f}$ of B at time t, and show that

$$\omega^2\mathbf{v} \cdot \mathbf{v}+\mathbf{f} \cdot \mathbf{f}$$

is constant.

Show that the scalar product of $\mathbf{i}$ with the velocity vector of the mid-point of AD is always zero.　　　　　　　　　　　　　　　　　　　　　　　　[L]

36 Two particles A and B have position vectors

$$[(2 \sin \omega t)\mathbf{i} + (2 \cos \omega t)\mathbf{j}] \text{ m}, \quad \text{where } \omega > 0, \quad \text{and } [2t\,\mathbf{i}+t^2\,\mathbf{j}] \text{ m}$$

respectively at time t s. Find the cartesian equations of the paths followed by A and B.

Find

(a) the magnitude of the velocity of A relative to B when $t = 0$

(b) the magnitude of the acceleration of A relative to B when $t = \pi/(2\omega)$.

Find also the values of t for which

(c) the accelerations of A and B are parallel and in the same sense.

(d) the accelerations of A and B are parallel and in opposite senses.　　[L]

13 Matrices

13.1 The order of a matrix

A matrix is a rectangular array of numbers or symbols, known as the elements of the matrix. The matrix itself does not have a numerical value. A matrix with m rows and n columns is said to be of order $m \times n$. Thus the matrix

$$\begin{pmatrix} 4 & 1 & -2 \\ 2 & 0 & 3 \end{pmatrix}$$

is of order 2×3. A matrix of order 1×2 such as (a, b) is a row vector, and a matrix of order 2×1 such as $\begin{pmatrix} p \\ q \end{pmatrix}$ is a column vector. Two matrices are equal only when they are identical, i.e. of the same order with corresponding elements equal.

13.2 Addition and subtraction of matrices

Two matrices P and Q of the same order can be added together. Their sum $P + Q$ is found by adding each element in P to the corresponding element in Q.

Matrix addition is commutative and associative, i.e. if P, Q, R are of the same order,

$$P + Q = Q + P,$$
$$(P + Q) + R = P + (Q + R).$$

The difference $P - Q$ is found by subtracting each element in Q from the corresponding element in P.

When $P = Q$, $P - Q$ will be a zero matrix (or null matrix) in which every element is zero. The product of a scalar k and a matrix of any order is found by multiplying each element in the matrix by k.

Example 1

$$\begin{pmatrix} 4 & 1 & -2 \\ 2 & 0 & 3 \end{pmatrix} + \begin{pmatrix} -1 & 2 & 3 \\ 3 & 1 & 1 \end{pmatrix} = \begin{pmatrix} 3 & 3 & 1 \\ 5 & 1 & 4 \end{pmatrix}.$$

Example 2

$$2\begin{pmatrix} 4 & 3 \\ 2 & 5 \end{pmatrix} - \begin{pmatrix} 4 & 4 \\ 3 & 3 \end{pmatrix} = \begin{pmatrix} 4 & 2 \\ 1 & 7 \end{pmatrix}.$$

Exercises 13.2

1 Simplify

(a) $\begin{pmatrix} 2 & 4 \\ 2 & 1 \end{pmatrix} + \begin{pmatrix} 1 & 3 \\ 4 & 2 \end{pmatrix}$

(b) $\begin{pmatrix} 6 & 4 & 2 \\ 5 & 3 & 1 \end{pmatrix} - \begin{pmatrix} 1 & 4 & 3 \\ 2 & 3 & 3 \end{pmatrix}$.

2 Find $A + B$, $B - A$, $2A + B$, $3A - 2B$ given that

$$A = \begin{pmatrix} 2 & 0 \\ 3 & -1 \end{pmatrix} \quad \text{and} \quad B = \begin{pmatrix} 3 & 1 \\ 0 & 2 \end{pmatrix}.$$

3 Find m and n if

$$m\begin{pmatrix} 2 & 0 \\ 1 & -1 \end{pmatrix} + n\begin{pmatrix} 1 & -1 \\ 2 & -2 \end{pmatrix} = \begin{pmatrix} 4 & 2 \\ -1 & 1 \end{pmatrix}.$$

4 Find $P + Q$, $P - Q$ and $P + 3Q$ given that

$$P = \begin{pmatrix} 1 & 3 \\ 4 & 0 \\ -3 & 6 \end{pmatrix} \quad \text{and} \quad Q = \begin{pmatrix} 2 & -1 \\ 0 & -2 \\ 1 & 0 \end{pmatrix}.$$

13.3 Matrix multiplication

The product $(a, b)\begin{pmatrix} p \\ q \end{pmatrix}$ of a row vector (a, b) and a column vector $\begin{pmatrix} p \\ q \end{pmatrix}$ is defined to be $ap + bq$. It equals the scalar product of the vectors $a\mathbf{i} + b\mathbf{j}$ and $p\mathbf{i} + q\mathbf{j}$. If $A = \begin{pmatrix} a & b \\ c & d \end{pmatrix}$ and $P = \begin{pmatrix} p & r \\ q & s \end{pmatrix}$ the product AP is given by

$$AP = \begin{pmatrix} ap + bq & ar + bs \\ cp + dq & cr + ds \end{pmatrix}.$$

Each element in the matrix AP is formed by taking the product of a row vector from A and a column vector from P. In this case A is said to be post-multiplied by P, and P is said to be pre-multiplied by A. The product PA is given by

$$PA = \begin{pmatrix} pa + rc & pb + rd \\ qa + sc & qb + sd \end{pmatrix}$$

so that, in general, AP and PA are not equal.

The product QR of the matrices Q and R will exist only if the number of columns of Q equals the number of rows of R. When this is so, each element in the product QR is the product of a row vector from Q and a column vector from R. The elements in the first row of QR are the products of the first row of Q with the columns of R taken in turn. The elements in the second row of QR are the products of the second row of Q with the columns of R taken in turn, and so on.

If the matrix $\mathbf{Q}$ is of order 2×3 and if $\mathbf{R}$ is of order 3×2, the product $\mathbf{QR}$ will be of order 2×2. The product $\mathbf{RQ}$ will also exist and it will be of order 3×3.

Example 1

Find the product $\mathbf{PQ}$ when $\mathbf{P} = \begin{pmatrix} 3 & -2 & 2 \\ 4 & -5 & 7 \end{pmatrix}$ and $\mathbf{Q} = \begin{pmatrix} 2 \\ 3 \\ 1 \end{pmatrix}$.

Note that $\mathbf{P}$ has 3 columns and $\mathbf{Q}$ has 3 rows.

$$\mathbf{PQ} = \begin{pmatrix} 3 \times 2 - 2 \times 3 + 2 \times 1 \\ 4 \times 2 - 5 \times 3 + 7 \times 1 \end{pmatrix} = \begin{pmatrix} 2 \\ 0 \end{pmatrix}.$$

$\mathbf{P}$ is of order 2×3; $\mathbf{Q}$ is of order 3×1. $\mathbf{PQ}$ is of order 2×1, and $\mathbf{QP}$ does not exist.

Example 2

Find the product $\mathbf{AB}$ when $\mathbf{A} = \begin{pmatrix} 2 & 3 \\ 1 & 4 \end{pmatrix}$ and $\mathbf{B} = \begin{pmatrix} -3 & 4 & 5 \\ 2 & -3 & -1 \end{pmatrix}$.

The elements in the first row of $\mathbf{AB}$ are given by

$$(2, 3)\begin{pmatrix} -3 \\ 2 \end{pmatrix} = 0, \quad (2, 3)\begin{pmatrix} 4 \\ -3 \end{pmatrix} = -1, \quad (2, 3)\begin{pmatrix} 5 \\ -1 \end{pmatrix} = 7.$$

The elements in the second row are given by

$$(1, 4)\begin{pmatrix} -3 \\ 2 \end{pmatrix} = 5, \quad (1, 4)\begin{pmatrix} 4 \\ -3 \end{pmatrix} = -8, \quad (1, 4)\begin{pmatrix} 5 \\ -1 \end{pmatrix} = 1.$$

Therefore $\mathbf{AB}$ is the matrix $\begin{pmatrix} 0 & -1 & 7 \\ 5 & -8 & 1 \end{pmatrix}$.

Exercises 13.3

1   The matrix $\mathbf{A}$ has 4 rows and 5 columns, while the matrix $\mathbf{B}$ has 5 rows and 4 columns. Find the order of $\mathbf{AB}$ and the order of $\mathbf{BA}$.

2   Find the following products.

(a) $(3, 2)\begin{pmatrix} 2 \\ 3 \end{pmatrix}$

(d) $(3, -1)\begin{pmatrix} 2 & 1 & 3 \\ 4 & 2 & 5 \end{pmatrix}$

(b) $(2, 4)\begin{pmatrix} -3 \\ 2 \end{pmatrix}$

(e) $\begin{pmatrix} 1 & 2 \\ 2 & 1 \end{pmatrix}\begin{pmatrix} 3 & 4 \\ 2 & 1 \end{pmatrix}$

(c) $(2, 3)\begin{pmatrix} 4 & 0 \\ -1 & 2 \end{pmatrix}$

(f) $\begin{pmatrix} 5 & 3 \\ 4 & 7 \end{pmatrix}\begin{pmatrix} 1 & -1 \\ -1 & 1 \end{pmatrix}$.

3   If $\mathbf{P} = \begin{pmatrix} 2 & 1 \\ 4 & 2 \end{pmatrix}$ and $\mathbf{Q} = \begin{pmatrix} 1 & -3 \\ -2 & 6 \end{pmatrix}$, show that $\mathbf{PQ}$ is a null matrix, and find the product $\mathbf{QP}$.

4   If $\mathbf{A} = \begin{pmatrix} 3 & 1 \\ x & 2 \end{pmatrix}$ and $\mathbf{B} = \begin{pmatrix} 6 & 2 \\ 4 & y \end{pmatrix}$, find the values of x and y given that $\mathbf{AB} = \mathbf{BA}$.

5 Prove that the equation $\mathbf{PQ} = \mathbf{PR}$ does not imply that $\mathbf{Q} = \mathbf{R}$ by showing that $\mathbf{PQ}$ and $\mathbf{PR}$ are equal when

$$\mathbf{P} = \begin{pmatrix} 6 & 3 \\ 2 & 1 \end{pmatrix}, \mathbf{Q} = \begin{pmatrix} 2 & -1 \\ -3 & 1 \end{pmatrix}, \mathbf{R} = \begin{pmatrix} 1 & 1 \\ -1 & -3 \end{pmatrix}.$$

6 Show that $\mathbf{PQ}$ and $\mathbf{QP}$ are equal when

$$\mathbf{P} = \begin{pmatrix} 1 & 1 & 4 \\ 1 & -1 & 2 \\ 1 & 1 & 1 \end{pmatrix} \quad \text{and} \quad \mathbf{Q} = \begin{pmatrix} -3 & 3 & 6 \\ 1 & -3 & 2 \\ 2 & 0 & -2 \end{pmatrix}.$$

13.4 Square matrices

A square matrix is one in which the number of rows equals the number of columns. A matrix with n rows and n columns is called a square matrix of order n.

If $\mathbf{A}, \mathbf{B}, \mathbf{C}$ are square matrices of the same order we have

$$(\mathbf{A} + \mathbf{B})\mathbf{C} = \mathbf{AC} + \mathbf{BC},$$
$$\mathbf{A}(\mathbf{B} + \mathbf{C}) = \mathbf{AB} + \mathbf{AC},$$
$$\mathbf{A}(\mathbf{BC}) = (\mathbf{AB})\mathbf{C}.$$

These relations are easily verified for square matrices of order 2.

A symmetric matrix is a square matrix which is unaltered if each row is interchanged with the corresponding column. For example,

$$\begin{pmatrix} 4 & 7 \\ 7 & 3 \end{pmatrix}, \begin{pmatrix} 1 & 2 & 3 \\ 2 & 0 & 5 \\ 3 & 5 & 4 \end{pmatrix}, \begin{pmatrix} a & f & g \\ f & b & h \\ g & h & c \end{pmatrix}.$$

Symmetric matrices of the form

$$\begin{pmatrix} 1 & 0 \\ 0 & 1 \end{pmatrix} \quad \text{or} \quad \begin{pmatrix} 1 & 0 & 0 \\ 0 & 1 & 0 \\ 0 & 0 & 1 \end{pmatrix}$$

are known as unit matrices, and are denoted by $\mathbf{I}$, the order being understood from the context. If $\mathbf{A}$ is a square matrix and $\mathbf{I}$ is the unit matrix of the same order, then $\mathbf{IA} = \mathbf{A}$ and $\mathbf{AI} = \mathbf{A}$.

Associated with each square matrix there is a number known as its determinant. A matrix with determinant equal to zero is said to be singular. The determinant of the matrix

$$\mathbf{A} = \begin{pmatrix} a & b \\ c & d \end{pmatrix}$$

is $(ad - bc)$, and is denoted by $\det \mathbf{A}$ or by $\begin{vmatrix} a & b \\ c & d \end{vmatrix}$.

The determinant of a square matrix of order 3 can be defined in terms of determinants of square matrices of order 2. Thus

$$\begin{vmatrix} a_1 & a_2 & a_3 \\ b_1 & b_2 & b_3 \\ c_1 & c_2 & c_3 \end{vmatrix} = a_1 \begin{vmatrix} b_2 & b_3 \\ c_2 & c_3 \end{vmatrix} - a_2 \begin{vmatrix} b_1 & b_3 \\ c_1 & c_3 \end{vmatrix} + a_3 \begin{vmatrix} b_1 & b_2 \\ c_1 & c_2 \end{vmatrix}$$

$$= a_1 b_2 c_3 - a_1 b_3 c_2 - a_2 b_1 c_3 + a_2 b_3 c_1 + a_3 b_1 c_2 - a_3 b_2 c_1.$$

Example 1

Evaluate the determinant of the matrix $\begin{pmatrix} 3 & 4 & 2 \\ 2 & 6 & 3 \\ 5 & 2 & 4 \end{pmatrix}$.

$$\begin{vmatrix} 3 & 4 & 2 \\ 2 & 6 & 3 \\ 5 & 2 & 4 \end{vmatrix} = 3 \begin{vmatrix} 6 & 3 \\ 2 & 4 \end{vmatrix} - 4 \begin{vmatrix} 2 & 3 \\ 5 & 4 \end{vmatrix} + 2 \begin{vmatrix} 2 & 6 \\ 5 & 2 \end{vmatrix}$$

$$= 3(24 - 6) - 4(8 - 15) + 2(4 - 30)$$
$$= 54 + 28 - 52 = 30.$$

Example 2

If the sum $(A + B)$ exists and the product AB exists, show that the product BA exists.

If the sum exists, and if A is of order $m \times n$, B must also be of order $m \times n$. If the product AB exists, the number of columns of A must equal the number of rows of B, so that $n = m$. Thus A and B are square matrices of the same order, and so the product BA exists.

Exercises 13.4

1 Find whether the equation $A^2 - B^2 = (A - B)(A + B)$ is true

(a) when $A = \begin{pmatrix} 4 & 2 \\ 2 & 3 \end{pmatrix}$ and $B = \begin{pmatrix} 5 & -3 \\ -3 & 2 \end{pmatrix}$

(b) when $A = \begin{pmatrix} 3 & 1 \\ 4 & 2 \end{pmatrix}$ and $B = \begin{pmatrix} 5 & 2 \\ 3 & 4 \end{pmatrix}$.

2 Given that $A = \begin{pmatrix} 3 & 4 \\ 1 & 2 \end{pmatrix}$, $B = \begin{pmatrix} -2 & 3 \\ 3 & -2 \end{pmatrix}$, $C = \begin{pmatrix} 1 & 0 \\ -2 & 3 \end{pmatrix}$, verify that $A(BC)$ equals $(AB)C$.

3 Evaluate the determinants of the matrices

(a) $\begin{pmatrix} 6 & 3 \\ 8 & 5 \end{pmatrix}$ (b) $\begin{pmatrix} 5 & 6 \\ -7 & -4 \end{pmatrix}$ (c) $\begin{pmatrix} 8 & 4 \\ 6 & 3 \end{pmatrix}$

$$(d) \begin{pmatrix} 2 & 5 & 1 \\ 3 & 8 & 3 \\ 0 & -1 & -3 \end{pmatrix} \qquad (e) \begin{pmatrix} 2 & 3 & 2 \\ 4 & 6 & 5 \\ 5 & 7 & 6 \end{pmatrix} \qquad (f) \begin{pmatrix} 2 & 4 & 3 \\ 4 & 6 & 5 \\ 7 & 9 & 8 \end{pmatrix}.$$

4 If $P = \begin{pmatrix} 1 & 0 \\ 0 & 1 \\ 0 & 0 \end{pmatrix}$ and $Q = \begin{pmatrix} 1 & 0 & 0 \\ 0 & 1 & 0 \end{pmatrix}$,

find the determinants of **PQ** and **QP**.

5 If **A** and **B** are square matrices of order 2, show that
   (a) $\det(k\mathbf{A}) = k^2 \det \mathbf{A}$, where k is a scalar
   (b) $\det(\mathbf{A}^2) = (\det \mathbf{A})^2$
   (c) $\det(\mathbf{AB}) = (\det \mathbf{A})(\det \mathbf{B})$.

13.5 The inverse of a matrix

If the square matrix **A** is not singular, it will possess an inverse, denoted by $\mathbf{A}^{-1}$, such that

$$\mathbf{AA}^{-1} = \mathbf{I} \quad \text{and} \quad \mathbf{A}^{-1}\mathbf{A} = \mathbf{I}.$$

For example, each of the matrices $\begin{pmatrix} 7 & 2 \\ 3 & 1 \end{pmatrix}, \begin{pmatrix} 1 & -2 \\ -3 & 7 \end{pmatrix}$ is the inverse of the other since

$$\begin{pmatrix} 7 & 2 \\ 3 & 1 \end{pmatrix}\begin{pmatrix} 1 & -2 \\ -3 & 7 \end{pmatrix} = \begin{pmatrix} 1 & 0 \\ 0 & 1 \end{pmatrix}.$$

To find the inverse of the matrix $\begin{pmatrix} a & b \\ c & d \end{pmatrix}$, put

$$\begin{pmatrix} a & b \\ c & d \end{pmatrix}\begin{pmatrix} p & q \\ r & s \end{pmatrix} = \begin{pmatrix} 1 & 0 \\ 0 & 1 \end{pmatrix}.$$

Then $ap + br = 1$ and $cp + dr = 0$,
giving $(ad - bc)p = d$ and $(bc - ad)r = c$.
Also $aq + bs = 0$ and $cq + ds = 1$,
giving $(bc - ad)q = b$ and $(ad - bc)s = a$.
Provided that $ad - bc$ is not zero, we have

$$p = d/(ad - bc), \qquad q = -b/(ad - bc),$$
$$r = -c/(ad - bc), \qquad s = a/(ad - bc).$$

Hence $\begin{pmatrix} a & b \\ c & d \end{pmatrix}^{-1} = \dfrac{1}{(ad - bc)}\begin{pmatrix} d & -b \\ -c & a \end{pmatrix}.$

Example 1

Solve the simultaneous equations

$$4x + 5y = 3,$$
$$2x + 3y = -1.$$

The equations can be written as a matrix equation

$$\begin{pmatrix} 4 & 5 \\ 2 & 3 \end{pmatrix} \begin{pmatrix} x \\ y \end{pmatrix} = \begin{pmatrix} 3 \\ -1 \end{pmatrix}.$$

Now multiply each side by the inverse of the matrix of the coefficients.

$$\tfrac{1}{2} \begin{pmatrix} 3 & -5 \\ -2 & 4 \end{pmatrix} \begin{pmatrix} 4 & 5 \\ 2 & 3 \end{pmatrix} \begin{pmatrix} x \\ y \end{pmatrix} = \tfrac{1}{2} \begin{pmatrix} 3 & -5 \\ -2 & 4 \end{pmatrix} \begin{pmatrix} 3 \\ -1 \end{pmatrix},$$

This simplifies to

$$\begin{pmatrix} x \\ y \end{pmatrix} = \begin{pmatrix} 7 \\ -5 \end{pmatrix},$$

so that $x = 7$, $y = -5$.

Example 2

Given that the square matrix $\mathbf{P}$ is not singular, show that if $\mathbf{PQ}$ equals $\mathbf{PR}$ the matrices $\mathbf{Q}$ and $\mathbf{R}$ are equal.

As $\mathbf{P}$ is not singular, it will have an inverse $\mathbf{P}^{-1}$. If $\mathbf{PQ} = \mathbf{PR}$, the matrix $\mathbf{P(Q-R)}$ is a null matrix and so $\mathbf{P}^{-1}\mathbf{P(Q-R)}$ is a null matrix. Since $\mathbf{P}^{-1}\mathbf{P} = \mathbf{I}$, this shows that $\mathbf{(Q-R)}$ is a null matrix, i.e. $\mathbf{Q} = \mathbf{R}$.

Exercises 13.5

1 Find the inverse of each of the matrices

(a) $\begin{pmatrix} 1 & 0 \\ 0 & -1 \end{pmatrix}$ (b) $\begin{pmatrix} 1 & 0 \\ 1 & -1 \end{pmatrix}$ (c) $\begin{pmatrix} 1 & 0 \\ 1 & 1 \end{pmatrix}$ (d) $\begin{pmatrix} 1 & 1 \\ 0 & 1 \end{pmatrix}$.

2 Find the inverse of each of the matrices

(a) $\begin{pmatrix} 4 & 5 \\ 3 & 4 \end{pmatrix}$ (b) $\begin{pmatrix} 4 & 2 \\ 7 & 3 \end{pmatrix}$ (c) $\begin{pmatrix} 3 & 4 \\ 2 & -3 \end{pmatrix}$ (d) $\begin{pmatrix} -3 & 4 \\ 4 & -6 \end{pmatrix}$.

3 Use an inverse matrix to solve the simultaneous equations

$$5x + 3y = 4,$$
$$8x + 6y = 10.$$

4 Given that $\mathbf{P} = \begin{pmatrix} 2 & 3 & 2 \\ 4 & 6 & 5 \\ 5 & 7 & 6 \end{pmatrix}$, and $\mathbf{Q} = \begin{pmatrix} 1 & -4 & 3 \\ 1 & 2 & -2 \\ -2 & 1 & 0 \end{pmatrix}$,

show that $\mathbf{PQ} = \mathbf{QP} = \mathbf{I}$, i.e. that $\mathbf{P}$ is the inverse of $\mathbf{Q}$.

5 Given that $A = \begin{pmatrix} 3 & 4 \\ 2 & 3 \end{pmatrix}$ and $B = \begin{pmatrix} 5 & 4 \\ 3 & 3 \end{pmatrix}$,

verify that the inverse of AB equals the product $B^{-1}A^{-1}$, and that the inverse of BA equals the product $A^{-1}B^{-1}$.

6 Find a matrix (other than I) which equals its inverse.

7 Find the product of the matrices

$$\begin{pmatrix} 3 & 2 & 2 \\ 1 & -3 & 1 \\ 2 & 1 & -4 \end{pmatrix}, \begin{pmatrix} 11 & 10 & 8 \\ 6 & -16 & -1 \\ 7 & 1 & -11 \end{pmatrix}.$$

Use these matrices to solve the equations

(a) $3x + 2y + 2z = 7,$
$\quad x - 3y + z = 10,$
$\quad 2x + y - 4z = 0.$

(b) $11x + 10y + 8z = 9,$
$\quad 6x - 16y - z = 21,$
$\quad 7x + y - 11z = -5.$

8 Find the matrices P^{-1}, Q^{-1} and R^{-1} where

$$P = \begin{pmatrix} 1 & x & y \\ 0 & 1 & 0 \\ 0 & 0 & 1 \end{pmatrix}, \quad Q = \begin{pmatrix} 1 & x & y \\ 0 & 1 & z \\ 0 & 0 & 1 \end{pmatrix}, \quad R = \begin{pmatrix} 1 & x & 0 \\ 0 & 1 & 0 \\ 0 & y & 1 \end{pmatrix}.$$

9 Find the inverse of the matrix

$$\begin{pmatrix} \cos\theta & -\sin\theta \\ \sin\theta & \cos\theta \end{pmatrix}.$$

10 Find the value of x, given that $A^2 = A^{-1}$, where

$$A = \begin{pmatrix} 1 & x & 1 \\ -1 & -1 & 0 \\ 1 & 0 & 0 \end{pmatrix}.$$

13.6 Linear transformations of the x–y plane

The equations

$$X = ax + by, \quad Y = cx + dy$$

define a linear transformation of the x–y plane, mapping the point (x, y) on the point (X, Y). The origin is mapped on itself, the point $A(1, 0)$ on $A'(a, c)$, the point $B(1, 1)$ on $B'(a+b, c+d)$ and the point $C(0, 1)$ on $C'(b, d)$ (Fig. 13.1).

The same transformation is defined by the equation

$$\begin{pmatrix} X \\ Y \end{pmatrix} = T\begin{pmatrix} x \\ y \end{pmatrix}, \quad \text{where } T = \begin{pmatrix} a & b \\ c & d \end{pmatrix},$$

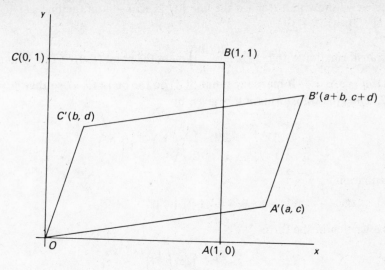

C(0, 1)

B(1, 1)

B'(a+b, c+d)

C'(b, d)

A'(a, c)

O

A(1, 0)

x

y

Fig. 13.1

mapping the vector $\mathbf{p} = \begin{pmatrix} x \\ y \end{pmatrix}$ on the vector $\mathbf{Tp} = \begin{pmatrix} X \\ Y \end{pmatrix}$.

For any two vectors $\mathbf{p}$ and $\mathbf{q}$ in the plane,

$$\mathbf{T(p+q)} = \mathbf{Tp} + \mathbf{Tq},$$

and
$$\mathbf{T}(k\mathbf{p}) = k\mathbf{Tp},$$

where k is a scalar. These two relationships are fundamental to a linear transformation.

Let the vectors $\mathbf{p}$ and $\mathbf{q}$ be mapped on the vectors $\mathbf{u}$ and $\mathbf{v}$ respectively. The line with equation $\mathbf{r} = \mathbf{p} + t\mathbf{q}$ is mapped on the line with equation $\mathbf{r} = \mathbf{u} + t\mathbf{v}$. All lines parallel to the vector $\mathbf{q}$ will be transformed into lines parallel to the vector $\mathbf{v}$. Any parallelogram wiil be mapped on a parallelogram.

If we know the points on which the points $(1, 0)$ and $(0, 1)$ are mapped, we can write down the matrix of the transformation. For if it is given that

$$\begin{pmatrix} 1 \\ 0 \end{pmatrix} \rightarrow \begin{pmatrix} a \\ c \end{pmatrix}$$

and
$$\begin{pmatrix} 0 \\ 1 \end{pmatrix} \rightarrow \begin{pmatrix} b \\ d \end{pmatrix},$$

the matrix must be $\begin{pmatrix} a & b \\ c & d \end{pmatrix}$.

Example 1

A linear transformation maps the points $(1, 0)$, $(0, 1)$ on the points $(3, 1)$, $(-1, 2)$

respectively. Show that it maps the line $\mathbf{r} = 2\mathbf{i} + t(\mathbf{i} + \mathbf{j})$ on the line
$\mathbf{r} = (6\mathbf{i} + 2\mathbf{j}) + t(2\mathbf{i} + 3\mathbf{j})$.

The transformation with matrix $\begin{pmatrix} a & b \\ c & d \end{pmatrix}$ maps the point $(1, 0)$ on the point (a, c),
so that $a = 3, c = 1$. It maps the point $(0, 1)$ on the point (b, d), so that $b = -1$,
$d = 2$. Thus the transformation is given by

$$\begin{pmatrix} X \\ Y \end{pmatrix} = \begin{pmatrix} 3 & -1 \\ 1 & 2 \end{pmatrix} \begin{pmatrix} x \\ y \end{pmatrix} = \mathbf{T} \begin{pmatrix} x \\ y \end{pmatrix}.$$

The equation

$$\mathbf{r} = 2\mathbf{i} + t(\mathbf{i} + \mathbf{j})$$

can be written in the form

$$\mathbf{r} = \begin{pmatrix} 2 \\ 0 \end{pmatrix} + t \begin{pmatrix} 1 \\ 1 \end{pmatrix}.$$

This becomes

$$\mathbf{r} = \mathbf{T} \begin{pmatrix} 2 \\ 0 \end{pmatrix} + t\mathbf{T} \begin{pmatrix} 1 \\ 1 \end{pmatrix} = \begin{pmatrix} 6 \\ 2 \end{pmatrix} + t \begin{pmatrix} 2 \\ 3 \end{pmatrix},$$

i.e. $\qquad\qquad \mathbf{r} = (6\mathbf{i} + 2\mathbf{j}) + t(2\mathbf{i} + 3\mathbf{j})$.

Example 2
Describe the transformations given by the matrices

(a) $\begin{pmatrix} 0 & 1 \\ 1 & 0 \end{pmatrix}$ (b) $\begin{pmatrix} 1 & 1 \\ 0 & 1 \end{pmatrix}$ (c) $\begin{pmatrix} 0 & 0 \\ 0 & 1 \end{pmatrix}$.

(a) We have $\begin{pmatrix} 0 & 1 \\ 1 & 0 \end{pmatrix} \begin{pmatrix} x \\ y \end{pmatrix} = \begin{pmatrix} y \\ x \end{pmatrix}$

so that the point $P(x, y)$ is mapped on the point $Q(y, x)$. The point Q is the mirror
image of P in the line $y = x$, and the transformation is a reflection in this line.

(b) Since $\begin{pmatrix} 1 & 1 \\ 0 & 1 \end{pmatrix} \begin{pmatrix} x \\ y \end{pmatrix} = \begin{pmatrix} x+y \\ y \end{pmatrix}$,

the point (x, y) is mapped on the point $(x + y, y)$. All points except those on the x-
axis, move parallel to the x-axis through a distance equal to their distance from
that axis.

This transformation is an example of a shear.

(c) In this transformation the point (x, y) is mapped on the point $(0, y)$. For each
value of c, every point on the line $y = c$ is mapped on the point $(0, c)$. The whole
plane is mapped on the y-axis.

Example 3

Show that the linear transformation given by

$$\begin{pmatrix} X \\ Y \end{pmatrix} = \begin{pmatrix} 3 & 4 \\ 6 & 8 \end{pmatrix} \begin{pmatrix} x \\ y \end{pmatrix}$$

maps all the points of the plane on a straight line.

Since

$$X = 3x + 4y, \quad Y = 6x + 8y,$$

any point (x, y) is mapped on a point on the line $Y = 2X$. All points on the line $3x + 4y = c$ are mapped on the single point $X = c$, $Y = 2c$. Note that the matrix of the transformation is singular.

Example 4

A triangle ABC is mapped by a linear transformation with matrix $\mathbf{T}$ on a triangle $A'B'C'$. Find the ratio of their areas.

Let the points $A(x_1, y_1)$, $B(x_2, y_2)$ be mapped on the points $A'(X_1, Y_1)$, $B'(X_2, Y_2)$. Then

$$\begin{pmatrix} X_1 & X_2 \\ Y_1 & Y_2 \end{pmatrix} = \mathbf{T} \begin{pmatrix} x_1 & x_2 \\ y_1 & y_2 \end{pmatrix}.$$

Since the determinant of the product of two matrices equals the product of their determinants, this gives

$$(X_1 Y_2 - X_2 Y_1) = (\det \mathbf{T})(x_1 y_2 - x_2 y_1),$$

i.e.
$$2|\text{area of } \triangle OA'B'| = |\det \mathbf{T}| \times 2|\text{area of } \triangle OAB|$$

where O is the origin. Therefore the ratio of the areas of the triangles $OA'B'$ and OAB is $|\det \mathbf{T}| : 1$. By considering the triangles OBC and OCA, we can deduce that this is the ratio of the areas of the triangles $A'B'C'$ and ABC.

Example 5

Find the matrix of the transformation in which all the points of the plane are rotated anticlockwise through an angle α about the origin. Find also the matrix of the inverse transformation.

The point P with polar coordinates (r, θ) will be mapped on the point Q with polar coordinates $(r, \theta + \alpha)$. The cartesian coordinates of P are $x = r \cos \theta$, $y = r \sin \theta$, and those of Q are $X = r \cos (\theta + \alpha)$, $Y = r \sin (\theta + \alpha)$.

The unit vectors $\begin{pmatrix} 1 \\ 0 \end{pmatrix}$, $\begin{pmatrix} 0 \\ 1 \end{pmatrix}$ become $\begin{pmatrix} \cos \alpha \\ \sin \alpha \end{pmatrix}$, $\begin{pmatrix} -\sin \alpha \\ \cos \alpha \end{pmatrix}$ respectively.

It follows that the transformation is given by

$$\begin{pmatrix} X \\ Y \end{pmatrix} = \begin{pmatrix} \cos \alpha & -\sin \alpha \\ \sin \alpha & \cos \alpha \end{pmatrix} \begin{pmatrix} x \\ y \end{pmatrix}.$$

The matrix of the inverse transformation is found by replacing α by $-\alpha$, giving

$$\begin{pmatrix} \cos \alpha & \sin \alpha \\ -\sin \alpha & \cos \alpha \end{pmatrix}.$$

The product of these two matrices (in any order) is easily shown to be **I**.

Example 4

The transformation with matrix **S** is followed by the transformation with matrix **T**. Show that the result is a transformation with matrix **TS**.

Let $\mathbf{S} = \begin{pmatrix} a & b \\ c & d \end{pmatrix}$ and $\mathbf{T} = \begin{pmatrix} p & q \\ r & s \end{pmatrix}$. If the first transformation maps the point (x_1, y_1) on the point (x_2, y_2), then

$$x_2 = ax_1 + by_1, \quad y_2 = cx_1 + dy_1.$$

If the second transformation maps the point (x_2, y_2) on the point (x_3, y_3), then

$$x_3 = px_2 + qy_2, \quad y_3 = rx_2 + sy_2.$$

By substituting for x_2 and y_2 we obtain

$$x_3 = (pa + qc)x_1 + (pb + qd)y_1,$$
$$y_3 = (ra + sc)x_1 + (rb + sd)y_1.$$

Hence

$$\begin{pmatrix} x_3 \\ y_3 \end{pmatrix} = \begin{pmatrix} pa + qc & pb + qd \\ ra + sc & rb + sd \end{pmatrix} \begin{pmatrix} x_1 \\ y_1 \end{pmatrix} = \mathbf{TS} \begin{pmatrix} x_1 \\ y_1 \end{pmatrix}.$$

If the order of the two transformations is reversed, the resulting transformation will be different, except when **TS** and **ST** are equal.

Exercises 13.6

1 Find the matrices of the transformations which map the points $(1, 0)$, $(0, 1)$ on the points
   (a) $(2, 0)$, $(3, 2)$ (b) $(2, 4)$, $(3, 6)$ (c) $(0, 3)$, $(2, 0)$ (d) $(3, 4)$, $(2, 3)$.
   Evaluate the determinant of each matrix.

2 Describe the transformations given by the following matrices.

   (a) $\begin{pmatrix} 2 & 0 \\ 0 & 2 \end{pmatrix}$ (c) $\begin{pmatrix} 1 & 0 \\ 0 & 0 \end{pmatrix}$

   (b) $\begin{pmatrix} -1 & 0 \\ 0 & 1 \end{pmatrix}$ (d) $\begin{pmatrix} 0 & 1 \\ -1 & 0 \end{pmatrix}$.

   Find the effect of each transformation on the square with vertices $(0, 0)$, $(1, 0)$, $(1, 1)$, $(0, 1)$.

3 Transform the triangle with vertices $(0, 0)$, $(1, 1)$, $(-1, 1)$ by each of the following matrices.

(a) $\begin{pmatrix} 1/\sqrt{2} & -1/\sqrt{2} \\ 1/\sqrt{2} & 1/\sqrt{2} \end{pmatrix}$ (c) $\begin{pmatrix} 1 & 2 \\ 2 & 1 \end{pmatrix}$

(b) $\begin{pmatrix} 1 & 0 \\ 2 & 1 \end{pmatrix}$ (d) $\begin{pmatrix} 1 & -1 \\ 1 & 1 \end{pmatrix}$.

Find the area of the transformed triangle in each case.

4   Find the matrices of the linear transformations giving
(a) a reflection in the x-axis
(b) a reflection in the y-axis
(c) a reflection in the line $y = x$
(d) a reflection in the line $x + y = 0$.

5   A linear transformation maps the points $(1, 0)$, $(0, 1)$ on the points $(2, 0)$, $(1, 1)$ respectively. Sketch in a diagram the lines on which the lines $x = 0$, $x = 1$, $y = 0$, $y = 1$ are mapped.

6   Find the matrix **A** of the transformation which rotates all vectors in the plane through 90° anticlockwise, and the matrix **B** of the transformation which rotates all vectors in the plane through 180° anticlockwise.

7   Find the matrix of the transformation which maps the points $(2, 0)$, $(0, 2)$ on the points $(0, 4)$, $(-4, 0)$ respectively. Show that this matrix is the product of a matrix **P** giving a magnification, and a matrix **Q** giving a rotation, and that **PQ = QP**.

8   Find the matrix of the transformation which reflects all points of the plane in the line $y = \sqrt{3}x$.

9   Find the effect on the square with vertices $(0, 0)$, $(1, 0)$, $(1, 1)$, $(0, 1)$ of the transformations with matrices **R**, **S**, **RS** and **SR** where $\mathbf{R} = \begin{pmatrix} 0 & 1 \\ -1 & 0 \end{pmatrix}$ and $\mathbf{S} = \begin{pmatrix} 1 & -1 \\ 0 & 1 \end{pmatrix}$.

10   Show that the line $y = x$ is mapped on itself by the transformation

$$\begin{pmatrix} X \\ Y \end{pmatrix} = \begin{pmatrix} 2 & 2 \\ 1 & 3 \end{pmatrix}\begin{pmatrix} x \\ y \end{pmatrix}.$$

Find the equation of another line which is mapped on itself.

Miscellaneous examples 13

1   Show that the inverse of the matrix **AB** is $\mathbf{B}^{-1}\mathbf{A}^{-1}$.

We have $(\mathbf{AB})(\mathbf{B}^{-1}\mathbf{A}^{-1}) = \mathbf{A}(\mathbf{BB}^{-1})\mathbf{A}^{-1}$
$= \mathbf{AIA}^{-1}$
$= \mathbf{AA}^{-1} = \mathbf{I}.$

Also $(\mathbf{B}^{-1}\mathbf{A}^{-1})(\mathbf{AB}) = \mathbf{B}^{-1}(\mathbf{A}^{-1}\mathbf{A})\mathbf{B}$
$= \mathbf{B}^{-1}\mathbf{IB}$
$= \mathbf{B}^{-1}\mathbf{B} = \mathbf{I}.$

Therefore $(\mathbf{AB})^{-1} = \mathbf{B}^{-1}\mathbf{A}^{-1}$.

2 Find the matrix of the transformation which reflects all points of the plane in the line $y = x \tan \alpha$.

First rotate the plane clockwise through the angle α, using the matrix

$$\mathbf{R} = \begin{pmatrix} \cos \alpha & \sin \alpha \\ -\sin \alpha & \cos \alpha \end{pmatrix}.$$

This will map the line $y = x \tan \alpha$ on the x-axis. Next reflect the plane in the x-axis, using the matrix

$$\mathbf{S} = \begin{pmatrix} 1 & 0 \\ 0 & -1 \end{pmatrix}.$$

Finally rotate the plane anticlockwise through the angle α, using the inverse of the matrix $\mathbf{R}$. The matrix required will be given by

$$\mathbf{R}^{-1}\mathbf{SR} = \begin{pmatrix} \cos \alpha & -\sin \alpha \\ \sin \alpha & \cos \alpha \end{pmatrix} \begin{pmatrix} 1 & 0 \\ 0 & -1 \end{pmatrix} \begin{pmatrix} \cos \alpha & \sin \alpha \\ -\sin \alpha & \cos \alpha \end{pmatrix}$$

$$= \begin{pmatrix} \cos \alpha & -\sin \alpha \\ \sin \alpha & \cos \alpha \end{pmatrix} \begin{pmatrix} \cos \alpha & \sin \alpha \\ \sin \alpha & -\cos \alpha \end{pmatrix}$$

$$= \begin{pmatrix} \cos 2\alpha & \sin 2\alpha \\ \sin 2\alpha & -\cos 2\alpha \end{pmatrix}.$$

3 Show that for any angles α and β
(i) $\cos (\alpha + \beta) = \cos \alpha \cos \beta - \sin \alpha \sin \beta$
(ii) $\sin (\alpha + \beta) = \sin \alpha \cos \beta + \cos \alpha \sin \beta$.

Rotate all the points of the plane about the origin through an anticlockwise angle β first and then through an anticlockwise angle α. The matrix of this transformation will be

$$\begin{pmatrix} \cos \alpha & -\sin \alpha \\ \sin \alpha & \cos \alpha \end{pmatrix} \begin{pmatrix} \cos \beta & -\sin \beta \\ \sin \beta & \cos \beta \end{pmatrix}$$

which equals

$$\begin{pmatrix} \cos \alpha \cos \beta - \sin \alpha \sin \beta & -\cos \alpha \sin \beta - \sin \alpha \cos \beta \\ \sin \alpha \cos \beta + \cos \alpha \sin \beta & -\sin \alpha \sin \beta + \cos \alpha \cos \beta \end{pmatrix}.$$

Since this transformation gives a rotation through the angle $(\alpha + \beta)$, its matrix is

$$\begin{pmatrix} \cos (\alpha + \beta) & -\sin (\alpha + \beta) \\ \sin (\alpha + \beta) & \cos (\alpha + \beta) \end{pmatrix}.$$

By equating each element of this matrix to the corresponding element in the preceding matrix, the required addition formulae are obtained.

4 Find the matrix of a linear transformation which maps the whole plane on the line $2y = 3x$, and which maps any line parallel to the vector $\begin{pmatrix} 2 \\ -1 \end{pmatrix}$ on a point.

Let $\begin{pmatrix} X \\ Y \end{pmatrix} = \mathbf{T}\begin{pmatrix} x \\ y \end{pmatrix} = \begin{pmatrix} a & b \\ c & d \end{pmatrix}\begin{pmatrix} x \\ y \end{pmatrix}$.

Then $X = ax + by$, $Y = cx + dy$. Since $2Y = 3X$ for all values of x and y, this gives $2c = 3a, 2d = 3b$. The points on the line with equation

$$\mathbf{r} = \begin{pmatrix} p \\ q \end{pmatrix} + t\begin{pmatrix} 2 \\ -1 \end{pmatrix}$$

will all be mapped on the single point with position vector $\mathbf{T}\begin{pmatrix} p \\ q \end{pmatrix}$ if $\mathbf{T}\begin{pmatrix} 2 \\ -1 \end{pmatrix}$ is zero. Hence

$$\begin{pmatrix} a & b \\ c & d \end{pmatrix}\begin{pmatrix} 2 \\ -1 \end{pmatrix} = \begin{pmatrix} 0 \\ 0 \end{pmatrix}.$$

This gives $2a = b$, $2c = d$. The four equations for the elements of the matrix $\mathbf{T}$ are satisfied by $a = 2k$, $b = 4k$, $c = 3k$, $d = 6k$ where k is any constant, and so

$$\mathbf{T} = k\begin{pmatrix} 2 & 4 \\ 3 & 6 \end{pmatrix}.$$

Miscellaneous exercises 13

1 If $A = \begin{pmatrix} -1 & -1 \\ 2 & 2 \end{pmatrix}$ and $B = \begin{pmatrix} 3 & 3 \\ -2 & -2 \end{pmatrix}$,
show that $AB = A$ and $BA = B$.

2 If $P = \begin{pmatrix} -1 & -2 \\ 1 & 1 \end{pmatrix}$ and $Q = \begin{pmatrix} 1 & 4 \\ 1 & -1 \end{pmatrix}$,
show that $(P - Q)^2 = P^2 + Q^2$.

3 The equation $\mathbf{M}^2 = a\mathbf{M} + b\mathbf{I}$, where a and b are scalars, is satisfied by the matrix $\mathbf{M}$ given by

$$\mathbf{M} = \begin{pmatrix} 1 & 2 & 2 \\ 2 & 1 & 2 \\ 2 & 2 & 1 \end{pmatrix}.$$

Find the values of a and b, and use the equation to find the inverse of the matrix $\mathbf{M}$.

4 Find the inverse of the matrix

$$\begin{pmatrix} 3 & 2 & 4 \\ 0 & 1 & 2 \\ 5 & 0 & 6 \end{pmatrix}$$

5 Find the matrix **A** such that $\mathbf{AB} = \mathbf{A} + \mathbf{B}$, where

$$\mathbf{B} = \begin{pmatrix} 3 & 0 & 0 \\ 0 & 5 & 0 \\ 1 & 0 & 2 \end{pmatrix}.$$

6 If **B** is the row vector $(1, 1, 1)$ and **C** is the matrix

$$\begin{pmatrix} 2 & 1 & 0 \\ 0 & 3 & 4 \end{pmatrix}$$

find the matrix **A** such that **BA** is the row vector $(3/2, -1/4)$ and **CA** is a unit matrix.

7 Find the matrix $\mathbf{M}^2$ where **M** is given by

$$\mathbf{M} = \begin{pmatrix} \cos\theta & k\sin\theta \\ (1/k)\sin\theta & -\cos\theta \end{pmatrix}$$

and k is non-zero.

Given that $\mathbf{MX} = \lambda\mathbf{X}$, where λ is a constant, and **X** is a non-null column matrix with elements x_1, x_2, show by premultiplying this equation by **M**, or otherwise, that $\lambda = \pm 1$.

Show that the values of **X** corresponding to these two values of λ are

$$\begin{pmatrix} \mu_1 k \cos\left(\tfrac{1}{2}\theta\right) \\ \mu_1 \sin\left(\tfrac{1}{2}\theta\right) \end{pmatrix} \quad \text{and} \quad \begin{pmatrix} -\mu_2 k \sin\left(\tfrac{1}{2}\theta\right) \\ \mu_2 \cos\left(\tfrac{1}{2}\theta\right) \end{pmatrix}$$

respectively, where μ_1 and μ_2 are arbitrary constants, and find the values of k if the vectors are to be orthogonal. [JMB]

8 Show that the column vectors

$$\mathbf{x}_1 = \begin{pmatrix} -1 \\ 2 \\ 2 \end{pmatrix}, \quad \mathbf{x}_2 = \begin{pmatrix} 2 \\ -1 \\ 2 \end{pmatrix}, \quad \mathbf{x}_3 = \begin{pmatrix} 2 \\ 2 \\ -1 \end{pmatrix}$$

form a linearly independent set. Express the vector

$$\mathbf{y} = \begin{pmatrix} 7 \\ 19 \\ 79 \end{pmatrix}$$

in terms of $\mathbf{x}_1, \mathbf{x}_2$ and $\mathbf{x}_3$.

9 If $\mathbf{A}(x)$ denotes the matrix

$$\begin{pmatrix} 2-x & 2x-2 \\ 1-x & 2x-1 \end{pmatrix},$$

prove that $A(x)$ is singular if and only if $x = 0$.

Prove also that $A(x)A(y) = A(xy)$ and hence show that the square of $A(-1)$ is the identity matrix. Find the inverse of $A(2)$.

10 (i) Find the value of the constant a, given that there is a matrix B such that $AB = C$ where

$$A = \begin{pmatrix} 2 & 1 \\ 4 & 3 \\ 3 & a \end{pmatrix}, C = \begin{pmatrix} 1 & 1 & 1 \\ 1 & 1 & 1 \\ 1 & 1 & 1 \end{pmatrix}.$$

(ii) Show, with the help of a diagram, that the matrix P of the linear transformation which rotates all points of the plane in the counter-clockwise sense through an angle θ about the origin is

$$\begin{pmatrix} \cos\theta & -\sin\theta \\ \sin\theta & \cos\theta \end{pmatrix}.$$

Find the matrix Q of the linear transformation which reflects the points of the plane in the line $y = x$. Find the values of θ for which $PQ = QP$.[L]

11 A transformation T of the plane is the product $T_3 T_2 T_1$ of the three transformations T_1, T_2, T_3, where

(a) T_1 is reflection in the line $y = x$

(b) T_2 is an anticlockwise rotation about the origin through $60°$

(c) T_3 is magnification, from the origin, by a factor 2.

Express each of T_1, T_2, T_3 in the matrix form

$$\begin{pmatrix} x' \\ y' \end{pmatrix} = \begin{pmatrix} a & b \\ c & d \end{pmatrix}\begin{pmatrix} x \\ y \end{pmatrix},$$

where a, b, c, d are constants, and deduce that the transformation T is given by

$$\begin{pmatrix} x' \\ y' \end{pmatrix} = \begin{pmatrix} -\sqrt{3} & 1 \\ 1 & \sqrt{3} \end{pmatrix}\begin{pmatrix} x \\ y \end{pmatrix}.$$

Find the gradients of the two lines through the origin each of which is transformed into itself by T. [L]

12 The matrices A and B are given by

$$A = \frac{1}{2}\begin{pmatrix} 1 & -\sqrt{3} \\ \sqrt{3} & 1 \end{pmatrix}, \quad B = \frac{1}{\sqrt{2}}\begin{pmatrix} 1 & -1 \\ 1 & 1 \end{pmatrix}.$$

Show that $A^3 = B^4 = -I$. Give a geometrical interpretation of this result.

13 Show that the transformation

$$\begin{pmatrix} x_2 \\ y_2 \end{pmatrix} = \begin{pmatrix} -3/5 & 4/5 \\ 4/5 & 3/5 \end{pmatrix}\begin{pmatrix} x_1 \\ y_1 \end{pmatrix}$$

represents a reflection in a fixed line through the origin.

14 A transformation **T** of the plane is the product $\mathbf{T}_3\mathbf{T}_2\mathbf{T}_1$ of the three transformations $\mathbf{T}_1, \mathbf{T}_2, \mathbf{T}_3$, where

(a) $\mathbf{T}_1$ halves the y-coordinate of every point, but leaves x-coordinates unaltered

(b) $\mathbf{T}_2$ is reflection in the line $y = -x$

(c) $\mathbf{T}_3$ is anticlockwise rotation about the origin through $135°$.

Express each of the transformations $\mathbf{T}_1, \mathbf{T}_2, \mathbf{T}_3$ in matrix form and hence show that the matrix of **T** is

$$\frac{1}{2\sqrt{2}}\begin{pmatrix} 2 & 1 \\ 2 & -1 \end{pmatrix}.$$

Find the gradients of the two lines through the origin each of which is transformed into itself by **T**. [L]

15 Find vectors **u** and **v** such that $\mathbf{Au} = \lambda\mathbf{u}$ and $\mathbf{Av} = \mu\mathbf{v}$, where **A** is the matrix

$$\begin{pmatrix} 1 & 2 \\ 1 & 1 \end{pmatrix}$$

and λ and μ are unequal constants.

Express the vector $\begin{pmatrix} 1 \\ 0 \end{pmatrix}$ as the sum of multiples of **u** and **v** .

A sequence of vectors is defined by the iteration $\mathbf{x}_{n+1} = \mathbf{Ax}_n$, with $\mathbf{x}_0 = \begin{pmatrix} 1 \\ 0 \end{pmatrix}$. Express $\mathbf{x}_n$ in terms of **u** and **v**, and show that the ratio of the two elements of $\mathbf{x}_n$ tends to a limit as n tends to infinity. [L]

16 Write down the 2×2 matrix **R** such that $\mathbf{R}\begin{pmatrix} x \\ y \end{pmatrix} = \begin{pmatrix} X \\ Y \end{pmatrix}$ represents a rotation of the coordinate axes through $\pi/4$ in an anticlockwise direction.

S is the curve whose equation in terms of x and y is

$$x^2 + y^2 - 2xy - 2x - 2y = 0.$$

Determine the equation of S in terms of X and Y. Sketch the two sets of axes and the curve S. [JMB]

Notation

Sets and functions

$\{\,\ldots\ldots\}$	the set of elements listed or described within the braces
$\{x:\ldots\ldots\}$	the set of values of x such that ...
$\in$	is an element of
$\notin$	is not an element of
$\subset$	is a subset of
$\cap$ as in $A \cap B$	the intersection of sets A and B
$\cup$ as in $A \cup B$	the union of sets A and B
$\varnothing$	the empty (or null) set
$\mathscr{E}$	the universal set
A'	the complement of set A
$A \times B$	the product set of A and B, i.e. the set of ordered pairs $\{(x, y): x \in A,\ y \in B\}$
$f: x \mapsto y$	the function f which maps the element x to the element y
$f(x)$	the image of x under the function f
fg	the composite function which maps x to $f[g(x)]$
f^{-1}	the inverse function of the function f
$\Rightarrow$	implies
$\Leftarrow$	is implied by
$\Leftrightarrow$	implies and is implied by

Algebra

$>$	is greater than
$<$	is less than
$\geqslant$	is greater than or equal to
$\leqslant$	is less than or equal to
$\approx$	is approximately equal to
$\equiv$	is identical to
$\displaystyle\sum_{r=1}^{n} u_r$ or $\displaystyle\sum_{1}^{n} u_r$	$u_1 + u_2 + u_3 + \ldots\ldots + u_n$
$\dbinom{n}{r}$	$\dfrac{n(n-1)(n-2)\ldots(n-r+1)}{(1)(2)(3)\ldots(r)}$ or $\dfrac{n!}{r!(n-r)!}$

Trigonometry

arcsin or $\sin^{-1}$ the inverse function of the function f defined by
$$f(x) = \sin x, \ -\tfrac{1}{2}\pi \leqslant x \leqslant \tfrac{1}{2}\pi,$$

arccos or $\cos^{-1}$ the inverse function of the function g defined by
$$g(x) = \cos x, 0 \leqslant x \leqslant \pi,$$

arctan or $\tan^{-1}$ the inverse function of the function h defined by
$$h(x) = \tan x, \ -\tfrac{1}{2}\pi < x < \tfrac{1}{2}\pi.$$

Differentiation

$\rightarrow$ tends to

$\lim_{x \to a}$ the limit as x tends to a of

f' the derived function of f

$\dfrac{dy}{dx}$ or $f'(x)$ first derivative

$\dfrac{d^2y}{dx^2}$ or $f''(x)$ second derivative

$\dfrac{d^ry}{dx^r}$ or $f^{(r)}(x)$ rth derivative

δx a small increment in x

δy the corresponding increment in y

Integration

$\int f(x)\,dx$ the indefinite integral of $f(x)$

$\left[F(x)\right]_a^b$ $F(b) - F(a)$

Real and complex numbers

$\mathbb{N}$ the set of positive integers and zero $\{0, 1, 2, \ldots\}$

$\mathbb{Z}$ the set of integers $\{0, \pm 1, \pm 2, \ldots\}$

$\mathbb{Q}$ the set of rational numbers

$\mathbb{R}$ the set of real numbers

$\mathbb{C}$ the set of complex numbers

$\mathbb{Z}^+$ the set of positive integers $\{1, 2, 3, \ldots\}$

$\mathbb{Q}^+$ the set of positive rational numbers

$\mathbb{R}^+$ the set of positive real numbers

$|z|$ the modulus of z

$\arg z$ the argument of z

z^* the conjugate of z

$\mathrm{Re}\,(z)$ x, when $z = x + iy$

$\mathrm{Im}\,(z)$ y, when $z = x + iy$

Vectors and matrices

a or $\lvert \mathbf{a} \rvert$	the magnitude of the vector $\mathbf{a}$
$\mathbf{i}, \mathbf{j}, \mathbf{k}$	unit vectors in the directions Ox, Oy, Oz
$\begin{pmatrix} a_1 \\ a_2 \end{pmatrix}$	the vector $a_1\mathbf{i} + a_2\mathbf{j}$
$\mathbf{a}.\mathbf{b}$	the scalar product of $\mathbf{a}$ and $\mathbf{b}$
$\mathbf{A}^{-1}$	the inverse of the matrix $\mathbf{A}$
det $\mathbf{A}$ or $\lvert \mathbf{A} \rvert$	the determinant of $\mathbf{A}$

Greek letters

α	alpha
β	beta
γ	gamma
δ	delta
λ	lambda
μ	mu
Π, π	pi
ρ	rho
Σ, σ	sigma
θ	theta
ϕ	phi
ψ	psi
ω	omega

Formulae

Algebra

$$ax^2 + bx + c = 0, \ (a \neq 0), \quad x = \frac{-b \pm \sqrt{(b^2 - 4ac)}}{2a}$$

$$\alpha + \beta = -\frac{b}{a}, \ \alpha\beta = \frac{c}{a}$$

$$ax^3 + bx^2 + cx + d = 0, \ (a \neq 0),$$

$$\alpha + \beta + \gamma = -\frac{b}{a}, \ \alpha\beta + \beta\gamma + \gamma\alpha = \frac{c}{a}, \ \alpha\beta\gamma = -\frac{d}{a}$$

$$1 + 2 + 3 + \ldots + n \ = \tfrac{1}{2}n(n+1)$$

$$1^2 + 2^2 + 3^2 + \ldots + n^2 = \tfrac{1}{6}n(n+1)(2n+1)$$

$$1^3 + 2^3 + 3^3 + \ldots + n^3 = \tfrac{1}{4}n^2(n+1)^2$$

Arithmetic progression

$$a + (a+d) + (a+2d) + \ldots + [a + (n-1)d]$$

$$S_n = \tfrac{1}{2}n(a+l) = \tfrac{1}{2}n[2a + (n-1)d]$$

Geometric progression

$$a + ar + ar^2 + \ldots + ar^{n-1}$$

$$S_n = \frac{a(1 - r^n)}{1 - r}$$

$$\text{Sum to infinity} = \frac{a}{1 - r}, \text{ for } |r| < 1$$

Binomial expansion

$$(1 + x)^n = 1 + nx + \frac{n(n-1)}{(1)(2)}x^2 + \ldots + \frac{n(n-1)\ldots(n-r+1)}{r!}x^r + \ldots$$

for $|x| < 1$

Trigonometry

General solution of equations

If $\tan x = \tan \alpha$, then $x = n\pi + \alpha$.

If $\cos x = \cos \alpha$, then $x = 2n\pi \pm \alpha$.

If $\sin x = \sin \alpha$, then $x = 2n\pi + \alpha \quad \text{or} \quad (2n+1)\pi - \alpha$

$$\text{i.e. } x = n\pi + (-1)^n \alpha$$

(π radians = 180 degrees)

Pythagorean identities

$$\cos^2 \theta + \sin^2 \theta = 1 \qquad 1 + \tan^2 \theta = \sec^2 \theta \qquad 1 + \cot^2 \theta = \csc^2 \theta$$

Sine formula

$$\frac{a}{\sin A} = \frac{b}{\sin B} = \frac{c}{\sin C} = 2R$$

Cosine formula

$$a^2 = b^2 + c^2 - 2bc \cos A$$

$$\cos A = \frac{b^2 + c^2 - a^2}{2bc}$$

Addition formulae

$$\sin (A + B) = \sin A \cos B + \cos A \sin B$$
$$\sin (A - B) = \sin A \cos B - \cos A \sin B$$
$$\cos (A + B) = \cos A \cos B - \sin A \sin B$$
$$\cos (A - B) = \cos A \cos B + \sin A \sin B$$

$$\tan (A + B) = \frac{\tan A + \tan B}{1 - \tan A \tan B}$$

$$\tan (A - B) = \frac{\tan A - \tan B}{1 + \tan A \tan B}$$

Double-angle formulae

$$\sin 2A = 2 \sin A \cos A$$
$$\cos 2A = \cos^2 A - \sin^2 A$$
$$\cos 2A = 2 \cos^2 A - 1 \quad \text{or} \quad \cos^2 A = \tfrac{1}{2}(1 + \cos 2A)$$
$$\cos 2A = 1 - 2 \sin^2 A \quad \text{or} \quad \sin^2 A = \tfrac{1}{2}(1 - \cos 2A)$$

$$\tan 2A = \frac{2 \tan A}{1 - \tan^2 A}$$

Half-angle formulae

$$\sin x = \frac{2t}{1 + t^2}$$

$$\cos x = \frac{1 - t^2}{1 + t^2} \qquad \text{where } t = \tan \tfrac{1}{2}x$$

$$\tan x = \frac{2t}{1 - t^2}$$

Factor formulae

$$\sin x + \sin y = 2 \sin \tfrac{1}{2}(x + y) \cos \tfrac{1}{2}(x - y)$$
$$\sin x - \sin y = 2 \cos \tfrac{1}{2}(x + y) \sin \tfrac{1}{2}(x - y)$$
$$\cos x + \cos y = 2 \cos \tfrac{1}{2}(x + y) \cos \tfrac{1}{2}(x - y)$$
$$\cos x - \cos y = -2 \sin \tfrac{1}{2}(x + y) \sin \tfrac{1}{2}(x - y)$$

Area of a triangle $= \tfrac{1}{2}bc \sin A$ or
$\sqrt{[s(s - a)(s - b)(s - c)]}$, where $s = \tfrac{1}{2}(a + b + c)$

Length of an arc of a circle $= r\theta$
Area of a sector of a circle $= \tfrac{1}{2}r^2\theta$

Coordinate geometry

Length of a line $= \sqrt{[(x_1 - x_2)^2 + (y_1 - y_2)^2]}$ 163

Gradient of a line $= \dfrac{y_1 - y_2}{x_1 - x_2}$ 163

Angle between two lines $\tan \theta = \dfrac{m_1 - m_2}{1 + m_1 m_2}$ 164

for parallel lines, $m_1 = m_2$

for perpendicular lines, $m_1 m_2 = -1$

Mid-point of a line $[\tfrac{1}{2}(x_1 + x_2), \tfrac{1}{2}(y_1 + y_2)]$ 166

Centroid of a triangle $[\tfrac{1}{3}(x_1 + x_2 + x_3), \tfrac{1}{3}(y_1 + y_2 + y_3)]$ 167

Area of a triangle $= \tfrac{1}{2}|(x_1 y_2 - x_2 y_1) + (x_2 y_3 - x_3 y_2) + (x_3 y_1 - x_1 y_3)|$ 168

$$
\begin{matrix}
x_1 & y_1 \\
x_2 & y_2 \\
x_3 & y_3 \\
x_1 & y_1
\end{matrix}
$$

Equation of the line through (x_1, y_1) with gradient m_1

$$y - y_1 = m_1(x - x_1)$$ 170

Distance from (x_1, y_1) to the line $ax + by + c = 0$

$$= \left| \frac{ax_1 + by_1 + c}{\sqrt{(a^2 + b^2)}} \right|$$ 175

Equation of circle with centre $(0, 0)$ and radius r is

$$x^2 + y^2 = r^2$$ 178

Equation of the tangent at (x_1, y_1) to the circle $x^2 + y^2 = r^2$ is

$$xx_1 + yy_1 = r^2$$ 179

Condition for the line $y = mx + c$ to touch the circle $x^2 + y^2 = r^2$ is

$$c^2 = r^2(1 + m^2)$$ 180

Equation of the circle with centre (a, b) and radius r is

$$(x - a)^2 + (y - b)^2 = r^2$$ 181

The circle $x^2 + y^2 + 2gx + 2fy + c = 0$ has centre $(-g, -f)$
and radius $\sqrt{(g^2 + f^2 - c)}$ 182

Equation of the circle with (x_1, y_1) and (x_2, y_2) as end-points of
a diameter is

$$(x - x_1)(x - x_2) + (y - y_1)(y - y_2) = 0$$ 183

The parabola $y^2 = 4ax$ has focus $(a, 0)$ and directrix $x = -a$ 185

Condition for the line $y = mx + c$ to touch the parabola $y^2 = 4ax$ is
$c = a/m$

Equation of the tangent at $(ap^2, 2ap)$ to the parabola $y^2 = 4ax$ is

$$py - x = ap^2$$ 189

Equation of the normal at $(ap^2, 2ap)$ to the parabola $y^2 = 4ax$ is

$$y + px = 2ap + ap^3 \qquad\qquad 189$$

Differentiation

$$f'(x) = \lim_{h \to 0} \left(\frac{f(x+h) - f(x)}{h} \right) \qquad\qquad 145$$

$$\frac{dy}{dx} = \lim_{\delta x \to 0} \left(\frac{\delta y}{\delta x} \right) \qquad\qquad 147$$

$$\delta y \approx \frac{dy}{dx} \delta x \qquad\qquad 243$$

$$\frac{dy}{dx} = \frac{dy}{dt} \bigg/ \frac{dx}{dt} = \frac{dy}{dt} \times \frac{dt}{dx} \qquad\qquad 147$$

$$\frac{d}{dx}(u \times v) = u\frac{dv}{dx} + v\frac{du}{dx} \qquad\qquad 151$$

$$\frac{d}{dx}(u/v) = \frac{v\dfrac{du}{dx} - u\dfrac{dv}{dx}}{v^2} \qquad\qquad 153$$

$f(x)$	$f'(x)$	
x^n	nx^{n-1}	145
$\sin x$	$\cos x$	150
$\cos x$	$-\sin x$	150
$\tan x$	$\sec^2 x$	154
$\sec x$	$\sec x \tan x$	155
$\csc x$	$-\csc x \cot x$	155
$\cot x$	$-\csc^2 x$	155
$\arcsin x$	$1/\sqrt{(1-x^2)}$	158
$\arccos x$	$-1/\sqrt{(1-x^2)}$	159
$\arctan x$	$1/(1+x^2)$	158
$\ln x$	$1/x$	281
$\exp x$	$\exp x$	292

Integration

$$\int f(x)dx = \int fg(t) \times g'(t)dt \qquad\qquad 252$$

$$\int u\frac{dv}{dx}dx = uv - \int v\frac{du}{dx}dx \qquad\qquad 255$$

$$\int x^n dx = x^{n+1}/(n+1) + c, \text{ if } n \neq -1 \qquad\qquad 250$$

Logarithms and the exponential function

Vectors and matrices

Answers

Chapter 1

1.1 3 (a) 2, 3, 5, 7, 11, 13, 17, 19, 23, 29
(b) A, E, F, H, I, K, L, M, N, T, V, W,
X, Y, Z   (c) 4, 9, 16, 25, 36, 49
5 (a) $\in$ (b) $\notin$ (c) A (d) B
(e) B (f) A (g) $\notin$ (h) $\notin$
(i) 8, 9 or 10 (j) 1, 2, 3, 4 or 5
1.2 3 (a) $\{4\}$, $\{1, 2, 3, 4, 5, 6, 8\}$
(b) $\{2, 4, 6\}$, $\{1, 2, 3, 4, 6, 8, 10, 12\}$
(c) $\{4, 5, 6\}$, $\mathbb{N}$
(d) $\{x:x \in \mathbb{Z}, x > 4\}$, $\mathbb{Z}^+$
5 8
7 (a) $\{1, 3, 5, 7, 9, 11, 12\}$ (b) $\{1, 2,$
$4, 5, 7, 8, 10, 11\}$ (c) $\{6\}$ (d) $\{1, 2,$
$3, 4, 5, 7, 8, 9, 10, 11, 12\}$ (e) $\{2, 3,$
$4, 6, 8, 9, 10, 12\}$ (f) $\{1, 5, 7, 11\}$
1.3 2 $(4, 1), (3, 2), (-1, 0), (2, -3)$
4 (a) mn (b) nm (c) m^2 (d) n^2
1.4 3 (a) odd (b) even (c) neither
(d) neither (e) even (f) even
(g) neither
1.5 1 $-4, -2, -5, -9, 2(x-3), 2x-3$
2 $1, 3, 4, 100, 2(x+1)^2+1, (2x+2)^2$
3 neither, even, neither, even
4 even, even, odd, even, even, even
1.6 1 (a) $\frac{1}{3}x$ (b) $\frac{1}{2}(x-1)$ (c) $\frac{1}{4}(1-x)$
2 (a) $\{x:x \in \mathbb{R}, x \neq 0\}$
(b) $\{x:x \in \mathbb{R}, x \neq 1\}$ (c) $\mathbb{R}$
(d) $\{x:x \in \mathbb{R}, x \neq 1, x \neq -1\}$
3 (a) $\mathbb{R}^+$, $+\frac{1}{2}\sqrt{x}$ (b) $\{x:x \in \mathbb{R},$
$x > 0\}$, $+\dfrac{1}{\sqrt{x}}$ (c) $\{x:x \in \mathbb{R}, x \neq -1\}$,
$\dfrac{1-x}{x}$ (d) $\mathbb{R}^+$, $\mathbb{R}$, $+\sqrt{(x+9)}$
1.7 1 (a) $\mathscr{E}$ (b) no inverse
2 (a) $\varnothing$ (b) no inverse
3 (a) yes (b) (i) no, (ii) no
(c) no (d) no
4 (a) yes (b) (i) no, (ii) no
(c) no (d) no
5 (a) yes (b) (i) no, (ii) yes
(c) yes, 0 (d) yes, self-inverse

Miscellaneous exercises 1

1 f even, g odd,

$$f^{-1} = \begin{cases} +\sqrt{x} & \text{for } x > 0 \\ -\sqrt{(-x)} & \text{for } x \leqslant 0 \end{cases}$$

2 $\{x:x \in \mathbb{R}^+, 0 < x < 1\}$,
$f^{-1}:x \mapsto \dfrac{x}{1-x}$, yes
3 $\mathbb{R}, f^{-1}:x \mapsto [x+\sqrt{(x^2+4)}]/2$,
$\{x:x \in \mathbb{R}, x \neq 0\}$, yes
4 $\{x:x \in \mathbb{R} -5 < x \leqslant 4\}$,
$\{x:x \in \mathbb{R}, -2 < x \leqslant 2\}$,
$$f^{-1}(x) = \begin{cases} \frac{1}{2}(1+x) & \text{for } -5 < x \leqslant 1 \\ +\sqrt{x} & \text{for } 1 < x \leqslant 4 \end{cases}$$
5 $\{x:x \in \mathbb{R}, -4 \leqslant x < 4, x \neq -2\}$,
$g:x \mapsto -5-x$ for $-4 \leqslant x < -2$
$g:x \mapsto (\frac{1}{2}x)^{\frac{1}{3}}$ for $-2 < x \leqslant 2$
$g:x \mapsto 5-x$ for $2 < x < 4$
g is a function with domain the range of f.
6 $\dfrac{a}{1+a^2}$, $\{x:x \in \mathbb{R}, 0 < x \leqslant \frac{1}{2}\}$
7 (i) yes (ii) $\{x:x \in \mathbb{R}_+, x \neq 1\}$
8 (i) odd, even, even
9 (i) odd, odd, odd (ii) no
10 (a) yes (b) (i) no, (ii) yes
(c) no (d) no
11 (a) (i) yes, (ii) yes, (iii) yes
(b) $\dfrac{1}{3}$ (c) $\dfrac{1}{9x}$

Chapter 2

2.1(i) 1 (a) $(0, 4)$ (b) $(0, 5)$ (c) $(0, 0)$
(d) $(0, -1\frac{1}{2})$ (e) $(0, -1\frac{1}{3})$
(f) $(0, 3)$ (g) $(0, -7)$
(h) $(0, 0)$ (i) $(0, 7)$ (j) $(0, -1)$
(k) $(0, 3)$
2 (a) 2, 1 (b) 3, -3 (c) -7, 2
(d) $-4\frac{1}{4}$, $-\frac{1}{2}$ (e) 0, 2
(f) 0, -4 (g) -1, 1 (h) -10, 1
(i) $-1\frac{1}{4}$, $\frac{3}{4}$ (j) $-\frac{17}{5}$, $-\frac{1}{5}$
3 (a) 6, 1 (b) 12, -2 (c) 5, 3
(d) $-\frac{3}{4}, \frac{1}{2}$ (e) 0, -3 (f) 0, 5
(g) 20, -2 (h) 11, 3
(i) $\frac{49}{8}$, $-\frac{1}{4}$ (j) $-\frac{11}{3}$, $\frac{2}{3}$
4 (a) least, -2 (b) greatest, 6

(c) least, 0 (d) least, 5
(e) greatest, 20 (f) least, -11
(g) greatest, $\frac{49}{8}$ (h) greatest, $-\frac{23}{16}$
(i) greatest, 0

2.1(ii) **1** (a), (d)
2 (a), (c), (f), (g), (h), (i)
3 (a), (e), (h)
4 (c), (e)
5 $a > 9/16$
6 $c < -\frac{1}{3}$
9 (a) $(1, 0), (3, 0), (0, 6), -2$
(b) $(4, 0), (-1, 0), (0, 4), \frac{25}{4}$
(c) $(2, 0), (0, 12), 0$ (d) $(\frac{1}{2}, 0),$
$(-1\frac{2}{3}, 0), (0, -5), -\frac{169}{24}$
(e) $(0, -6), -\frac{23}{4}$

2.2 **1** (a) $-1\frac{1}{2}, -2$ (b) $\frac{1}{3}, -2$
(c) $-2\frac{2}{3}, 16/9$ (d) $-\frac{2}{5}, -\frac{3}{5},$
(e) $1, 3$ (f) $1, -1$
2 (a) $\frac{1}{2}$ (b) 2 (c) 1 (d) $\frac{1}{4}$
(e) $-3\frac{3}{4}$
3 (a) 4 (b) $13\frac{1}{3}$ (c) $4\sqrt{(10/3)}$
4 (a) -27 (b) 24 (c) -72
5 (a) $14\frac{1}{4}$ (b) $\dfrac{57^{3/2}}{8}$ (c) $\dfrac{\pm 33\sqrt{57}}{8}$
6 (a) $x^2 + x - 14 = 0$
(b) $2x^2 + 3x - 63 = 0$
(c) $7x^2 - x - 2 = 0$
(d) $14x^2 + 29x + 14 = 0$
(e) $2x^2 - 3x - 6 = 0$
7 (a) $9x^2 + 2x + 1 = 0$
(b) $x^2 + 2x + 9 = 0$
(c) $81x^2 + 36x - 32 = 0$
8 (a) $r^2 x^2 - (q^2 - 2pr)x + p^2 = 0$
(b) $p^2 x^2 - (q^2 - 2pr)x + r^2 = 0$
(c) $p^3 x^2 + q(q^2 - 3pr)x + r^3 = 0$
10 $\pm\sqrt{17}$
11 (a) 1 (b) 2 (c) -3
(d) -3 (e) $-\frac{2}{3}$
12 (a) $4x^3 - 13x^2 + 25x - 16 = 0$
(b) $4x^3 + x^2 - 3x - 2 = 0$
(c) $2x^3 + 15x^2 + 35x + 22 = 0$

2.3 **1** (a) $x > 2$ (b) $x > -8$
(c) $x < 1$ (d) $x < 7$
(e) $x < -5$ (f) $x < -9$
(g) $x > 2$ (h) $x > 8$ (i) $x > 3$
(j) $x < \frac{7}{3}$ (k) $x < -\frac{1}{4}$
(l) $x < \frac{1}{4}$ (m) $x < 1$
(n) $x > -\frac{1}{3}$ (o) $x < \frac{7}{3}$
2 (a) $\emptyset$ (b) $-2 < x < 0,$
$-2 < x < 0$
3 (a) $0 < x < \frac{4}{3}$ (b) $\emptyset, 0 < x < \frac{4}{3}$
4 (a) $x > 1$ (b) $x < -11, x > 1,$
$x < -11$
5 (a) $-3 < x < 0$
(b) $-\frac{1}{2} < x < 2$ (c) $x < -3, x > 2$

(d) $x < -1, x > 1$ (e) $-2 < x < -\frac{7}{8}$
(f) $x < \frac{3}{7}, x > \frac{1}{2}$
2.4(i) **1** (a) $2 < x < 3$ (b) $-4 < x < -3$
(c) $-2 < x < 3$ (d) $x < 0,$
$x > 3$ (e) $x < -1 - \sqrt{2},$
$x > -1 + \sqrt{2}$ (f) $\mathbb{R}$ (g) $x \leqslant -4,$
$x \geqslant 3$ (h) $-2 \leqslant x \leqslant 5$ (i) $\mathbb{R}$
2 (a) $x < -6, x > 2$ (b) $x < -3,$
$x > -1$ (c) $\mathbb{R},$
(d) $-2 < x < 2$ (e) $0 < x < 4$
(f) $1 < x < 2,$ (g) $\mathbb{R}$
(h) $x \leqslant -3, x \geqslant 5$ (i) $-3 \leqslant x \leqslant 3$
3 (a) $\frac{3}{2} < x < 4$ (b) $x < -\frac{1}{2}, x > -\frac{1}{3}$
(c) $-1 \leqslant x \leqslant \frac{2}{3}$ (d) $x \leqslant -\frac{3}{2}, x \geqslant \frac{3}{2}$
(e) $0 \leqslant x \leqslant \frac{2}{3}$ (f) $\mathbb{R}$
(g) $x < -\frac{1}{2}, x > \frac{1}{3}$
(h) $-\frac{1}{2} < x < 3$ (i) $\mathbb{R}$
(j) $-1 \leqslant x \leqslant \frac{1}{4}$ (k) $x \leqslant -\frac{1}{2}, x \geqslant \frac{1}{2}$
(l) $-\frac{2}{3} \leqslant x \leqslant \frac{3}{4}$
4 (a) $x < -1, 0 < x < 2$
(b) $x < -3, 0 < x < 2$
(c) $x < 1, x > 2$
(d) $-3 < x < 1$
(e) $x < -6, x > -1$ (f) $0 < x < \frac{3}{2}$
2.4(ii) **1** (a) $y \leqslant 2, y \geqslant 6$ (b) $y \leqslant -6,$
$y \geqslant 2$ (c) $y \leqslant 8, y \geqslant 16$
2 (a) $-\frac{1}{6} \leqslant y \leqslant \frac{1}{2}$
(b) $-\frac{1}{10} \leqslant y \leqslant \frac{1}{2}$ (c) $-\frac{1}{4} \leqslant y \leqslant \frac{1}{2}$
3 (a) $\mathbb{R}$ (b) $y \leqslant -3, y \geqslant 1$
(c) $y \leqslant -3, y \geqslant 1$ (d) $\frac{1}{2} \leqslant y \leqslant \frac{3}{2}$
(e) $-\frac{1}{2} \leqslant y \leqslant \frac{5}{2}$ (f) $y \leqslant -\frac{1}{2}, y \geqslant \frac{1}{2}$
2.5(i) **1** (a) $x < -1, x > 7$
(b) $-1 < x < 7$ (c) $x \leqslant -3, x \geqslant -1$
(d) $-3 \leqslant x \leqslant -1$ (e) $x < -2,$
$x > 4$ (f) $3 \leqslant x \leqslant 7$
(g) $x \leqslant -4, x \geqslant -1$ (h) $0 < x < \frac{3}{2}$
2 (a) $-1 < x < 1$ (b) $x > 2$
(c) $-3 \leqslant x \leqslant \frac{1}{3}$ (d) $1 \leqslant x \leqslant 3$
(e) $x < -5, x > \frac{1}{3}$
(f) $-2 < x < \frac{4}{9}$ (g) $x \leqslant 0, x \geqslant \frac{2}{5}$
(h) $x \leqslant 0, x \geqslant \frac{4}{7}$
2.5(ii) **5** (a) $x < -2, x > 6$
(b) $-5 < x < 3$ (c) $x \leqslant -8, x \geqslant 2$
(d) $2 \leqslant x \leqslant 8$
6 (a) $x < -\frac{9}{2}, x > \frac{3}{2}$
(b) $-\frac{1}{3} < x < 3$ (c) $-\frac{1}{2} \leqslant x \leqslant \frac{7}{2}$
(d) $x \leqslant -1, x \geqslant 2$
7 (a) $2 < x < 4$
(b) $-\frac{4}{3} < x < 2$ (c) $x \leqslant \frac{1}{3}, x \geqslant 1$
(d) $-8 \leqslant x \leqslant \frac{2}{3}$
2.6 **1** $1\cdot 67$
2 $-1\cdot 56$
3 $1\cdot 47$
4 $2\cdot 23$
5 $1\cdot 61$
6 $-2\cdot 19$

Miscellaneous exercises 2

1 $x^2 + 5x - 12 = 0$
3 $2x^3 + 3x^2 + 8 = 0$
4 $2x^2 + 3x - 45 = 0$
5 $x^2 - 7x + 1 = 0$
7 $3y^2 - (p + 8)y + 3 = 0$, $p = -2$ or -14
8 $243x^2 + 30x + 1 = 0$
9 $q = \frac{1}{2}(b + 6)$,
$r = \frac{3b}{4} + \frac{c}{2} + 2 - \frac{1}{4}\sqrt{(b^2 - 8c)}$
10 $y^2 + pqy + q^3 = 0$
11 (a) $x^3 - 2x^2 - 16x + 40 = 0$
(b) $5x^3 - 4x^2 - x + 1 = 0$
(c) $x^3 - 2x^2 - 3x - 1 = 0$, 10
12 -38, $10x^3 + 3x^2 - x + 1 = 0$
13 $2(p^2 - 3q)$
14 $1, 0, x^3 - 2x^2 + x = 0$
15 $x^3 - 6x^2 - x + 30 = 0$, $-2, 3, 5$
16 $\alpha + \dfrac{1}{\beta}$, $\beta + \dfrac{1}{\alpha}$
17 $a - b + c = 0$
18 (i) $x^3 + 4(1 - k)x - k^2 = 0$
(ii) $a = 2$; $2, 1 \pm \sqrt{3}$
19 $-4 < x < 2$, $a = 3$
21 $k = -\frac{1}{48}$, (a) $k > 1$, $k \leqslant \frac{1}{2}$
22 $p = \frac{b}{2a}$, $q = (4ac - b^2)/4a$; $-3 < k < 5$
23 $-2 < k < 6$; $0 < k < 6$
24 (a) $k < 2$ and $k > 6$
(b) (i) $p > 1$, (ii) $p < -4$
25 $k < -\frac{37}{12}$, $k = -25/12$
27 (i) $-\dfrac{1}{2} \leqslant \dfrac{2x + 1}{x^2 + 2} \leqslant 1$
(ii) $x^2 - (2p^2 - 4q)x + p^2(p^2 - 4q) = 0$
29 $-1 \leqslant \dfrac{4(x - 2)}{4x^2 + 9} \leqslant \dfrac{1}{9}$
31 $6 - \sqrt{20} < y < 6 + \sqrt{20}$
32 $-1 \leqslant k \leqslant 1$
34 $x > -1$, $-3 < x < -2$, $2 \leqslant \lambda \leqslant 3$
36 $\{x : 2 < x < 2\frac{1}{2}\}$
37 (i) $-\frac{2}{3} < x < \frac{4}{3}$
(ii) $-2 < x < -1$, $1 < x < 2$
38 (i) $\{x : x < -9, x > 5\}$
(ii) $\{x : \frac{1}{3} < x < 6\}$
39 $qy^2 + 2py + 4 = 0$
40 $0 < x < 3$, $x > 4$
41 $0 < x < 2$, $x > 3$
42 $\{x : -3 < x < 3, x > 5\}$
43 (a) (i) $-4 < x < \frac{3}{2}$, (ii) $x < 1$, $2 < x < 3$; (b) real, $a = 1$
44 (a) $\{x : x \in \mathbb{R}, -4 < x < -\frac{2}{3}\}$
(b) $\{x : x \in \mathbb{R}, -4 < x < 1\}$
45 $x > 0$
46 $x \leqslant 0$, $\frac{3}{2} \leqslant x \leqslant \frac{5}{2}$, $4 \leqslant x$

47 $-1 \leqslant x \leqslant 2$, $4 \leqslant x \leqslant 7$
48 $-4 < x < -2$, $1 < x < 2$, $x > 3$

Chapter 3

3.1 **1** (a) 2, 3, 4, 5 (b) $-1, 1, 3, 5$
(c) 6, 10, 14, 18 (d) $-2, 1, 4, 7$
(e) 4, 9, 16, 25 (f) $1, \frac{1}{2}, \frac{1}{3}, \frac{1}{4}$
(g) 0, 3, 8, 15 (h) 1, 2, 6, 24
(i) 6, 12, 20, 30 (j) 15, 105, 315, 693
(k) $\frac{1}{3}, \frac{1}{4}, \frac{1}{5}, \frac{1}{6}$ (l) $\frac{1}{2}, \frac{2}{3}, \frac{3}{4}, \frac{4}{5}$
(m) $\frac{1}{28}, \frac{1}{70}, \frac{1}{130}, \frac{1}{208}$ (n) 2, 4, 8, 16
(o) $\frac{1}{3}, \frac{1}{9}, \frac{1}{27}, \frac{1}{81}$ (p) $-1, 4, -9, 16$
(q) $1, 1, -3, 5$
2 (a) 0 1, 2, 3, 4 (b) 1, 4, 7, 10
(c) 5, 7, 9, 11 (d) $\frac{1}{3}, \frac{2}{4}, \frac{3}{5}, \frac{4}{6}, \frac{5}{7}$
(e) 4, 7, 12, 19 (f) $\frac{1}{2}, \frac{1}{6}, \frac{1}{12}, \frac{1}{20}, \frac{1}{30}, \frac{1}{42}$
(g) 1, 3, 9, 27, 81 (h) $0\cdot1, 0\cdot01$,
$0\cdot001, 0\cdot0001$ (i) $\frac{1}{4}, \frac{7}{7}, \frac{9}{70}, \frac{8}{65}$
(j) $\frac{1}{2}, \frac{1}{4}, \frac{1}{8}, \frac{1}{16}, \frac{1}{32}, \frac{1}{64}$ (k) $-\frac{1}{2}, \frac{2}{3}, -\frac{3}{4}, \frac{4}{5}$

3.2 **1** (a) 10 (b) 55 (c) 24 (d) 4
(e) $\frac{1}{6} + \frac{1}{12} + \frac{1}{20} = \frac{3}{10}$
(f) $-24 + 35 - 48 = -37$
(g) $1 + 2 + 6 + 24 + 120 = 153$
(h) $\frac{1}{4} + 1 + \frac{9}{4} + 4 = 7\frac{1}{2}$
(i) $3 + 9 + 27 = 39$
(j) $1 - \frac{2}{3} + \frac{1}{2} = \frac{5}{6}$
(k) $-1 + 0 + 4 + 16 = 19$
(l) $2 - 6 + 12 - 20 = -12$

2 (a) $\displaystyle\sum_{r=1}^{5} r$ (b) $\displaystyle\sum_{r=1}^{4} (4r + 3)$
(c) $\displaystyle\sum_{r=1}^{5} (5 - 2r)$ (d) $\displaystyle\sum_{r=1}^{4} \frac{1}{2r + 1}$
(e) $\displaystyle\sum_{r=1}^{5} (-1)^r r^2$
(f) $\displaystyle\sum_{r=1}^{5} (-1)^{r+1} \frac{1}{r^2}$
(g) $\displaystyle\sum_{r=1}^{5} r(r + 1)$
(h) $\displaystyle\sum_{r=1}^{4} (3r - 2)(3r + 1)$
(i) $\displaystyle\sum_{r=1}^{5} \left(\frac{1}{2}\right)^{r-2}$ (j) $\displaystyle\sum_{r=1}^{4} rx^r$
(k) $\displaystyle\sum_{r=1}^{5} (-1)^{r+1} a^{r-1}$

3.3 **1** (a) n^2 (b) $\frac{1}{2}n(3n + 7)$ (c) $4n(n - 1)$
(d) $\frac{1}{6}n(n + 1)(2n + 7)$ (e) $\frac{1}{3}n(n + 1)(n - 1)$
(f) $\frac{1}{3}n(n^2 + 6n + 11)$
(g) $\frac{1}{6}(n + 1)(n + 2)(2n + 3)$ (h) n
2 (a) $\frac{1}{3}n(2n + 1)(4n + 1)$

(b) $\frac{2}{3}n(n+1)(2n+1)$

(c) $\frac{1}{3}n(2n-1)(2n+1)$

4 (a) $\frac{1}{12}n(n+1)(n+2)(3n+1)$

(b) $\frac{1}{4}n(n+1)(n+2)(n+3)$

(c) $\frac{1}{4}n(n+1)(n+4)(n+5)$

3.5 1 (a) 189 (b) $\frac{1023}{256}$ (c) $\frac{4^{12}-1}{5}$

(d) $\frac{12[1+(\frac{3}{4})^{15}]}{7}$ (e) $\frac{1-x^{20}}{x^2(1-x)}$

(f) $\frac{a(1-a^{12})}{1+a}$

2 (a) $\frac{3}{2}, \frac{1}{2}\left(\frac{3}{2}\right)^5$ (b) $-3, -2(3)^7$

(c) $\frac{1}{4}, \frac{1}{2^9}$ (d) $-\frac{3}{2}, \frac{81}{4}$

(e) $-2x, 64x^7$ (f) $\frac{b}{a}, b^9/a^{10}$

3 $\frac{x^{2n}-1}{x^{2n-1}(x^2-1)}$

4 $\frac{1-x^{2n}}{x(1-x^2)}$

5 $0\cdot01+0\cdot03+0\cdot09+0\cdot27+0\cdot81$

6 (a) $\pm1\cdot2$ (b) $\pm\frac{1}{2}$

7 $(0\cdot8)^{10}$

8 £88.85

3.6 1 (a) yes, 6 (b) no (c) yes, 200

(d) no (e) yes, $\frac{a}{a-1}$ (f) $\frac{1}{a-1}$

2 4/33

3 1600/17

4 $10+6+3\cdot6+2\cdot16$

5 $2+1\cdot8+1\cdot62+1\cdot458$,

$18+1\cdot8+0\cdot18+0\cdot018$

3.7 1 (a) $1-5x+10x^2-10x^3+5x^4-x^5$

(b) $1+8x+24x^2+32x^3+16x^4$

(c) $27+54x+36x^2+8x^3$ (d) 1024

$-1280x+640x^2-160x^3+20x^4-x^5$

2 (a) $1-\frac{1}{2}x-\frac{1}{8}x^2-\frac{1}{16}x^3$

(b) $1-2x+3x^2-4x^3$ (c) $1+x-x^2$

$+\frac{5}{3}x^3$ (d) $2+\frac{1}{4}x-\frac{1}{64}x^2+\frac{1}{512}x^3$

(e) $\frac{1}{3}+\frac{1}{9}x+\frac{1}{27}x^2+\frac{1}{81}x^3$

(f) $\frac{1}{2}-\frac{1}{48}x+\frac{1}{576}x^2-\frac{7}{41472}x^3$

3 (a) $1-5x+30x^3$

(b) $\frac{1}{4}+\frac{1}{4}x+\frac{7}{16}x^2+\frac{1}{2}x^3$

4 (a) $0\cdot9899$ (b) $3\cdot045$ (c) $0\cdot49835$

5 $+1\%, +4\%$

3.8 1 (a) 26 (b) 2 (c) 49/8

(d) 70/27 (e) 4

2 (a) -1 (b) -7

(c) $-77/27$ (d) -5

3 (a) $(x-1)(x-2)(x-3)$

(b) $(x+2)(2x+1)(2x-1)$

(c) $(x+2)(x^2+x+1)$

(d) $(x-a)(x^2+ax+a^2)$

(e) $(x+a)(x^2-ax+a^2)$

(f) $(x+y)(x-3y)(x-5y)$

4 (a) $x=-2,-1,3$

(b) $x=-3,-2,-1,1$

(c) $x=\frac{2}{3}, \frac{1}{2}(-1\pm\sqrt{5})$

3.9 1 $\frac{2}{x+2}-\frac{1}{x+1}$

2 $\frac{5}{x-1}+\frac{2}{x-2}$

3 $\frac{3}{2x+1}-\frac{2}{3x+2}$

4 $\frac{1}{2x-3}-\frac{1}{1+4x}$

5 $\frac{1}{x+1}+\frac{2}{x+2}-\frac{3}{x+3}$

6 $\frac{1}{x-1}-\frac{1}{x+1}+\frac{2}{x+2}$

7 $\frac{2}{x+2}+\frac{1}{(x+2)^2}$

8 $\frac{3}{(2x-1)^2}-\frac{4}{2x-1}$

9 $\frac{4}{x-1}+\frac{x-1}{x^2+1}$

10 $\frac{1}{4x-1}-\frac{1}{x^2+2}$

11 $\frac{2}{x}-\frac{x-1}{x^2-x+1}$

12 $\frac{1}{x+1}+\frac{2}{(x+1)^2}+\frac{x+2}{x^2+1}$

13 $\frac{2}{x-1}-\frac{3}{(x-1)^2}+\frac{1-x}{x^2+x+1}$

14 $1+\frac{2}{x+1}+\frac{3}{x+2}$

15 $-x+2-\frac{1}{x-2}+\frac{1}{(x-2)^2}$

16 $\quad 1-\dfrac{3}{2(x+1)}+\dfrac{3x+7}{2(x^2+1)}$

3.10 1 $\quad x-x^2+x^3,\ |x|<1$

2 $\quad \dfrac{1}{3}-\dfrac{1}{9}x+\dfrac{1}{27}x^2,\ |x|<3$

3 $\quad 2+3x+3x^2,\ |x|<1$

4 $\quad x-2x^2+4x^3,\ |x|<\tfrac{1}{2}$

5 $\quad -3-9x-27x^2,\ |x|<\tfrac{1}{3}$

6 $\quad -1-4x-8x^2,\ |x|<\tfrac{1}{2}$

7 $\quad \tfrac{1}{2}x-\tfrac{3}{4}x^2+\tfrac{7}{8}x^3,\ |x|<1$

8 $\quad \tfrac{5}{4}-\tfrac{3}{4}x+\tfrac{7}{16}x^2,\ |x|<2$

9 $\quad -3-3x-5x^2,\ |x|<1$

3.11 1 (a) $\dfrac{n}{n+1}$ (b) 1

2 (a) $\dfrac{n}{3n+1}$ (b) $\tfrac{1}{3}$

3 (a) $\dfrac{1}{6}\left[\dfrac{7}{6}-\dfrac{2}{n+1}+\dfrac{1}{n+2}+\dfrac{1}{n+3}\right]$

(b) $7/36$

4 (a) $\dfrac{n(n+2)}{3(2n+1)(2n+3)},\ 1/12$

Miscellaneous exercises 3

1 $\quad \tfrac{1}{2}n(3n+1),\ n=7$

2 $\quad -2\tfrac{1}{2},\ \tfrac{1}{2},\ 205$

3 $\quad 50(r+1),\ 257\,500$

4 (a) $(n+1)(2n+1)$ (b) -610

5 (a) $21,\ 67$ (b) $|r|<1,\ \dfrac{a}{1-r},\ 21$

6 (a) n^3+6n^2+11n

(b) $(14n^2+15n+1)/6n$

7 (ii) $n(n+1)(2n^2+2n-1)$

8 $\quad 2n^2(n+1)^2,$

$\tfrac{1}{4}n^2(n+1)^2-2n(n+1)+3$

9 $\quad n^2(2n^2-1)$

10 (i) $-\tfrac{1}{2}n(n+1)$ (ii) $\tfrac{1}{2}n(n+1)$

15 (a) $n(2n^3+8n^2+7n-2)$

16 (b) $\dfrac{n(n+2)}{3(2n+1)(2n+3)}$

19 $\quad \tfrac{1}{6}n(n+1)(2n+1)$

20 $\quad x<0,\ 48\cdot6,\ 21$

21 $\quad \tfrac{1}{3}$

22 $\quad \dfrac{(1-x^n)}{(1-x)^2}-\dfrac{nx^n}{1-x}$

23 $\quad \dfrac{(b-1)^n}{b^{n-1}},\ b(b-1)\left[1-\left(\dfrac{b-1}{b}\right)^n\right]$

24 $\quad 39$

25 $\quad 5$

26 $\quad p=3,\ q=-2$

27 $\quad \tfrac{55}{53}$

28 $\quad 1+x+\tfrac{1}{2}x^2,\ 3\cdot315$

29 $\quad a=\dfrac{1}{2},\ b=\dfrac{9}{16},\ c=-\dfrac{27}{64}$

30 $\quad 3\cdot1\%$

31 (a) $-\tfrac{3}{2},\ 14$ (b) $\tfrac{3121}{3200}$

32 (b) $35,\ 27$

33 $\quad 0\cdot969,\ 0\cdot031$

34 $\quad 1+4x-6x^2-32x^3,\ a=6,\ b=4,$ $|x|<\tfrac{1}{4}$

35 $\quad p=-1,\ q=5$

36 $\quad a=3,\ b=-4$

37 $\quad a=2,\ b=1,\ c=2;\ x>-2$

38 $\quad 2x^2-5x-3,$

$(x+1)^2(2x^2-5x-3)$

39 $\quad 3x-2,\ x-1$

40 $\quad 3x^3-7x^2-2x+8$

41 (i) 0 (ii) $a^{n-1}x-a^n$

42 (a) $\dfrac{1}{x}-\dfrac{x}{x^2+2}$

(b) $\dfrac{1}{x}-\dfrac{1}{x+2}-\dfrac{2}{(x+2)^2}$

43 $\quad \dfrac{1}{3(1+x)}+\dfrac{2}{3(1-2x)},\ 1+x+3x^2,$

$\tfrac{1}{3}\left[(-1)^n+2^{n+1}\right],\ |x|<\tfrac{1}{2}$

44 $\quad \dfrac{1}{4(1+x)}+\dfrac{3}{4(1-3x)},$

$1+2x+7x^2+20x^3,\ \dfrac{x^n}{4}\left[3^{n+1}+(-1)^n\right]$

45 (a) $-\dfrac{5}{3(x-2)}-\dfrac{4}{(x-2)^2}$

$-\dfrac{1}{3(1-2x)}$ (b) $\dfrac{3}{1+x}+\dfrac{4-3x}{1+x^2}$

$a_0=7,\ a_1=-6,\ a_2=-1,\ a_3=0,$

$a_4=7,\ a_5=-6$

46 $\quad \dfrac{2}{1-x}-\dfrac{2}{2-x},\ 3+\tfrac{3}{2}x+\tfrac{7}{4}x^2+\tfrac{15}{8}x^3,$

$|x|<1,\ 2-(\tfrac{1}{2})^n$

47 $\quad \dfrac{2}{1+x}-\dfrac{1}{1-3x},\ 2(-1)^r-3^r$

48 $\quad \dfrac{3}{2(x-1)}+\dfrac{1}{2(x+3)},$

(i) $-\tfrac{4}{3}-\tfrac{14}{9}x-\tfrac{40}{27}x^2$

(ii) $\dfrac{2}{x}+\dfrac{6}{x^3}$

49 (i) $\dfrac{n}{4(n+1)}$ (ii) $\dfrac{n}{2n+1},$

$\dfrac{n(6n+5)}{4(n+1)(2n+1)}$

50 $\quad \dfrac{1}{4}-\dfrac{1}{2(n+1)}+\dfrac{1}{2(n+2)};\ \dfrac{1}{4},\ 21$

Chapter 4

4.1(i) **1** (a) $0.9848, -0.1736, -5.6713$
(b) $-0.3420, -0.9397, 0.3640$
(c) $-0.8660, 0.5, -1.7321$
(d) $0.9397, -0.3420, -2.7475$
(e) $-0.8660, -0.5, 1.7321$
(f) $-0.6428, 0.7660, -0.8391$
(g) $0.6820, -0.7314, -0.9325$
(h) $-0.9563, -0.2924, 3.2709$
(i) $-0.8829, 0.4695, -1.8807$
(j) $0.6225, -0.7826, -0.7954$
(k) $-0.4446, -0.8957, 0.4964$
(l) $-0.7804, 0.6252, -1.2482$
(m) $0.3239, -0.9461, -0.3423$
(n) $-0.8462, -0.5329, 1.5880$
(o) $-0.6884, 0.7254, -0.9490$
2 (a) $0.6428, 0.7660, 0.8391$
(b) $0.9848, 0.1736, 5.6713$
(c) $0.3420, 0.9397, 0.3640$
(d) $0.6428, -0.7660, -0.8391$
(e) $-0.8660, -0.5, 1.7321,$
(f) $0.8660, -0.5, -1.7321$
(g) $0.9397, 0.3420, 2.7475$
(h) $0.3420, -0.9397, -0.3640$
(i) $-0.6428, -0.7660, 0.8391$
(j) $0.9063, -0.4226, -2.1445$
(k) $-0.6820, -0.7314, 0.9325$
(l) $-0.7771, -0.6293, 1.2349$
(m) $0.5906, -0.8070, -0.7319$
(n) $-0.9622, -0.2723, 3.5339$
(o) $-0.1495, -0.9888, 0.1512$
(p) $0.1874, -0.9823, -0.1908$
(q) $-0.3322, 0.9432, -0.3522$
(r) $-0.4648, -0.8854, 0.5250$
3 (a) $1, 0,$ undefined
(b) $0, -1, 0$ (c) $-1, 0,$ undefined
(d) $0, 1, 0$ (e) $1, 0,$ undefined
(f) $0, -1, 0$ (g) $-1, 0,$ undefined
(h) $0, 1, 0$ (i) $0, -1, 0$

4.1(ii) **1** (a) $-0.7660, 0.6428, -1.1918$
(b) $-0.9511, 0.3090, -3.0777$
(c) $-0.8788, 0.4772, -1.8418$
(d) $-0.6428, -0.7660, 0.8391$
(e) $-0.9744, -0.2250, 4.3315$
(f) $-0.8934, -0.4493, 1.9883$
(g) $0.5, -0.8660, -0.5774$
(h) $0.9877, -0.1564, -6.3138$
(i) $0.8181, -0.5750, -1.4229$
(j) $0.8660, 0.5, 1.7321$ (k) $0.7193,$
$0.6947, 1.0355$ (l) $0.8721, 0.4894,$
1.7820 (m) $-0.6428, 0.7660,$
-0.8391 (n) $0.8660, -0.5,$
-1.7321 (o) $0.3420, 0.9397,$
0.3640 (p) $-0.9205, -0.3907,$
2.3559 (q) $-0.4772, -0.8788,$

0.5430 (r) $0.0645, -0.9979,$
-0.0647 (s) $-0.9912, -0.1323,$
7.4947 (t) $0.9806, 0.1959, 5.0045$
(u) $-0.1409, -0.9900, 0.1423$

4.2   **1** (a) $23.6°, 156.4°$
(b) $-36.9°, -143.1°$
2 (a) $53.1°, 306.9°, -53.1°$
$-306.9°$ (b) $101.5°, 258.5°$
$-101.5°, -258.5°$
3 (a) $63.4°, 243.4°$
(b) $108.4°, 288.4°$

4.3   **1** (a) $30°, 150°$ (b) $21.9°,$
$158.1°$ (c) $200.5°, 339.5°$
(d) $200.2°, 339.8°$ (e) $0°, 180°$
(f) $90°$ (g) $270°$
(h) $41.8°, 138.2°$ (i) $210°,$
$330°$ (j) $228.6°, 311.4°$
2 (a) $60°, 300°$ (b) $41.8°,$
$318.2°$ (c) $150.0°, 210.0°$
(d) $97.1°, 262.9°$ (e) $90°, 270°$
(f) $0, 360°$ (g) $180°$ (h) $75.5°,$
$284.5°$ (i) $120°, 240°$
(j) $131.8°, 228.2°$
3 (a) $45°, 225°$ (b) $68.2°$
$248.2°$ (c) $123.7°, 303.7°$
(d) $166.8°, 346.8°,$ (e) $0°$
$180°$ (f) $135°, 315°$ (g) $36.9°$
$216.9°$ (h) $146.3°, 326.3°$
(i) $107.1°, 287.1°$ (j) $120°, 300°$

4.4(i) **1** (a) $180n° + 66.5°$ (b) $180n°$
$+139.6°$ (c) $180n°$ (d) $360n°$
$\pm 90°$ (e) $360n°$ (f) $360n°$
$+180°$ (g) $360n° \pm 60°$
(h) $360n° \pm 120°$ (i) $360n°$
$\pm 138.9°$ (j) $180n°$ (k) $360n°$
$+90°$ (l) $360n° - 90°$
(m) $180n° + (-1)^n(210°)$
(n) $180n° + (-1)^n(14.5°)$
(o) $180n° + (-1)^n 221.8°$
2 (a) $180n° + (-1)^n(5.7°)$; $180n°$
$+ (-1)^n(191.5°)$. $5.7°, 174.3°$
$191.5°, 348.5°$ (b) $360n°$
$\pm 70.5°$; $360n° \pm 120°$. $70.5°,$
$289.5°$; $120°, 240°$ (c) $180n°$
$+ (-1)^n(30°)$; $180n° + (-1)^n(210°)$.
$30°, 150°$; $210°, 330°$ (d) $180n°$
$+ 63.4°$; $180n° + 135°$. $63.4°,$
$243.4°$; $135°, 315°$ (e) $360n°$
$\pm 60°$; $360n° \pm 109.5°$. $60°, 300°$;
$109.5°, 250.5°$

4.4(ii) **1** (a) $180n°$ (b) $60n° + 30°$
(c) $90n°$ (d) $60n°$ (e) $360n°$ or
$72n° + 36°$ (f) $120n°$ or $72n°$
$+ 36°$ (g) $90n° + 22\frac{1}{2}°$ or
$180n° + 45°$ (h) $180n° + 15°$

364   Advanced mathematics 1

4.5   **1** (a) $180n° + 45°$ or $180n°$
63·4°, 45°, 225°; 63·4°
243·4° (b) $180n° + 71·6°$ or
$180n° + 116·6°$, 71·6°, 251·6°
116·6°, 296·6° (c) $180n°$ or
$180n° + 45°$, 0°, 180°, 360°; 45°,
225° (d) $360n°$ or $360n° \pm 90°$, 0°,
360°; 90°, 270° (e) $360n° \pm 70·5°$;
70·5°, 289·5° or $360n° \pm 120°$; 120°,
240° (f) $180n° + 45°$ or $180n°$
$+ 63·4°$, 45°, 225°; 63·4°
243·4° (g) $180n° + 116·6°$ or
$180n° + 18·4°$, 116·6°, 296·6°;
18·4°, 198·4°

4.6(i) **1** $\pi/6$, $\pi/4$, $\pi/2$, $2\pi/3$, $3\pi/4$, $5\pi/6$,
$4\pi/3$, $3\pi/2$, $5\pi/3$
    2 45°, 60°, 210°, 36°, 300°
    3 0·4363, 1·9199, 3·4907, 0·7068,
0·9472, 2·5261, 4·9428
    4 114·6°, 802·1°, 80·2°, 143·2°
100·8°, 70·7°

4.6(ii) **1** (a) $n\pi + 0·8761$
    (b) $2n\pi \pm \pi/3$ (c) $2n\pi + 3\pi/2$
    (d) $n\pi - 1·107$ (e) $2n\pi \pm 1·8235$
    (f) $n\pi + (-1)^n(3·3783)$
    2 (a)$n\pi + 1·249$ or $n\pi - 1·107$
    (b) $n\pi + (-1)^n(\pi/6)$ or $n\pi +$
$(-1)^n(3·481)$ (c) $2n\pi \pm 0·8411$ or
$2n\pi \pm 2·4189$ (d) $\frac{1}{2}n\pi$
    (e) $\frac{2}{3}n\pi$ or $\frac{2}{7}n\pi$

4.7   **1** (a)$\pi/3$, $5\pi/6$ (b) $\pi/12$
    (c) 1·37, 2·30
    2 (a) $-3\pi/2$ (b) $-\pi/6$, $7\pi/6$
    3 (a) $-\frac{1}{4}\pi$, $\frac{1}{4}\pi$ (b) $-0·2$, 1·37

4.8(i) **1** (a) 60°, 120°, 240°, 300°
    (b) 60°, 300°
    2 (a) 0·386, 1·708, 2·481, 3·802
4·575, 5·897 (b) π
    3 (a) $45n° + (-1)^n(7·5°)$
    (b) $1440n° \pm 212·5°$
    4 (a) $\frac{1}{3}n\pi + (-1)^n(\pi/9)$
    (b) $6n\pi \pm 2·3862$

4.8(ii) **1** (a) $360n° + 100·5°$ or $360n°$
$-40·5°$ (b) $180n°$
$+ (-1)^n(48·6°) - 60°$
    2 (a) $2n\pi + \pi/6$ or $2n\pi - \pi/2$
    (b) $n\pi + \frac{1}{4}\pi + (-1)^n\ (\pi/3)$

4.8(iii) **1** 0·824
    2 1·274
    3 0·424

Miscellaneous exercises 4
    1 $\pm 35·3°$
    2 $\pi/4$, $5\pi/4$
    3 $180n° + 135°$, $180n° + 18·4°$

    4 $n\pi \pm \pi/6$
    5 $\pi/6$, $5\pi/6$
    6 (a) 70·5°, 120° (b) $\pi/4$
    7 $n\pi$ or $(2n \pm \frac{1}{6})\pi$
    8 143·7° or 336·3°
    9 $2\pi - \alpha$, $\alpha/3$, $\frac{2}{3}\pi + \frac{1}{3}\alpha$, $\dfrac{4\pi}{3} + \frac{1}{3}\alpha$
    10 $\pi/12$, $5\pi/12$, $13\pi/12$, $17\pi/12$
    11 (a) $n\pi + \pi/3$ (i) 30°, 120°, 210°,
300°, (ii) 120° (b) $-148·8°$
$-31·2°$, 31·2°, 148·8°
    12 $n\pi + \frac{1}{4}\pi$, $\frac{1}{2}n\pi + \frac{1}{8}\pi$
    13 $x = 52·2°$, $y = 23·3°$
    14 0·86
    15 0°, 90°, 360°; $+\sqrt{2}$, $-\sqrt{2}$
    16 1·12
    17 1·3

Chapter 5
5.1   **1** (a) $a = 10·7$, $b = 11·6$, $C = 65°$
(b) $a = 15·0$, $b = 20·6$, $C = 66·3°$
(c) $A = 70·7°$, $b = 14·0$, $c = 11·8$
(d) $A = 116·7°$, $b = 5·05$,
$c = 8·85°$ (e) $a = 2·75$,
$B = 125·5°$, $c = 4·80$
    2 (a) $B = 31·1°$, $C = 111·9°$
$c = 21·6$ (b) $A = 77·7°$, $C = 41·1°$,
$a = 8·47$ (c) $A = 47·4°$,
$B = 58·9°$, $b = 55·2$ (d) $B = 50·5°$
$C = 92·2°$, $c = 18·1$ or
$B = 129·5°$, $C = 13·2°$, $c = 4·13$
(e) $A = 134·4°$, $C = 28·8°$
$a = 37·1$ or $A = 12·0°$,
$C = 151·2°$, $a = 10·8$ (f) $A = 59·1°$,
$B = 67·7°$, $a = 17·3°$ or
$A = 14·5°$, $B = 112·3°$, $a = 5·03$
(g) $B = 40·5°$, $C = 36·3°$, $c = 142$
5.2   **1** (a) $A = 48·2°$, $B = 58·4°$
$C = 73·4°$ (b) $A = 136·0°$,
$B = 25·3°$, $C = 18·7°$ (c) $A = 31·8°$,
$B = 24·8°$, $C = 123·5°$
    2 (a) $A = 65·1°$, $B = 42·9°$,
$c = 8·39$ (b) $a = 6·49$, $B = 76·5°$,
$C = 54·0°$ (c) $A = 69·8°$
$b = 9·66$, $C = 49·0$ (d) $A = 33·1°$
$B = 43·1°$, $c = 11·2$
5.3   **1** 56/65, 33/65, 56/33, 16/65, 63/65,
16/63
    2 416/425, 87/425, 416/87, $-304/425$,
$-297/425$, 304/297
    3 13/85, $-84/85$, $-13/84$, $-77/85$,
36/85, $-77/36$
    4 24/25, 7/25, 24/7, 240/289,
$-161/289$, $-240/161$

5 0.96, −0.28, −24/7, 120/169, −119/169, −120/119

6 120/169, 119/169, 120/119, 24/25, −7/25, −24/7

7 $\cos A$, $-\sin A$, $-\cos A$, $-\cos A$, $-\sin A$, $-\cos A$, $\sin A$, $-\sin A$, $-\cot A$, $\tan A$, $-\cot A$, $\cot A$

8 (a) $\frac{1}{2}, \frac{1}{2}\sqrt{3}, 1/\sqrt{3}$ (b) $1/\sqrt{2}$,

$1/\sqrt{2}, 1$ (c) $\frac{1}{2}\sqrt{3}, \frac{1}{2}, \sqrt{3}$

(d) $\frac{1+\sqrt{3}}{2\sqrt{2}}, \frac{\sqrt{3}-1}{2\sqrt{2}}, \frac{\sqrt{3}+1}{\sqrt{3}-1}, \frac{\sqrt{3}-1}{1+\sqrt{3}},$

$\frac{2\sqrt{2}}{\sqrt{3}-1}, \frac{2\sqrt{2}}{1+\sqrt{3}}$ (e) $\frac{1+\sqrt{3}}{2\sqrt{2}},$

$\frac{1-\sqrt{3}}{2\sqrt{2}}, \frac{1+\sqrt{3}}{1-\sqrt{3}}, \frac{1-\sqrt{3}}{1+\sqrt{3}}, \frac{2\sqrt{2}}{1-\sqrt{3}},$

$\frac{2\sqrt{2}}{1+\sqrt{3}}$ (f) $\frac{\sqrt{3}-1}{2\sqrt{2}}, \frac{1+\sqrt{3}}{2\sqrt{2}},$

$\frac{\sqrt{3}-1}{1+\sqrt{3}}, \frac{1+\sqrt{3}}{\sqrt{3}-1}, \frac{2\sqrt{2}}{1+\sqrt{3}}, \frac{2\sqrt{2}}{\sqrt{3}-1}$

9 $4/5, \pm 3/5, \pm 4/3$

10 12/13, 5/13, 12/5

11 $\frac{1}{3}$ or -3

12 $\frac{1}{2}$ or -2

13 $\sqrt{2}-1, \frac{1}{4}(2-\sqrt{2}), \frac{1}{4}(2+\sqrt{2})$

14 $4\cos^3 x - 3\cos x$

15 $\dfrac{3\tan x - \tan^3 x}{1 - 3\tan^2 x}$

16 $4\sin x \cos x - 8\sin^3 x \cos x$

17 $8\cos^4 x - 8\cos^2 x + 1$

18 (a) 90°, 270°, 194·5°, 345·5°

(b) 26·6°, 90°, 206·6°, 270°

(c) 0°, 60°, 180°, 300°   (d) 120°, 240°

(e) 70·5°, 180°, 289·5°

19 (a) $90(2n+1)°$ (b) $n\pi$ or $n\pi - 1·1072$ (c) $180n° + (-1)^n(30°)$

(d) $n\pi + (-1)^n 0·2527$, $n\pi - (-1)^n 0·5236$

(e) $2n\pi \pm 1·2310$

20 (a) 90°, 157·4°   (b) 208·1°, 270°

21 (a) $2n\pi - 0·9274$ or $2n\pi - 2·4982$

(b) $2n\pi + 2·2144$ or $2n\pi - 1·9658$

5.4 **1** (a) $+10, -10$ (b) $+13, -13$

(c) $+17, -17$ (d) $+\sqrt{10}, -\sqrt{10}$

(e) $+\sqrt{2}, -\sqrt{2}$

2 (a) $360n° + 96·9°$ or $360n° - 23·1°$ (b) $360n° + 4·7°$ or $360n° - 139·5°$

(c) $360n°$ or $360n° + 126·9°$

(d) $180n° + 56·3° + (-1)^n(33·7°)$

(e) $180n°$ or $180n° - 63·4°$

5.5 **1** (a) $2\sin 4x \cos x$

(b) $2\sin 6x \cos 2x$

(c) $2\sin(x + 45°)\cos 15°$

(d) $\sqrt{2}\cos 5°$ (e) $2\sin\alpha\cos\beta$

2 (a) $2\cos 2x \sin x$

(b) $2\cos 3x \sin x$ (c) $2\cos\frac{3x}{2}\sin\frac{x}{2}$

(d) $\sqrt{2}\sin 15°$

(e) $2\cos(45° - x)\sin 15°$

(f) $2\cos\alpha\sin\beta$

3 (a) $2\cos 3x \cos 2x$

(b) $2\cos\frac{3x}{2}\cos\frac{x}{2}$ (c) $\sqrt{2}\cos x$

(d) $\sqrt{2}\cos 15°$

4 (a) $2\sin 3x \sin 2x$

(b) $2\sin 4x \sin 2x$ (c) $-\sqrt{3}\sin x$

(d) $-2\sin\frac{5x}{2}\sin\frac{x}{2}$

5 (a) 30°, 90°, 150°, 210°, 270°, 330°   (b) 45°, 60°, 135°, 225°, 300°, 315°   (c) 0°, 60°, 90°, 180°, 270°, 300°, 360°   (d) 20°, 45°, 100°, 135°, 140°, 220°, 225°, 260°, 315°, 340°

6 (a) $120n° \pm 30°$ or $180n° \pm 60°$

(b) $n\pi$ or $\frac{1}{3}n\pi + (-1)^n(\pi/18)$

(c) $90n° \pm 22\frac{1}{2}°$, $360n° \pm 90°$, $180n° + (-1)^n 30°$

5.6 **1** (a) 21·4   (b) 149   (c) 23·0

(d) 4·01

2 (a) 23·4   (b) 1298   (c) 30·1

(d) 7010   (e) 116

5.7 **1** 432·5 m

2 34 cm

3 37·2 cm

5.8 **1** (a) 36 cm   (b) 105 cm

(c) 9·006 cm   (d) 237·4 cm

2 (a) 216 cm²   (b) 1312·5 cm²

(c) 38·73 cm²   (d) 16 140 cm²

3 (a) 0·9375, 53·7°   (b) 1·3889, 79·6°   (c) 0·725, 41·5°

4 (a) 4 cm   (b) 8·229 cm

(c) 2·566 m

5 (a) 19·05 cm   (b) 44·76 cm

(c) 3·167   (d) 135·7°

6 (a) 25·64 cm², 681·2 cm²

(b) 1·645 m², 4·513 m²   (c) 226·2 cm², 2343 cm²

5.9 **1** (a) 0·8481   (b) 0·3469

(c) −0·2527   (d) −0·3473

(e) 0·6435   (f) 1·244   (g) 3·614

(h) 1·176   (i) −0·2227

2 (a) −8   (b) 7/22

3 (a) 63/65   (b) 36/85

4 (a) 13/85   (b) $(\sqrt{8}+\sqrt{3})/6$

Miscellaneous exercises 5

1 120°, 60°

2 $2\sqrt{7}$, $\sqrt{3}$, 2, $\sqrt{(3/7)}$

4 342·4, 56·6°

5 150·6°

6 (i) $(3\sqrt{3}d^2)/14$

(ii) $(3\sqrt{3}d^3)/(28\sqrt{7})$ (iii) $\sqrt{(7/12)}$

7 $EP_1 = \sqrt{5}$, $EP_2 = \sqrt{8}$

8 78·5°

9 $(\sqrt{3})r \arcsin [2/\sqrt{3}) \sin (\tfrac{1}{2}\theta)]$

11 (a) 5/3 (b) 126·9°, 323·1°

12 (i) $-63/65$, 56/33 (ii) $\theta = \pi/6$ or $5\pi/6$

14 0°, 48·6°, 131·4°, 180°, 270°, 360°

15 (a) 199·5°, 340·5°

16 (a) $n\pi + \tfrac{1}{4}\pi$, $\tfrac{1}{2}\pi(n + \tfrac{1}{4})$

(b) $\tfrac{1}{4}\pi(n + \tfrac{1}{2})$, $\tfrac{1}{3}(2n\pi \pm \tfrac{1}{3}\pi)$

(c) $(2n+1)\pi$, $(2n + \tfrac{1}{2})\pi$

17 $n\pi - 68\cdot2°$, $360n° \pm 42\cdot0° - 21\cdot8°$

18 (i) 0°, 90°, 180° (ii) 22·6°, 90°

19 (i) 0°, 126·9° (ii) 0°, 90°, 180°, 0°, 30°, 70°,

20 (a) 32·8°, 147·2° (b) $2n\pi \pm \tfrac{1}{2}\pi$, $\dfrac{4n\pi}{3} \pm \dfrac{1}{3}\pi$, $\dfrac{2}{3}n\pi$

21 (i) 45°, 90°, 135°, 180°, 225°, 270°, 315° (ii) 4·5 or -3

22 $360n° + 90°$, $360n° - 36\cdot9°$

23 23·2°, 135°, 203·2°, 315°

24 (ii) 0°, 30°, 60°, 120°, 150°, 180° (iii) $360n° + 119\cdot6°$, $360n° - 13\cdot3°$

25 (i) $n\pi$, $n\pi - \tfrac{1}{4}\pi$ (ii) $x = 3\cdot759$, $-3\cdot064$, $-0\cdot695$

26 (a) (i) 0°, 120°, 240°, 360° (ii) 40·9°, 220·9°

(b) $(r-p)t^2 + 2qt + (p+r) = 0$

27 $\pm\tfrac{1}{2}\pi$, $\pi/6$, $5\pi/6$; $-\pi \leqslant \theta \leqslant -\tfrac{1}{2}\pi$, $\pi/6 \leqslant \theta \leqslant \tfrac{1}{2}\pi$, $5\pi/6 \leqslant \theta \leqslant \pi$

28 110·6°, 216·9°

29 70°, 190°, 310°, 330°

30 (i) $n\pi$ or $(2n \pm \tfrac{1}{6})\pi$ (ii) (a) 73·7° or 180° (b) 1, 1/11

32 (i) (a)26·6°, 63·4°, 206·6°, 243·4°, (b) 60°, 300° (ii) 34·6°, 163·8°

33 (a) 1, $\tfrac{1}{2}$ (b) 8/15

34 (a) 14·5°, 165·5° (b) 248·7° 344·4°

35 115·3°, 318·4°

36 (a) $180n° + 26\cdot6°$

(b) $360n° - 36\cdot9°$

37 (a) 18°, 30°, 90°, 150°, 162°

38 (a) 0°, 20°, 45°, 90°, 100°, 135°, 140°, 180° (b) $180n° + (-1)^n(30°)$, $180n° - (-1)^n(51\cdot3°)$

39 $\pi/2$, $3\pi/2$, $\pi/8$, $5\pi/8$, $9\pi/8$, $13\pi/8$

40 (b) $(3 + 3t^2 - 2t)/(1 - t^2)$

41 (i) 45°, 161·6°, 341·6°

(ii) $\tfrac{1}{2}(2n-1)\pi$, $n\pi + (-1)^n(\pi/6)$

(iii) $a = 7$, $b = 11$

$\dfrac{\cot A \cot B - 1}{\cot A + \cot B}$, $\cot 22\tfrac{1}{2}° = 1 + \sqrt{2}$,

$\cot 67\tfrac{1}{2} = \sqrt{2} - 1$

43 (b) (i) 27°, 63°, 90° (ii) 210°, 330°

44 (a) $1/\sqrt{50}$

45 (a) 14·9°, 28·1°

Chapter 6

6.1 **1** (a) $10^6 + 1$ (b) 20 (c) 35

2 99

3 1

4 (a) $\tfrac{2}{3}$ (b) $\tfrac{2}{3}$ (c) $\tfrac{1}{2}$ (d) 2

5 (a) 2 (b) 8 (c) 2

6 (a) 3 (b) 3 (c) $-\tfrac{1}{2}$ (d) $\tfrac{1}{2}$

6.2 **1** $\tfrac{2}{3}$, $-\tfrac{2}{3}$, -2

2 (a) 6 (b) -4 (c) 10

3 (a) (2, 1) (b) $(2\tfrac{1}{2}, 1\tfrac{1}{4})$

(c) $(1\tfrac{1}{2}, 1\tfrac{1}{4})$

4 (a) (1, 4) (b) $(1\tfrac{1}{2}, 4\tfrac{1}{2})$ (c) (2, 4)

5 $y = 7x - 14$

6 3

7 -10

8 $y = 2x - 1$

9 (c)

6.3 **1** (a) $2x$ (b) $4x$ (c) $4x^3$ (d) $5x^4$

5 (a) $6x^5$, $30x^4$ (b) $6x$, 6

(c) $10x^9$, $90x^8$ (d) $6x^2 - 3$, $12x$

7 5, -1, 5   **9** -2 and 3

10 (a) 1, 0 (b) $2t$, 2 (c) $3t^2$, $6t$

(d) $2t + 2$, 2

6.4 **1** (a) $5x^4 + 4$ (b) $8x^3 + 8x$

(c) $18x^5 - 6x^2 + 1$ (d) $3x^2 + 6x + 2$

3 (a) $12\,\mathrm{m\,s^{-1}}$, $12\,\mathrm{m\,s^{-2}}$ (b) $13\,\mathrm{m\,s^{-1}}$, $6\,\mathrm{m\,s^{-2}}$ (c) $48\,\mathrm{m\,s^{-1}}$, $96\,\mathrm{m\,s^{-2}}$

4 40 m

5 (4, 1), $(-4, -1)$

6 $1/3\,\mathrm{m\,s^{-1}}$

7 $2\pi rh$, $6\pi h$

8 (a) 4, 12 (b) 12, 12 (c) -1, 2

6.5 **2** (a) $-\sin x - \cos x$, $-\cos x + \sin x$

(b) $2\cos x - 3\sin x$, $-2\sin x - 3\cos x$

6.6 **1** (a) $\sin x + x \cos x$ (b) $\cos^2 x - \sin^2 x$ (c) $2x \cos x - x^2 \sin x$

2 (a) $-\sin 2t$ (b) $2t \sin t + t^2 \cos t$

(c) $\tfrac{1}{2}\sin 2t + t \cos 2t$

3   $2 + 8x - 4x^3 - 10x^4$
4   (a) $(3x+2)/2\sqrt{(x+1)}$
   (b) $2x(\sin x + \cos x) + x^2(\cos x - \sin x)$

6.7   1   (a) $1/(x+1)^2$ (b) $-5/(x-1)^2$
   (c) $2/(x+1)^2$ (d) $2x/(x^2+1)^2$
   3   (a) $(t\cos t - \sin t)/t^2$
   (b) $-(t\sin t + \cos t)/t^2$ (c) $(\sin t - t\cos t)/\sin^2 t$ (d) $1/(\sin t + \cos t)^2$
   6   (a) $-2/x^3$ (b) $-4/x^5$
   (c) $-3/x^4$ (d) $-3/(x+1)^4$

6.8   1   (a) $6(2x+1)^2$ (b) $6(3x-4)$
   (c) $3\cos 3x$ (d) $-2x\sin(x^2)$
   (e) $\frac{1}{2}\cos(\frac{1}{2}x+1)$ (f) $-6(3-2x)^2$
   2   (a) $\frac{1}{2}x^{-1/2}$ (b) $(3/2)x^{1/2}$
   c) $-\frac{1}{2}x^{-3/2}$ (d) $(-3/2)x^{-5/2}$
   3   (a) $(2t+1)^{-1/2}$ (b) $2(3t-1)^{-1/3}$
   (c) $-4(t+1)/(t^2+2t)^3$
   4   (a) $3\sin^2 x\cos x$
   (b) $2\tan x\sec^2 x$ (c) $-2\sin 4x$
   (d) $2\sec 2x\tan 2x$
   5   (a) $-2/(2x+3)^2$ (b) $2x\sin 3x$
   $+3x^2\cos 3x$ (c) $\sec^2(\frac{1}{2}x)$
   (d) $6x(x^2+1)^2$
   6   $\pi\,m^2 s^{-1}$
   7   $1.5\times10^{-4}\,m^3 s^{-1}$
   8   $2u$
   9   $6\,m^2 s^{-1}$
   10   $-0.12\,m^3 s^{-1}$
   15   $1/t^3$

6.9   2   (a) $-\pi/6$ (b) $\pi/4$ (c) $-\pi/4$
   (d) $\pi/3$
   3   (a) $2/\sqrt{(1-4x^2)}$ (b) $3/(1+9x^2)$
   (c) $2\arcsin x/\sqrt{(1-x^2)}$
   4   $A=\frac{1}{2},\,B=2$
   5   (a) $\arcsin x + x/\sqrt{(1-x^2)}$
   (b) $2x/\sqrt{(1-x^4)}$ (c) $2x\arcsin 2x + 2x^2/\sqrt{(1-4x^2)}$ (d) 1

Miscellaneous exercises 6

1   (a) 4/3   (b) 1/2
2   (a) 0   (b) 1
3   (a) 1/4   (b) 2
4   $x<1/3,\,x>1$
5   $(1/\sqrt{2}, -1/\sqrt{2}),\,(-1/\sqrt{2}, 1/\sqrt{2})$
6   $-1<x<0$
7   $0.01\,cm\,s^{-1}$
8   $(2/\pi)\,cm\,s^{-1}$
9   $18.4°$
10   $4\pi av$
11   $2\,m,\,-2\,m\,s^{-2}$
12   $\pi/4,\,5\pi/4$
13   (a) $2x\cos(x^2)$ (b) $-4\sin 2x$
   (c) $4\tan(2x)\sec^2(2x)$
14   (a) $2x\tan x + x^2\sec^2 x$

(b) $x\cos x$ (c) $\sin 2x\cos 2x$
(d) $-x\sin x$
15   (a) $(x+1)/(2x\sqrt{x})$
   (b) $\frac{1}{2}(2x+3)(x^2+3x)^{-1/2}$
   (c) $(2x^2+2)(x^3+3x)^{-1/3}$
   (d) $-\frac{2}{3}(x+1)^{-5/3}$
16   (a) $8(x^2+1)(x^2-1)^7/x^9$
   (b) $40x(3-4x^2)^{-6}$
   (c) $-3x^2(x^2+3)^{-2}$
   (d) $4(2x+1)(x^2+x+1)^3$
17   (a) $-2x/(x^2-2)^2$
   (b) $-5x/(1+2x^2)^2$
   (c) $-2x/(x+3)^3$
18   $-1/(1+x^2),\,2x/(1+x^2)^2$
21   1
22   $3+\cos x - 4\cos(\frac{1}{2}x)$
23   $(1/30)\,m\,s^{-1}$
24   $2.2\times10^{-5}\,m/degree$
27   $-1/3,\,19/3$
28   $a=\frac{1}{2}\sqrt{3},\,b=\frac{1}{2}(1+\sqrt{3})$
30   $a=9/8,\,b=-1/24$
31   $x=r(\sec\theta-1),$
   $y=h+r(1-\operatorname{cosec}\theta)$
32   (a) $(py+qx)\,cm^2 s^{-1}$
   (b) $(px+qy)/\sqrt{(x^2+y^2)}\,cm\,s^{-1}$
   (c) $(qx-py)/(x^2+y^2)\,radians\,s^{-1}$
37   $\pi/2,\,2\pi/3$

Chapter 7

7.1   1 $AB=\sqrt{13},\,BC=\sqrt{26},\,CA=\sqrt{13}$
   2   $PQ\sqrt{17},\,-\frac{1}{4},\,QR\sqrt{37},\,-6,$
   $RS\sqrt{20},\,\frac{1}{2},\,PS\sqrt{10},\,-3$

7.2   1   (a) $(5,2),\,(6\frac{1}{2},5),\,(3\frac{1}{2},\frac{1}{2}),\,(2,-2\frac{1}{2})$
   (b) $(4,3),\,(5,5);\,(5,0),\,(3,-3);$
   $(2\frac{1}{3},-1\frac{1}{3}),\,(1\frac{2}{3},-3\frac{2}{3})$ (c) $(5\frac{1}{3},3\frac{2}{3}),$
   $(4\frac{2}{3},1\frac{1}{3}),\,(3\frac{1}{3},\frac{2}{3})$
   2   $(-7,-13),\,(5,11)$

7.3   1   9   2   5   3   17   4   40

7.4   1   (a) $y=2x+1$ (b) $y=3x-6$
   (c) $y=-4x+14$ (d) $y=-5x-7$
   2   (a) $2y=x+3$ (b) $3y=4x-6$
   (c) $4y=-3x+12$ (d) $7y=2x-19$
   3   (a) $y=\frac{2}{3}x+\frac{4}{3},\,\frac{2}{3},\,(-2,0),\,(0,4/3)$
   (b) $y=\frac{3}{2}x-\frac{1}{2},\,3/2,\,(\frac{1}{3},0),\,(0,-\frac{1}{2})$
   (c) $y=-4x+6,\,-4,\,(3/2,0),\,(0,6)$
   (d) $y=-5x-5,\,-5,\,(-1,0),\,(0,-5)$
   4   $a=2,\,b=3$
   5   $a=-1.5,\,b=4$
   6   $p=-1.3,\,q=2.2$
   7   (a) $2x+3y-16=0$
   (b) $x-4y+17=0$ (c) $2x+y-6=0$
   (d) $7x+3y-28=0$ (e) $8x+2y+45=0$

7.5   1 3.6
   2   $36/13$

3 $1/5$

4 $14/\sqrt{17}$

5 $8/\sqrt{10}$

6 $1/\sqrt{2}$

7 (a) $x+3y-16=0$, $3x-y-8=0$

(b) $14x-112y+71=0$,

$64x+8y-19=0$

7.6 **1** $x-y\sqrt{3}-4=0$

2 $5x-12y+169=0$

3 $3\sqrt{2}$

4 $\sqrt{3}$

5 $3/\sqrt{5}$

6 ±10

7 $\pm4/3$

7.7 **1** (a) $(2,5)$, 4   (b) $(-1,3)$, 2

(c) $(-2,\frac{1}{2})$, 3

2 (a) $4x+3y-36=0$ (b) $2x-y$

$+13=0$ (c) $2x-3y-17=0$

(d) $15x+8y+38=0$

3 $x-y-3=0$, $x+y+1=0$, $90°$

4 (a) $3\sqrt{2}$ (b) $\sqrt{6}$ (c) $2\sqrt{14}$

5 (a) $(x-3)^2+(y+2)^2=25$

(b) $x^2+y^2-x+4y-12=0$

(c) $(x-7)^2+(y+5)^2=169$

6 (a) $3y=4x$, $y=0$ (b) $y=3x$,

$3y+x=0$

7 (a) $x^2+y^2-x+y-18=0$

(b) $x^2+y^2-4x+4y-17=0$

(c) $x^2+y^2+x+2y-3=0$

(d) $x^2+y^2-x+4y-53=0$

7.8 **1** (a) $x-2y+4=0$ (b) $x+4y+12$

$=0$ (c) $x+y-2=0$

2 (a) $x-2y+4=0$, $x+2y+4=0$

(b) $x-2y+3=0$, $3x+2y+1=0$

(c) $3x-2y-4=0$, $x+y-3=0$

4 (a) $(1,0)$, $x=-1$ (b) $(3,0)$,

$x=-3$ (c) $(0,2)$, $y=-2$

(d) $(0,-\frac{1}{2})$, $2y=1$

7.9 **1** $(0,ap)$, $(-ap^2,0)$, $(0,2ap+ap^3)$,

$(2a+ap^2,0)$

2 (a) $[\frac{1}{3}a(2+p^2),\frac{2}{3}ap]$

(b) $(\frac{1}{3}ap^2,\frac{2}{3}ap)$

3 $2ap\sqrt{(1+p^2)}$, ap^3, ap, $2a\sqrt{(1+p^2)}$

4 $\frac{1}{2}a^2p^3$, $\frac{1}{2}a^2p(2+p^2)^2$, $a^2p(1+p^2)$,

$a^2p(1+p^2)$

5 $\frac{1}{2}a^2p(3p^2+4)$

6 $[\frac{1}{3}ap^2, \frac{1}{3}(5ap+ap^3)]$,

$[\frac{1}{3}a(3+2p^2), \frac{2}{3}ap]$

7 $90°$, p

8 (a) $y^2+ax=0$

(b) $3y^2=2a(x-a)$

9 (a) $x\cos\alpha+y\sin\alpha=a$, $y=$

$x\tan\alpha$ (b) $bx\cos\alpha+ay\sin\alpha=ab$,

$ax\sin\alpha-by\cos\alpha=(a^2-b^2)\sin\alpha\cos\alpha$

(c) $x\cot\frac{1}{2}\alpha-y = a(\alpha\cot\frac{1}{2}\alpha-2)$,

$y+x\tan\frac{1}{2}\alpha = a\alpha\tan\frac{1}{2}\alpha$

10 (a) $x+p^2y=2cp$, $p^3x-py=$

$c(p^4-1)$ (b) $3px-y=p^3$,

$x+3py=p^2(6p^2+1)$

7.10 **1** $r=\pm d\sec\theta$

2 $r=\pm d\,\text{cosec}\,\theta$

3 (a) $\theta=\frac{1}{4}\pi$ (b) $r(\cos\theta-\sin\theta)=3$

4 $r=\pm2\sqrt{3}\sec(30°-\theta)$

5 $r=-2a\cos\theta$

6 (a) $r=4a\cot\theta\,\text{cosec}\,\theta$

(b) $r=a\,\text{cosec}^2\frac{1}{2}\theta$

Miscellaneous exercises 7

1 $(5,15)$, $x^2+y^2-5x-15y=0$

2 $(-7,4)$, $62\cdot5$, $(\frac{1}{2},6\frac{1}{2})$, $(-4\frac{1}{2},-3\frac{1}{2})$

3 $2:1:2$

4 $a=0\cdot35$, $b=-0\cdot5$

5 $x+2y=9$, $(x+1)^2+(y-5)^2=25$

6 $(x-a)^2+(y-b)^2=a^2$, $x=0$ and

$y=\dfrac{(b^2-a^2)x}{2ab}$, $A(0,b)$

$B\left(\dfrac{2ab^2}{a^2+b^2}, \dfrac{b(b^2-a^2)}{a^2+b^2}\right)$

7 $(x-3)^2+y^2=25$, $10(1+\sqrt{5})$

8 $2x-y=4$, $P(4,4)$, $Q(16,28)$,

$(x-10)^2+(y-16)^2=180$, $(2,20)$

9 (a) $x^2+y^2-6x-6y+5=0$

(b) $3x^2+3y^2-6x-26y+3=0$

10 $(x-12)^2+y^2=90$

11 $(x+3)^2+(y-6)^2=36$,

$4x-3y=0$

12 $x^2+(y-7)^2=50$, $x^2+(y+1)^2=2$

13 8, $y=0$, $3y=4x$

14 $4x-3y+10=0$, $x=2$, $(-\frac{8}{5},\frac{6}{5})$,

$(2,0)$

15 $x(x-4)+y(y-2)=0$, $2x+y=5$

16 $x^2+y^2-6x-10y+9=0$, $(0,9)$,

$(55\pm10\sqrt{10})/9$, 45

19 $[apq, a(p+q)]$, $y^2=4a(x+a)$

20 $y^2=2ax-8a^2$, parabola

22 (a) $y=2a+2x$ (b) $y^2=8ax$

24 vertices: $(2a,0)$, $(-a,0)$, foci: $(a,0)$,

$(0,0)$, $(\frac{1}{2}a,\pm\sqrt{6}a)$, $78\cdot5°$

26 (i) $ax=a^2+y^2$ (iii) $(3a,2a\sqrt{3})$

28 $(0,a)$ and $y=-a$, $y=tx-at^2$,

$Q(at,0)$, $R(0,-at^2)$, $2x^2+ay=0$

29 $[apq, a(p+q)]$

30 $my=2a$, $y(y+a)=2ax$

31 $y=tx-at-at^2$, $x+ty=a+2at$

$+at^3$, $x=a-2at-4a/t$, $y=a(t+2/t)^2$

32 $(y-4p)(p+q)=2(x-2p^2)$, $yp=$

$x+2p^2$, $[2pq, 2(p+q)]$, $\dfrac{4pq}{\sqrt{[(p+q)^2+4]}}$

$y^2=24x$ or $y^2=-8x$

33 $[\frac{1}{3}a(\lambda^2+\lambda\mu+\mu^2), a(\lambda+\mu)]$,
$3ax = y^2 + a^2$
34 $(8, 2)$
35 $x - 2y + 4 = 0$
36 $-\sec^2 t$ (b) $1\cdot5$
37 $2y = 3\sqrt{3}x - 3\sqrt{3}$
39 $t = \arccos(\pm\frac{3}{4})$

Chapter 8
8.5 1 (a) $\mathbb{Q}, \mathbb{R}$ (b) $\mathbb{R}$ (c) $\mathbb{Z}, \mathbb{Q}, \mathbb{R}$
(d) $\mathbb{R}$ (e) $\mathbb{R}$ (f) $\mathbb{N}, \mathbb{Z}, \mathbb{Q}, \mathbb{R}$
(g) $\mathbb{Q}, \mathbb{R}$ (h) $\mathbb{Q}, \mathbb{R}$ (i) $\mathbb{Q}, \mathbb{R}$
2 (a) $\mathbb{N}, \mathbb{Z}, \mathbb{Q}, \mathbb{R}$ (b) $\mathbb{Q}, \mathbb{R}$
(c) none (d) $\mathbb{Z}, \mathbb{Q}, \mathbb{R}$ (e) $\mathbb{Q}, \mathbb{R}$
(f) $\mathbb{R}$
8.6 1 $8 + 10i$
2 $-1 + 5i$
3 $4 + 32i$
4 $-8 + 36i$
5 $0\cdot7 - 0\cdot6i$
6 $2 + 6i$
7 $5 + i$
8 $16 + 30i$
9 $-1 + 2\sqrt{2}i$
10 $-7 - 24i$
11 $-1 - 5i$
12 $-9 - 2i$
13 10
14 $2i$
15 i
16 $-i$
17 $3 + 4i$
18 $-8 + 6i$
19 $-2 + 2\sqrt{3}i$
20 -8
21 1
22 $(5 + i)/26$
23 $-2i/5$
26 $3 + 5i, 3 - 5i$
27 $2 + i, 2 - i$
8.9 1 (i) $13 \angle 67°$ (ii) $13 \angle -67°$
(iii) $13 \angle -23°$ (iv) $13 \angle 157°$
(v) $13 \angle -157°$
2 (i) $1 + i$ (ii) $1 + \sqrt{3}i$ (iii) -1
$+ \sqrt{3}i$ (iv) $-3\sqrt{-i}$ (v) $5i$
(vi) $-5i$ (vii) $\frac{1}{\sqrt{2}}(1 - i)$ (viii) $4i$
3 (i) 1 (ii) $10 \cos(\pi/12 - i \sin \pi/12)$
(iii) -2 (iv) 2
(v) 8 (vi) $2(\cos 11\pi/12 - i \sin 11\pi/12)$
4 (i) $\cos\theta - i\sin\theta$ (ii) $2\cos\theta$
(iii) $2\cos 2\theta$ (iv) $2\cos 3\theta$
(v) $2i \sin\theta$ (vi) $2i \sin 2\theta$
7 (i) 1 (ii) α, where $\tan\alpha = \frac{4}{3}$,

$0 < \alpha < \frac{\pi}{2}$ (iii) $8/17$ (iv) $-1/5$
(v) $11/37$

Miscellaneous exercises 8
2 $2 \angle 60°, n = 3k \ (k \in \mathbb{Z})$
3 $2 \cos \frac{1}{2}\theta, \frac{1}{2}\theta$
4 $2\cos\dfrac{\phi-\theta}{2} \angle \left(\dfrac{\phi+\theta}{2}\right)$
8 $-1024\sqrt{3}i$
9 $a = 3, b = 1; a = -3, b = -1$
10 (i) $\pm(4 - 3i)$ (ii) $\pm(4 - i)$
(iii) $\pm(8 + i)/13$ (iv) $\pm(1 + i)$
(v) $\pm(1 - i)$
12 (i) $(x + 3i)(x - 3i)$
(ii) $(x + ia)(x - ia)$
(iii) $(x + 1 + 2i)(x + 1 - 2i)$
(iv) $(x + a + ib)(x + a - ib)$
13 $\dfrac{i}{2a}\left[\dfrac{1}{x+ia} - \dfrac{1}{x-ia}\right]$
14 $\dfrac{1}{2}\left[\dfrac{1}{x+ia} + \dfrac{1}{x-ia}\right]$
17 $x^4 - 8x^3 + 32x^2 - 80x - 100 = 0$
18 $2 + i$
20 $(a, b), r$
21 (b) $|z - 2|z = 2$ (c) $2\angle \pm\frac{2}{3}\pi$
(d) $2 \angle 11\pi/12, 2\angle -5\pi/12$
22 $(\cos 5\theta + 5\cos 3\theta + 10\cos\theta)/16$,
23 $3 + i, 2i, 2 - 2i$
25 $34 + 6\cos\theta - 24\cos^2\theta$
26 (i) $2\sqrt{10}, 10, \sqrt{10}/5$
(ii) $2(-1 + 3i)/5$
27 $\pm(3 + i), \pm(1 - 3i)$
28 (i) $-\frac{3}{4}(1 + 5i)$
29 $p = \sqrt{2}, q = -2 - 2\sqrt{2}$;
$p = -\sqrt{2}, q = -2 + 2\sqrt{2}$
31 $1 + i, 2 - i$,
$z^2 - 4(1 + i)z + (8 + 2i) = 0$
32 $1\cdot27 \pm 0\cdot79i - 1\cdot27 \pm 0\cdot79i$
33 $1 + 4i, -1, 3 - 2i$
35 $(1 - \frac{1}{2}i)z_1 + \frac{1}{2}iz_2$

Chapter 9
9.1 1 (a) $(3, -7)$ min (b) $(1, 6)$ max
(c) $(3, 4)$ max (d) $(-1, -4)$ min
(e) $(0, 2)$ neither (f) $(-\frac{1}{3}, \frac{2}{3})$ min
2 (a) $(1, 0)$ min (b) none
(c) $(0, 0)$ max, $(2/3, -4/27)$ min
(d) $(2, 12)$ min (e) $(0, -4)$ min,
$(-2, 0)$ max (f) $(0, 0)$ min,
$(-4, -8)$ max
3 (a) $32, 0$ (b) $12, -15$ (c) $20, 16$
4 100

5 2
6 $1/(6\sqrt{3})\,\text{m}^3$
7 a^2
8 $1/2$
9 $200\,\text{m}$
12 $2\pi a^2/3$
9.2 1 all false
2 (a) yes (b) no (c) yes
5 (a) $(-1, 0)$ (b) $(-1, 2)$
(c) $(1\pm 1/\sqrt{3}, 4/9)$ (d) $(\pm 1, 1/4)$
6 $(2,1)$
9.3 1 -2
2 $y = \pm 4$
3 $x + y = \pm 4$
4 $x/y;\ -1/y^3$
9.5 1 (a) $0{\cdot}1$ (b) $0{\cdot}2$ (c) $0{\cdot}32$
2 (a) $0{\cdot}24$ (b) $0{\cdot}54$
3 $2\,\text{cm}$
4 $\pi/3\sqrt{3}$
6 (a) $3x^2\delta x$ (b) $(-1/x^2)\delta x$
(c) $-\sin x\,\delta x$ (d) $\delta x/(1+x^2)$

Miscellaneous exercises 9

4 $1/8,$
5 30°
6 (i) $2:1$ (ii) $2:1$
7 $(-1/2, 1/12)$ maximum;
$(1/2, -1/12)$ minimum
9 $a = 5, b = 6$
11 $(3\sqrt{3})/2,\ -(3\sqrt{3})/2$
12 $4/3;\ -1/3$
15 maxima at $x = 0, \pi, 2\pi$; minima at
$x = 2\pi/3, 4\pi/3$
16 (a) $0, 2;\ -1 < x < 0, 2 < x < 3$
17 $(-1, 0);\ 1, -3$; minimum when
$x = (1/2)^{\frac{1}{3}}$
18 $(2,0), (0, -2/3), (2,0),$
$(-22, 144/147);\ x = 6, x = -1, y = 1$
20 $x = 1/\sqrt{3}, y = 2/3;\ 4/(3\sqrt{3})$
21 $a = 2, b = -3/2;$
maximum value 0
22 $(1,1), (-1, -1)$
24 $y = 22x - 7, 32x + y = 128;$
$(7,147), (-2,192)$
25 $a/3$
27 $32\pi a^3/81$
28 $1/2$
29 $(1 + \sqrt{3})a/2u$
30 $2:1$; yes

Chapter 10

10.1 1 (a) $x^7/7 + c$ (b) $x^4 + c$
(c) $x + x^2 + c$ (d) $x^6 + c$
(e) $x^2 - 3x + c$ (f) $\frac{1}{2}x^2 + x^3 + c$
2 (a) $2x^{3/2} + c$ (b) $-(1/x) + c$

(c) $x - (1/x) + c$ (d) $-(2/x^6) + c$
(e) $(4x^{7/4}/7) + c$
(f) $x - \frac{2}{3}x^3 + (x^5/5) + c$
3 (a) $\frac{3}{4}x^{4/3} + c$ (b) $2x^{5/2} + c$
(c) $2x^{1/2} + c$
4 (a) $2t^5 + c$ (b) $t - t^2 + (t^3/3) + c$
(c) $(3t^{5/3}/5) + c$
5 (a) $-(1/2u^2) + c$ (b) $2u^4 + 4u^3$
$+3u^2 + u + c$ (c) $(4u^{5/4}/5) + c$
6 (a) $3\sin x + 2\cos x + c$
(b) $x - \cos x + c$
(c) $(2x^3/3) + (x^2/2) - x + c$
(d) $(x^5/5) - x^4 + (4x^3/3) + c$
(e) $\frac{2}{3}x^{3/2} + 2x^{1/2} + c$
(f) $(x^5/5) + x^3 + 3x - (1/x) + c$
7 (a) $y = 2x^2 - 3x + 3$ (b) $y = 2x^2 - (x^4/4) - 4$ (c) $y = x^2 + x^3 - 9$
(d) $y = \sin x - \cos x + 2$
8 (a) 100 (b) 40 (c) 35
10.2 1 $\{(2x+3)^3/6\} + c$
2 $\{(3x-2)^4/12\} + c$
3 $\{(x+1)^5(5x-1)/30\} + c$
4 $\{(4x-3)^{3/2}/6\} + c$
5 $\frac{1}{3}\sin 3\theta + c$
6 $\{(1+x^3)^3/9\} + c$
7 $\frac{1}{3}\sin^3\theta + c$
8 $\{(2x+1)^5/10\} + c$
9 $\frac{1}{2}\tan 2x + c$
10 $-\frac{1}{3}(1-x^2)^{3/2} + c$
11 $\arctan 2x + c$
12 $2\sqrt{(x+2)} + c$
10.3 2 (a) $\frac{1}{2}x\sin 2x + \frac{1}{4}\cos 2x + c$
(b) $-\frac{1}{2}x\cos 2x + \frac{1}{4}\sin 2x + c$
3 $x(2x+3)^5/10 - (2x+3)^6/120 + c$
4 $\frac{1}{4}x\sin 4x + (1/16)\cos 4x + c$
5 $-(1/5)x\cos 5x + (1/25)\sin 5x + c$
9 $-(2x+1)/(x+1)^2 + c$
10 (a) $2x\sin(x/2) + 4\cos(x/2) + c$
(b) $-2x\cos(x/2) + 4\sin(x/2) + c$
10.4 1 4
2 21
3 60
4 $16/3$
5 2
6 2
7 10
8 -2
9 0
10 $2 - \sqrt{2}$
11 1
12 $\pi + 2$
10.5 3 $14/3$
4 $15/4$
5 $\pi/4$
10.6 1 76

2 $10\frac{1}{2}$
3 125/6
4 33
5 $84a^2$
6 $1\frac{1}{2}$
7 18
8 (a) 1   (b) 1
9 $3\pi/2$
10 36
11 1/6
12 1/3
13 4/3
14 4
15 4π

10.7 1 (a) $2/\pi$ (b) $2/\pi$
2 30
3 1/5
4 $-2/\pi$
5 $4/\pi$
6 $\pi/(3\sqrt{3})$
7 $\pi/6 + \sqrt{3} - 2$
8 -2
9 $2/3\pi$
10 3/25
11 $1/\pi$
12 $4/\pi$

10.9 1 $112\pi/5$
2 $\frac{1}{3}\pi h^3 \tan^2\alpha$
3 $\pi/105$
4 $\pi^2/4$
5 $3\pi c^3/2$
6 $4\pi/3$
8 $160\pi a^3$
9 $96\pi/5$
10 $16\pi a^3/5$

Miscellaneous exercises 10

1 $\frac{1}{4}; \frac{1}{4}$
4 8/15
5 7/3
8 (a) $\pi/2 + 1$ (b) $23\frac{33}{40}$ (c) $\frac{1}{3}$
(d) $14\frac{2}{3}$
9 $(5, 9/2)$; 18
10 $-2, 1, 4$
11 $1\frac{1}{4}$
12 $4\frac{2}{3}$
13 $32\pi/45$
14 $(\pi/2 - 1, 1)$; $(\pi/2 - 3/2)$
15 4/3; $16\pi/15$
16 $(1 \pm k, 1 - k^2)$; $y = 1 - (\frac{1}{3})^{2/3}$,
$y = 1 - (\frac{2}{3})^{2/3}$
17 (a) $-2\cos 2x + c$
(b) $\sqrt{(1 + x^2)} + c$ (c) $\tan x + c$
(d) $\tan x - x + c$ (e) $\frac{1}{2}(x^4 + 1)^{3/2} + c$
(f) $x - \frac{1}{4}\sin 4x + c$

18 (a) 4/15   (b) $2 - \sqrt{2}$
19 (a) 1/16   (b) 7/6
20 256/15   (i) $64\pi/3$ (ii) $1024\pi/35$
21 1/6; $\pi/30$
22 $3\pi/4$
23 $y = 12x - 16$; $256\pi/63$
24 $5\pi^2/12 + \pi(\sqrt{3})/8$
25 120π
26 $\pi/3$
28 $2\pi/3$
29 $9\pi/2$
30 a
31 $1; -3; -1$
32 $2p^5/5$; $\pi p^8/4$ (i) 3π (ii) $15/2\sqrt{2}$
(iii) 3/4

Chapter 11

11.1 1 (a) $2\frac{1}{2}$ (b) -2 (c) 6
2 (a) $2, -1$ (b) $0, -1$ (c) $1, 2$
3 (a) $4, 6$ (c) $-2, -1/3$
(c) $9, 1/3$
4 $x = 9, y = 27$
5 $x = 2, y = 8$
6 $(\log_y x)^2$
8 $x = 4, y = 2$; $x = 10, y = 4$
9 $x = 64, y = 16$
10 $x = 9, y = 3$
11 $k = 12{\cdot}7, n = 0{\cdot}5$
12 $n = 0{\cdot}71, C = 110$

11.2 2 $0{\cdot}208$
3 (a) $1/x$ (b) $2/x$ (c) $-1/x$
(d) $1/(x + 2)$
4 (a) $x > 0$ (b) $x > -1$
(c) $-1 < x < 1$
5 (a) $3/(3x + 4)$ (b) $-1/(2 - x)$
(c) 0   (d) $1 + \ln x$
7 (a) $1/e$ (b) $1/(2e)$
8 (a) $2\cot x$ (b) $(4x^3 + 1)/(x^4 + x)$
(c) $1/(x \ln 2)$ (d) $1/(x \ln x)$
10 (a) $\frac{1}{2}x^2\ln x - \frac{1}{4}x^2 + c$
(b) $\frac{1}{2}(\ln x)^2 + c$

11.3 2 (a) $\ln 2$ (b) $\frac{1}{2}\ln 3$
3 (a) $x + \ln(x - 1) + c$
(b) $x + \frac{1}{2}\ln(2x - 1) + c$
(c) $2x - 2\ln(x + 1) + c$
(d) $2x - \ln(2x + 1) + c$
5 (a) $2 - \ln 2$ (b) $3\ln(3/2) - 1$
6 (a) $\frac{1}{2}x^2 - x + \ln(x + 1) + c$
(b) $\frac{1}{3}x^3 + \frac{1}{2}x^2 + x + \ln(x - 1) + c$
8 (a) $\frac{1}{2}x^2 + x + 2\ln(x - 1) + c$
(b) $\frac{1}{2}x^2 - x + 2\ln(x + 2) + c$

11.4 1 (a) $\ln(x - 2) - \ln(x + 2) + c$
(b) $3\ln(x + 1) - 2\ln(x + 2) + c$
2 (a) $2\ln x - \ln(2x + 1) + c$
(b) $3\ln x - 2\ln(x + 1) + c$

3 (a) $2\ln(x+2)+\{1/(x+2)\}+c$
(b) $\ln(x-1)-[1/(x-1)]+c$
4 (a) $\ln(2x+1)-\ln(x+1)+c$
(b) $\ln(3x+1)-\ln(2x+1)+c$
8 $\frac{1}{2}\ln 3$

11.5 **1** (a) $\ln(x^3+1)+c$ (b) $\ln(x^2+2x)$
$+c$ (c) $\frac{1}{4}\ln(2x^2+1)+c$
(d) $2\ln(x^2+4)+c$
2 (a) $\frac{1}{2}\ln 2$ (b) $\ln 3$
3 (a) $\frac{1}{2}\ln 2$ (b) $\frac{1}{2}\ln 2$
4 $\ln 2$
5 (a) $\ln 2$ (b) $\ln 6$
6 $\ln(3/2)$
9 $\ln 2-\pi/4$
10 $\pi/4+2\ln 2$

11.6 **1** $2^x\ln 2;\ 2^x(\ln 2)^2$
2 20
4 $-2x(1+x^2)^{-1}(1-x^4)^{-1/2}$
5 1

11.7 **1** (a) x^2 (b) 2^x (c) $x+\ln 2$
3 (a) $e^x(\cos x+\sin x)$
(b) $5e^{2x}\cos x$ (c) $2(e^{2x}-e^{-2x})$
(d) $e^x(1+x)$ (e) $-xe^{-x}$
(f) $e^{3x}(3\cos 2x-2\sin 2x)$
5 (a) $-e^{-x}+c$ (b) $\frac{1}{4}e^{4x}+c$
(c) $-e^{2-x}+c$ (d) $2e^{x/2}+c$
6 (a) $\frac{1}{2}(1-e^{-2})$ (b) e^2-1
(c) 1 (d) 1
7 (a) $\ln(1+e)-\ln 2$ (b) $3/\ln 4$
9 (a) $(x-1)e^x$ (b) $-(x+1)e^{-x}$
10 2 or 3

11.9 **1** $5+7x^2+x^3$
2 (a) e^2-1 (b) $e^{-2}-1$
(c) $2(e^4-1)$
3 (a) $1/3$ (b) $-5/6$ (c) 2
5 $x^n(\ln 2)^n/n!$
7 $a=3,\ b=-2$
9 $1{\cdot}6487$

11.10 **1** (a) $2x-2x^2+(8x^3/3)-4x^4;$
$-1<2x\leqslant 1$ (b) $\ln 2+(x/2)$
$-(x^2/8)+(x^3/24);-2<x\leqslant 2$
(c) $x^2-(x^4/2)+(x^6/3)-(x^8/4);$
$-1\leqslant x\leqslant 1$
(d) $2x-x^2+(2x^3/3)-(x^4/2);$
$-1<x\leqslant 1$
2 $0{\cdot}6930$
7 (a) $(-1)^{n-1}3^n/(n2^n)$
(b) $-(1+(-\frac{1}{2})^n)/n$
8 $-1/3$
10 (a) $\ln(3/2)$ (b) $\ln(3/2)$
(c) $\ln 4-1$

11.11 **1** (a) $y\,dy/dx=x$ (b) $y+x\,dy/dx$
$=0$ (c) $y\,dy/dx=2$ (d) $x\,dy/dx$
$=y-1$

2 (a) $x^2-y^2=c$
(b) $y^2=2\ln x+c$
3 $y^2+y=x^2+x+6$
4 $\arctan y=x^3/3+\pi/4$
5 $x=4e^{2t}$
6 $y=e^{4-x}$
7 $u=2\exp(1-t^2)$
8 $(2xy)\,dy/dx=y^2-x^2$
10 $2x\,dy/dx=y$

Miscellaneous exercises 11
1 (a) $\ln(\sec\theta+\tan\theta)+c$
(b) $\ln\tan(\theta/2)+c$
2 (a) $2e-1$ (b) $e-1$ (c) e^2-1
3 $x+y=2$
4 (a) $2\operatorname{cosec}2x$ (b) $\sec x$
(c) $(1+\ln x)x^x$
6 4
8 (a) $\ln(x-1)-\ln(x+1)-2\arctan x+c$
(b) $\ln x-\frac{1}{2}\ln(4-x^2)+c$
11 $a=1/2,\ b=1/12$
12 $-3(x/2+x^2/4+x^3/2);\ -\frac{1}{2}\leqslant x<\frac{1}{2}$
13 $1/8$
14 $a=-1,\ b=2$
15 $x=2y^2,\ 4x+2y=3$
16 $y^2=4x-x^2+21,\ 4y=\pm(3x+19)$
17 $\ln(y^2+1)=2x/(x+1)$
18 $0{\cdot}080\,043$
19 (i) $a=1,\ b=2$ (ii) $-2<x\leqslant 2;$
$\ln(1+x/2);\ -3\leqslant x<3;\ -\ln(1-x/3);$ 0 or 1
20 (i) $2^{2n}/(2n)!$ (ii) $3x-3x^2/2+3x^3$
$-15x^4/4;\ -\frac{1}{2}<x\leqslant\frac{1}{2}$
21 $xy=3y-x-1,\ y=3$
23 $\frac{1}{2}\ln(45/17)$
26 $34{\cdot}3$
27 $a=3,\ b=4$ (i) $2{\cdot}214$ (ii) $0{\cdot}927$
29 $0,\ \ln 2$
31 e^2-e^{-2}
33 (a) $1-e^\pi$ (b) $(3/2)\ln 3-1/4$
35 $x\geqslant 2$ or $x<-2;\ 0{\cdot}278$
37 (i) $0,\ n/2,\ -n/3$
(ii) $x^{2n-1}/(2n-1),\ -1<x<1$
38 $y^3=x^4;\ 4x+3y+1=0,\ 3x-4y+7=0$
39 20 min.
40 (i) $x=\ln 2,\ y=4$ (ii) $1,\ -6,\ 9$
41 $20/9$ max., $16/9$ min.
42 $(1/b)\ln(3/2);\ a/b$
43 50

Chapter 12
12.2 **1** $5,5$
2 13
4 $\sqrt 3$
6 $(\mathbf{q}-\mathbf{p}),\ \frac{1}{2}(\mathbf{q}-\mathbf{p}),\ \frac{1}{2}(\mathbf{p}+\mathbf{q})$
7 $(\mathbf{q}-\mathbf{p}),\ \frac{1}{2}(\mathbf{p}+\mathbf{q}),\ \mathbf{q}-\frac{1}{2}\mathbf{p},\ -\frac{1}{2}(\mathbf{p}+\mathbf{q})$

10 (a) $12\sqrt{3}$, N60°E
(b) 12, S30°E
12.3 2 $\frac{1}{2}(2\mathbf{q}+\mathbf{p})$, $\frac{1}{2}(3\mathbf{q}-\mathbf{p})$
3 $\mathbf{b}-\mathbf{a}$, $-\mathbf{a}$, $-\mathbf{b}$, $\mathbf{a}-\mathbf{b}$
4 $3\mathbf{p}-2\mathbf{q}$, $2\mathbf{p}-\mathbf{q}$, $2\mathbf{q}-\mathbf{p}$, $3\mathbf{q}-2\mathbf{p}$
12.4 1 $(2\mathbf{b}+\mathbf{a})/3$, $(4\mathbf{a}-\mathbf{b})/3$, $\mathbf{a}$
5 $10\mathbf{q}-6\mathbf{p}$
12.5 1 $\mathbf{r}=2\mathbf{a}+t\mathbf{b}$, $\mathbf{r}=3\mathbf{b}+s\mathbf{a}$, $2\mathbf{a}+3\mathbf{b}$
2 (a) $\mathbf{r}=\mathbf{p}+t\mathbf{q}$ (b) $\mathbf{r}=\mathbf{p}+$
$t(\mathbf{q}-\mathbf{p})$ (c) $\mathbf{r}=t(\mathbf{p}+\mathbf{q})$
(d) $\mathbf{r}=(\mathbf{p}+\mathbf{q})+t(\mathbf{q}-\mathbf{p})$
3 $\mathbf{r}=\mathbf{b}+t(\mathbf{c}-\mathbf{a})$, $\mathbf{r}=\mathbf{c}+s(\mathbf{a}-\mathbf{b})$,
$(\mathbf{b}+\mathbf{c}-\mathbf{a})$
4 $\mathbf{r}=\frac{1}{3}(\mathbf{p}+\mathbf{q})+t(\mathbf{p}-\mathbf{q})$,
$\mathbf{r}=\mathbf{p}+s(\mathbf{p}+\mathbf{q})$, $(5\mathbf{p}-\mathbf{q})/6$
5 $(\mathbf{a}-\mathbf{b})$
12.6 1 (a) $\begin{pmatrix}5/13\\12/13\end{pmatrix}$ (b) $\begin{pmatrix}3/5\\-4/5\end{pmatrix}$
(c) $\begin{pmatrix}20/29\\21/29\end{pmatrix}$ (d) $\begin{pmatrix}-1/\sqrt{2}\\1/\sqrt{2}\end{pmatrix}$
2 (a) $3\mathbf{a}+2\mathbf{b}$ (b) $4\mathbf{a}+3\mathbf{b}$
(c) $\frac{1}{2}(\mathbf{a}+3\mathbf{b})$ (d) $(\mathbf{a}+2\mathbf{b})/7$
3 5, $\sqrt{2}$, $\sqrt{2}$
4 (a) $\begin{pmatrix}11/5\\27/5\end{pmatrix}$ (b) $\begin{pmatrix}-11/3\\1\end{pmatrix}$
5 (a) $\mathbf{r}=\mathbf{i}+t\mathbf{j}$ (b) $\mathbf{r}=t\mathbf{i}+(1-t)\mathbf{j}$
7 $6\mathbf{i}-8\mathbf{j}$, $5\mathbf{i}-4\mathbf{j}$
12.7 2 (a) 7 (b) -1 (c) -16 (d) 0
3 (a) $45°$ (b) $90°$ (c) $30°$
(d) $60°$ (e) $135°$
5 $\mathbf{r}=(5-2t)\mathbf{i}+(1-2t)\mathbf{j}$; $2\mathbf{i}-2\mathbf{j}$
12.8 1 (a) $\mathbf{i}-\mathbf{j}$ (b) $6t\mathbf{j}$
(c) $-\mathbf{i}\sin t-\mathbf{j}\cos t$ (d) $4t\mathbf{i}+3t^2\mathbf{j}$
2 (a) π (b) 5 (c) 13 (d) 0
3 $(1+2t^2)/\sqrt{(1+t^2)}$, $\sqrt{(1+4t^2)}$
12.9 1 (a) $(3\mathbf{i}+4\mathbf{j}+12\mathbf{k})/13$ (b) $(2\mathbf{i}-5\mathbf{j}$
$-14\mathbf{k})/15$ (c) $(2\mathbf{i}-6\mathbf{j}+3\mathbf{k})/7$
2 $(\mathbf{i}+\mathbf{j}+\mathbf{k})/\sqrt{3}$, $54\cdot7°$
3 $11{:}\mathbf{r}=(8\mathbf{i}+\mathbf{j}+3\mathbf{k})+t(6\mathbf{i}-7\mathbf{j}+6\mathbf{k})$
4 $3\mathbf{i}+3\mathbf{j}+4\mathbf{k}$, $4\mathbf{i}+\mathbf{j}+5\mathbf{k}$, $5\mathbf{i}-\mathbf{j}+6\mathbf{k}$
5 $(\mathbf{i}+2\mathbf{j}+\mathbf{k})/\sqrt{6}$
6 -2
8 $3{:}-1{:}-3$

Miscellaneous exercises 12
1 $(4\mathbf{i}+3\mathbf{k})/5$; $2\mathbf{i}+4\mathbf{j}+3\mathbf{k}$
2 $\lambda=-1$, $\mu=3$
3 $(-2\mathbf{i}+2\mathbf{j}+5\mathbf{k})/\sqrt{33}$
4 $\mathbf{j}+2\pi(\mathbf{i}+\mathbf{j})$
6 $60°$
7 $60°{:}(\mathbf{i}-\mathbf{j}+\mathbf{k})/\sqrt{3}$;
$\mathbf{r}=\mathbf{i}+\mathbf{j}+t(\mathbf{i}-\mathbf{k})$;
8 (i) $-60(6\mathbf{i}+5\mathbf{j}+\mathbf{k})\mathrm{km\,h^{-1}}$

(ii) $[30\mathbf{i}+30\mathbf{j}+2\mathbf{k}-t(6\mathbf{i}+5\mathbf{j}+\mathbf{k})]$ km
(iii) 5 min 20 s
9 $\mathbf{i}+2\mathbf{j}+\mathbf{k}$
10 $\pm(4\mathbf{i}-12\mathbf{j}+3\mathbf{k})/13$; $54\cdot7°$,
$125\cdot3°$
11 $ACBD$ is a parallelogram
13 (ii) $-4\mathbf{i}-3\mathbf{j}$
14 $90°$; $220/3$; $-7\mathbf{i}+22\mathbf{j}-\mathbf{k}$
15 $\mathbf{r}=\mathbf{i}+\mathbf{j}+t(\mathbf{i}+\mathbf{k})$; $1/\sqrt{3}$
16 $2\sqrt{5}$, $4\sqrt{2}$
17 $\mathbf{r}=t(2\mathbf{i}+\mathbf{j})$
18 $4\mathbf{j}+2\mathbf{k}$
21 $\sqrt{2}$
22 $9\mathbf{i}+2\mathbf{j}$
23 $\mathbf{r}=\mathbf{i}+t(\mathbf{j}+\mathbf{k})$
24 $3\mathbf{i}+3\mathbf{j}+4\mathbf{k}$
25 $\mathbf{i}-\mathbf{j}+\mathbf{k}$
26 $y+z=0$; $\sqrt{6}$
27 $(2t+1)\mathbf{i}+\mathbf{j}$; $(t^2+t)\mathbf{i}+jt$;
speed $=6\,\mathrm{m\,s^{-1}}$, $t=2$
28 $\alpha=1/3$, $\beta=2/3$, $\gamma=0$; $1/3$
29 $\mathbf{r}=\pm2\mathbf{j}+t\mathbf{i}$
30 $(\mathbf{i}+2\mathbf{j}+2\mathbf{k})/3$
31 $6\mathbf{i}+2\mathbf{j}$, $6\mathbf{j}$
33 $(\mathbf{a}+3\mathbf{b})/4$
34 3; $\sqrt{134}\,\mathrm{m}$
35 $(a\sin\omega t+2a\cos\omega t)\mathbf{i}-(2a\sin\omega t)\mathbf{j}$
36 $x^2+y^2=4$, $x^2=4y$;
(a) $|2\omega-2|$ (b) $2\sqrt{(\omega^4+1)}$
(c) $(2n-1)\pi/\omega$ (d) $2n\pi/\omega$

Chapter 13
13.2 1 (a) $\begin{pmatrix}3&7\\6&3\end{pmatrix}$ (b) $\begin{pmatrix}5&0&-1\\3&0&-2\end{pmatrix}$
2 $\begin{pmatrix}5&1\\3&1\end{pmatrix}$, $\begin{pmatrix}1&1\\-3&3\end{pmatrix}$, $\begin{pmatrix}7&1\\6&0\end{pmatrix}$,
$\begin{pmatrix}0&-2\\9&-7\end{pmatrix}$
3 $m=3$, $n=-2$
4 $\begin{pmatrix}3&2\\4&-2\\-2&6\end{pmatrix}$, $\begin{pmatrix}-1&4\\4&2\\-4&6\end{pmatrix}$,
$\begin{pmatrix}7&0\\4&-6\\0&6\end{pmatrix}$
13.3 1 $\mathbf{AB}{:}4\times4$; $\mathbf{BA}{:}5\times5$
2 (a) 12 (b) 2 (c) $(5,6)$
(d) $(2,1,4)$ (e) $\begin{pmatrix}7&6\\8&9\end{pmatrix}$

(f) $\begin{pmatrix} 2 & -2 \\ -3 & 3 \end{pmatrix}$

3 $\begin{pmatrix} -10 & -5 \\ 20 & 10 \end{pmatrix}$

4 $x = 2, y = 4$

13.4 1 (a) no (b) no
3 (a) 6 (b) 22 (c) 0 (d) 0
(e) 1 (f) 0
4 $\det \mathbf{PQ} = 0$, $\det \mathbf{QP} = 1$

13.5 1 (a) $\begin{pmatrix} 1 & 0 \\ 0 & -1 \end{pmatrix}$ (b) $\begin{pmatrix} 1 & 0 \\ 1 & -1 \end{pmatrix}$
(c) $\begin{pmatrix} 1 & 0 \\ -1 & 1 \end{pmatrix}$ (d) $\begin{pmatrix} 1 & -1 \\ 0 & 1 \end{pmatrix}$

2 (a) $\begin{pmatrix} 4 & -5 \\ -3 & 4 \end{pmatrix}$
(b) $\begin{pmatrix} -3/2 & 1 \\ 7/2 & -2 \end{pmatrix}$
(c) $\dfrac{1}{17}\begin{pmatrix} 3 & 4 \\ 2 & -3 \end{pmatrix}$ (d) $\begin{pmatrix} -3 & -2 \\ -2 & -3/2 \end{pmatrix}$

3 $x = -1, y = 3$

6 $\begin{pmatrix} 1 & 0 \\ 0 & -1 \end{pmatrix}$

7 $59\mathbf{I}$ (a) $x = 3, y = -2, z = 1$
(b) $x = 1, y = -1, z = 1$

8 $\begin{pmatrix} 1 & -x & -y \\ 0 & 1 & 0 \\ 0 & 0 & 1 \end{pmatrix}$,

$\begin{pmatrix} 1 & -x & xz-y \\ 0 & 1 & -z \\ 0 & 0 & 1 \end{pmatrix}$, $\begin{pmatrix} 1 & -x & 0 \\ 0 & 1 & 0 \\ 0 & -y & 1 \end{pmatrix}$

9 $\begin{pmatrix} \cos\theta & \sin\theta \\ -\sin\theta & \cos\theta \end{pmatrix}$

10 $x = 2$

13.6 1 (a) $\begin{pmatrix} 2 & 3 \\ 0 & 2 \end{pmatrix}$ (b) $\begin{pmatrix} 2 & 3 \\ 4 & 6 \end{pmatrix}$
(c) $\begin{pmatrix} 0 & 2 \\ 3 & 0 \end{pmatrix}$ (d) $\begin{pmatrix} 3 & 2 \\ 4 & 3 \end{pmatrix}$

3 (a) 1 (b) 1 (c) 3 (d) 2

4 (a) $\begin{pmatrix} 1 & 0 \\ 0 & -1 \end{pmatrix}$ (b) $\begin{pmatrix} -1 & 0 \\ 0 & 1 \end{pmatrix}$
(c) $\begin{pmatrix} 0 & 1 \\ 1 & 0 \end{pmatrix}$ (d) $\begin{pmatrix} 0 & -1 \\ -1 & 0 \end{pmatrix}$

6 $\mathbf{A} = \begin{pmatrix} 0 & -1 \\ 1 & 0 \end{pmatrix}$, $\mathbf{B} = \begin{pmatrix} -1 & 0 \\ 0 & -1 \end{pmatrix}$

7 $\begin{pmatrix} 0 & -2 \\ 2 & 0 \end{pmatrix}$, $\mathbf{P} = \begin{pmatrix} 2 & 0 \\ 0 & 2 \end{pmatrix}$,
$\mathbf{Q} = \begin{pmatrix} 0 & -1 \\ 1 & 0 \end{pmatrix}$

8 $\dfrac{1}{2}\begin{pmatrix} -1 & \sqrt{3} \\ \sqrt{3} & 1 \end{pmatrix}$

10 $x + 2y = 0$

Miscellaneous exercises 13

3 $a = 4, b = 5;$

$\mathbf{M}^{-1} = \dfrac{1}{5}\begin{pmatrix} -3 & 2 & 2 \\ 2 & -3 & 2 \\ 2 & 2 & -3 \end{pmatrix}$

4 $\dfrac{1}{18}\begin{pmatrix} 6 & -12 & 0 \\ 10 & -2 & -6 \\ -5 & 10 & 3 \end{pmatrix}$

5 $\begin{pmatrix} 3/2 & 0 & 0 \\ 0 & 5/4 & 0 \\ -1/2 & 0 & 2 \end{pmatrix}$

6 $\begin{pmatrix} 5/2 & -1 \\ -4 & 2 \\ 3 & -5/4 \end{pmatrix}$

7 ± 1

8 $21x_1 + 17x_2 - 3x_3$

9 $\begin{pmatrix} 3/2 & -1 \\ 1/2 & 0 \end{pmatrix}$

10 (i) 2 (ii) $\begin{pmatrix} 0 & 1 \\ 1 & 0 \end{pmatrix}$, $\theta = n\pi$

11 $\pm 2 + \sqrt{3}$

14 $(-3 \pm \sqrt{17})/2$

15 $\begin{pmatrix} \sqrt{2} \\ 1 \end{pmatrix}$, $\begin{pmatrix} -\sqrt{2} \\ 1 \end{pmatrix}$

16 $\dfrac{1}{\sqrt{2}}\begin{pmatrix} 1 & 1 \\ -1 & 1 \end{pmatrix}$; $Y^2 = \sqrt{2}X$